AF412506

Special Topics in the Theory of Piezoelectricity

Jiashi Yang

Editor

Special Topics in the Theory of Piezoelectricity

 Springer

Editor
Jiashi Yang
Department of Engineering Mechanics
University of Nebraska, Lincoln
Nebraska Hall
Lincoln, NE 68588-0526
USA
jyang1@unl.edu

ISBN 978-0-387-89497-3   e-ISBN 978-0-387-89498-0
DOI 10.1007/978-0-387-89498-0
Springer Dordrecht Heidelberg London New York

Library of Congress Control Number: 2009929638

Mathematics Subject Classification (2000): 74-xx

© Springer Science+Business Media, LLC 2009
All rights reserved. This work may not be translated or copied in whole or in part without the written
permission of the publisher (Springer Science+Business Media, LLC, 233 Spring Street, New York, NY
10013, USA), except for brief excerpts in connection with reviews or scholarly analysis. Use in connection
with any form of information storage and retrieval, electronic adaptation, computer software, or by similar
or dissimilar methodology now known or hereafter developed is forbidden.
The use in this publication of trade names, trademarks, service marks, and similar terms, even if they are
not identified as such, is not to be taken as an expression of opinion as to whether or not they are subject
to proprietary rights.

Printed on acid-free paper

Springer is part of Springer Science+Business Media (www.springer.com)

Preface

Piezoelectric or, more generally, electroelastic materials, exhibit electromechanical coupling. They experience mechanical deformations when placed in an electric field and become electrically polarized under mechanical loads. These materials have been used to make various electromechanical devices. Examples include transducers for converting electric energy to mechanical energy or vice versa, resonators and filters for telecommunication and timekeeping, and sensors for information collection.

Piezoelectricity has been a steadily growing field for more than a century, progressed mainly by researchers from applied physics, acoustics, materials science and engineering, and electrical engineering. After World War II, piezoelectricity research has gradually concentrated in the IEEE Society of Ultrasonics, Ferroelectrics, and Frequency Control. The two major research focuses have always been the development of new piezoelectric materials and devices. All piezoelectric devices for applications in the electronics industry require two phases of design. One aspect is the device operation principle and optimal operation which can usually be established from linear analyses; the other is the device operation stability against environmental effects such as a temperature change or stress, which is usually involved with nonlinearity. Both facets of design usually present complicated electromechanical problems.

Due to the application of piezoelectric sensors and actuators in civil, mechanical, and aerospace engineering structures for control purposes, piezoelectricity has also become a topic for mechanics researchers. Mechanics can provide effective tools for piezoelectric device and material modeling. For example, the finite element and boundary element methods for numerical analysis and the one- and two-dimensional theories of piezoelectric beams, plates, and shells are effective tools for the design and optimization of piezoelectric devices. Mechanics theories of composites are useful for predicting material behaviors.

In spite of the wide and growing applications of piezoelectric devices, books published on the topic of piezoelectricity are relatively few. Following the editor's previous book, *An Introduction to the Theory of Piezoelectricity*, Springer ©2005, this book addresses more advanced topics that require a collective effort. Each self-contained chapter has been written by a group of international experts and includes quite a few advanced topics in the theory of piezoelectricity. Each chapter attempts to present a basic picture of the subject area addressed.

Piezoelectricity is a broad field and, practically speaking, this volume can only cover a fraction of the many relatively advanced topics. Following a brief summary of the three-dimensional theory of linear piezoelectricity, Chapters 2 through 5 discuss selected topics within the linear theory. The linear theory of piezoelectricity assumes a reference state free of deformations and fields. When initial deformations and/or fields are present, the theory for small incremental fields superimposed on a bias is needed, which is the subject of Chapter 6. The theory for incremental fields needs to be obtained from the fully nonlinear theory by linearization about an initial state, and, therefore, is a subject that is inherently nonlinear. Chapter 7 covers the fully dynamic effects due to electromagnetic coupling. Chapter 8 addresses nonlocal and gradient effects of electric field variables.

I would like to take this opportunity to thank all chapter contributors. My thanks also go to Patricia A. Worster and Ziguang Chen of the College of Engineering at the University of Nebraska-Lincoln for their editing assistance on Chapters 1, 7, and 8.

Jiashi Yang
January 2009

Contents

Chapter 1
Basic Equations

Jiashi Yang

1.1 Introduction

This chapter presents a brief summary of the basic theory of linear piezo-electricity based mainly on the *IEEE Standard on Piezoelectricity* [1] and the classical book on piezoelectricity [2] by H. F. Tiersten who also wrote the theoretical part of [1]. The organization of this chapter is essentially a short-ened version of Chapter 2 of *An Introduction to the Theory of Piezoelectricity* [3]. This chapter uses Cartesian tensor notation, the summation convention for repeated tensor indices, and the convention that a comma followed by an index denotes partial differentiation with respect to the coordinate associated with the index. A superimposed dot represents a time derivative.

1.2 Basic Equations

The equations of linear piezoelectricity can be obtained by linearizing the nonlinear electroelastic equations [4, 5] under the assumption of infinitesimal deformation and fields. The equations of motion and the charge equation are

$$T_{ji,j} + \rho_0 f_i = \rho_0 \ddot{u}_i, \qquad D_{i,i} = q, \tag{1.1}$$

where $\mathbf{T}$ is the stress tensor, ρ_0 is the reference mass density, $\mathbf{f}$ is the body force per unit mass, $\mathbf{u}$ is the displacement vector, $\mathbf{D}$ is the electric displace-ment vector, and q is the body free charge density which is usually zero. Within the linear theory, the conservation of mass that determines the present

Jiashi Yang
Department of Engineering Mechanics University of Nebraska, Lincoln, NE 68588-0526, USA, e-mail: Jyang1@unl.edu

J. Yang (ed.), *Special Topics in the Theory of Piezoelectricity*, DOI: 10.1007/978-0-387-89498-0_1, 1
© Springer Science + Business Media, LLC 2009

mass density ρ takes the following form,

$$\rho_0 \cong \rho(1 + u_{k,k}), \tag{1.2}$$

which can be treated separately once the displacement has been obtained. Constitutive relations are given by an electric enthalpy function H,

$$H(S_{kl}, E_k) = \frac{1}{2}\, c^E_{ijkl} S_{ij} S_{kl} - e_{ijk} E_i S_{jk} - \frac{1}{2}\, \varepsilon^S_{ij} E_i E_j \tag{1.3}$$

through

$$T_{ij} = \frac{\partial H}{\partial S_{ij}} = c^E_{ijkl} S_{kl} - e_{kij} E_k,$$

$$D_i = -\frac{\partial H}{\partial E_i} = e_{ikl} S_{kl} + \varepsilon^S_{ik} E_k, \tag{1.4}$$

where the strain tensor $\mathbf{S}$ and the electric field vector $\mathbf{E}$ are related to the displacement $\mathbf{u}$ and the electric potential, ϕ, by

$$S_{ij} = (u_{i,j} + u_{j,i})/2, \qquad E_i = -\phi_{,i}. \tag{1.5}$$

c^E_{ijkl}, e_{ijk}, and ε^S_{ij} are the elastic, piezoelectric, and dielectric constants. The superscript E in c^E_{ijkl} indicates that the independent electric constitutive variable is the electric field $\mathbf{E}$. The superscript S in ε^S_{ij} indicates that the mechanical constitutive variable is the strain tensor $\mathbf{S}$. The material constants have the following symmetries.

$$c^E_{ijkl} = c^E_{jikl} = c^E_{klij}, \qquad e_{kij} = e_{kji}, \qquad \varepsilon^S_{ij} = \varepsilon^S_{ji}. \tag{1.6}$$

We also assume that the elastic and dielectric tensors are positive definite in the following sense.

$$c^E_{ijkl} S_{ij} S_{kl} \geq 0 \quad \text{for any } S_{ij} = S_{ji}, \quad \text{and} \quad c^E_{ijkl} S_{ij} S_{kl} = 0 \Rightarrow S_{ij} = 0,$$

$$\varepsilon^S_{ij} E_i E_j \geq 0 \quad \text{for any } E_i, \quad \text{and} \quad \varepsilon^S_{ij} E_i E_j = 0 \Rightarrow E_i = 0. \tag{1.7}$$

The internal energy density per unit volume can be obtained from H through a Legendre transform by

$$U(\mathbf{S}, \mathbf{D}) = H(\mathbf{S}, \mathbf{E}(\mathbf{S}, \mathbf{D})) + \mathbf{E}(\mathbf{S}, \mathbf{D}) \cdot \mathbf{D}. \tag{1.8}$$

Constitutive relations in the following form then follow.

$$\mathbf{T} = \frac{\partial U}{\partial \mathbf{S}}, \qquad \mathbf{E} = \frac{\partial U}{\partial \mathbf{D}}, \tag{1.9}$$

or

$$T_{ij} = c^D_{ijkl} S_{kl} - h_{kij} D_k, \qquad E_i = -h_{ikl} S_{kl} + \beta^S_{ik} D_k. \qquad (1.10)$$

It can be shown that U is positive definite:

$$\begin{aligned}
U &= H + E_i D_i \\
&= \frac{1}{2}\, c^E_{ijkl} S_{ij} S_{kl} - e_{ijk} E_i S_{jk} - \frac{1}{2}\, \varepsilon^S_{ij} E_i E_j + E_i(e_{ikl} S_{kl} + \varepsilon^S_{ik} E_k) \\
&= \frac{1}{2}\, c^E_{ijkl} S_{ij} S_{kl} + \frac{1}{2}\, \varepsilon^S_{ij} E_i E_j \\
&\geq 0. \qquad (1.11)
\end{aligned}$$

Similar to Equations (1.4) and (1.10), linear constitutive relations can also be written as

$$S_{ij} = s^E_{ijkl} T_{kl} + d_{kij} E_k, \qquad D_i = d_{ikl} T_{kl} + \varepsilon^T_{ik} E_k, \qquad (1.12)$$

and

$$S_{ij} = s^D_{ijkl} T_{kl} + g_{kij} D_k, \qquad E_i = -g_{ikl} T_{kl} + \beta^T_{ik} D_k. \qquad (1.13)$$

With successive substitutions from Equations (1.4) and (1.5), Equation (1.1) can be written as four equations for $\mathbf{u}$ and ϕ

$$\begin{aligned}
c_{ijkl} u_{k,lj} + e_{kij} \phi_{,kj} + \rho f_i &= \rho \ddot{u}_i, \\
e_{ikl} u_{k,li} - \varepsilon_{ij} \phi_{,ij} &= q, \qquad (1.14)
\end{aligned}$$

where we have neglected the superscripts of the material constants and the subscript of the reference mass density.

Let the region occupied by a piezoelectric body be V and its boundary surface be S, as shown in Figure 1.1. Let the unit outward normal of S be $\mathbf{n}$.

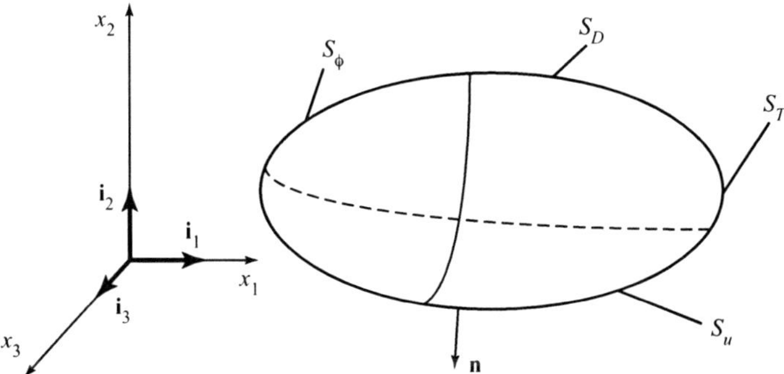

Fig. 1.1 A piezoelectric body and partitions of its boundary surface.

For boundary conditions, we consider the following partitions of S,

$$S_u \cup S_T = S_\phi \cup S_D = S, \qquad S_u \cap S_T = S_\phi \cap S_D = 0, \qquad (1.15)$$

where S_u is the part of S on which the mechanical displacement is prescribed, and S_T is the part of S where the traction vector is prescribed. S_ϕ represents the part of S which is electroded where the electric potential is no more than a function of time, and S_D is the unelectroded part. For mechanical boundary conditions, we have prescribed displacement $\bar{u}_i$

$$u_i = \bar{u}_i \quad \text{on} \quad S_u, \qquad (1.16)$$

and prescribed traction $\bar{t}_j$

$$T_{ij} n_i = \bar{t}_j \quad \text{on} \quad S_T. \qquad (1.17)$$

Electrically, on the electroded portion of S,

$$\phi = \bar{\phi} \quad \text{on} \quad S_\phi, \qquad (1.18)$$

where $\bar{\phi}$ does not vary spatially. On the unelectroded part of S, the charge condition can be written as

$$D_j n_j = -\bar{\sigma} \quad \text{on} \quad S_D, \qquad (1.19)$$

where $\bar{\sigma}$ is the free charge density per unit surface area. In the above formulation, we assume very thin electrodes, and the mechanical effects, such as inertia and stiffness, of the electrodes are neglected. On an electrode S_ϕ, the total free electric charge Q can be represented by

$$Q = \int_{S_\phi} -n_i D_i dS. \qquad (1.20)$$

The electric current flowing out of the electrode is given by

$$i = -\dot{Q}. \qquad (1.21)$$

Sometimes there are two (or more) electrodes on a body, and the electrodes are connected to an electric circuit. In this case, circuit equation(s) need to be considered.

1.3 Principle of Superposition

The linearity of Equation (1.14) allows the superposition of solutions. Suppose the solutions under two different sets of loads $\{\mathbf{f}^{(1)}, q^{(1)}\}$ and $\{\mathbf{f}^{(2)}, q^{(2)}\}$

are $\{\mathbf{u}^{(1)}, \phi^{(1)}\}$ and $\{\mathbf{u}^{(2)}, \phi^{(2)}\}$, respectively. Then, under the combined load of $\{\mathbf{f}^{(1)} + \mathbf{f}^{(2)}, q^{(1)} + q^{(2)}\}$, the solution to Equation (1.14) is $\{\mathbf{u}^{(1)} + \mathbf{u}^{(2)}, \phi^{(1)} + \phi^{(2)}\}$. This is called the principle of superposition and can be shown as

$$
\begin{aligned}
c_{ijkl}&(u_k^{(1)} + u_k^{(2)})_{,lj} + e_{kij}(\phi^{(1)} + \phi^{(2)})_{,kj} + \rho(f_i^{(1)} + f_i^{(2)}) - \rho\frac{\partial^2}{\partial t^2}(u_i^{(1)} + u_i^{(2)}) \\
&= c_{ijkl}u_{k,lj}^{(1)} + c_{ijkl}u_{k,lj}^{(2)} + e_{kij}\phi_{,kj}^{(1)} + e_{kij}\phi_{,kj}^{(2)} + \rho f_i^{(1)} + \rho f_i^{(2)} - \rho\ddot{u}_i^{(1)} - \rho\ddot{u}_i^{(2)} \\
&= (c_{ijkl}u_{k,lj}^{(1)} + e_{kij}\phi_{,kj}^{(1)} + \rho f_i^{(1)} - \rho\ddot{u}_i^{(1)}) + (c_{ijkl}u_{k,lj}^{(2)} + e_{kij}\phi_{,kj}^{(2)} + \rho f_i^{(2)} - \rho\ddot{u}_i^{(2)}) \\
&= 0 + 0 \\
&= 0,
\end{aligned}
\tag{1.22}
$$

and

$$
\begin{aligned}
e_{ikl}&(u_k^{(1)} + u_k^{(2)})_{,li} - \varepsilon_{ij}(\phi^{(1)} + \phi^{(2)})_{,ij} - (q^{(1)} + q^{(2)}) \\
&= e_{ikl}u_{k,li}^{(1)} + e_{ikl}u_{k,li}^{(2)} - \varepsilon_{ij}\phi_{,ij}^{(1)} - \varepsilon_{ij}\phi_{,ij}^{(2)} - q^{(1)} - q^{(2)} \\
&= (e_{ikl}u_{k,li}^{(1)} - \varepsilon_{ij}\phi_{,ij}^{(1)} - q^{(1)}) + (e_{ikl}u_{k,li}^{(2)} - \varepsilon_{ij}\phi_{,ij}^{(2)} - q^{(2)}) \\
&= 0 + 0 \\
&= 0.
\end{aligned}
\tag{1.23}
$$

The principle of superposition can be generalized to include boundary loads.

1.4 Hamilton's Principle

The equations and boundary conditions of linear piezoelectricity can be derived from a variational principle [2]. Consider

$$
\Pi(\mathbf{u}, \phi) = \int_{t_0}^{t_1} dt \int_V \left[\frac{1}{2}\rho\dot{u}_i\dot{u}_i - H(\mathbf{S}, \mathbf{E}) + \rho f_i u_i - q\phi \right] dV
$$
$$
+ \int_{t_0}^{t_1} dt \int_{S_T} \bar{t}_i u_i \, dS - \int_{t_0}^{t_1} dt \int_{S_D} \bar{\sigma}\phi \, dS,
\tag{1.24}
$$

where $\mathbf{S}$ and $\mathbf{E}$ are considered as functions of the displacement and potential through

$$
S_{ij} = \frac{(u_{i,j} + u_{j,i})}{2}, \qquad E_i = -\phi_{,i}.
\tag{1.25}
$$

$\mathbf{u}$ and ϕ are variationally admissible if they are smooth enough and satisfy

$$
\begin{aligned}
\delta u_i|_{t_0} = \delta u_i|_{t_1} &= 0 \quad \text{in} \quad V, \\
u_i = \bar{u}_i \quad &\text{on} \quad S_u, \quad t_0 < t < t_1, \\
\phi = \bar{\phi} \quad &\text{on} \quad S_\phi, \quad t_0 < t < t_1.
\end{aligned}
\tag{1.26}
$$

The first variation of Π is

$$\delta\Pi = \int_{t_0}^{t_1} dt \int_V \left[(T_{ji,j} + \rho f_i - \rho\ddot{u}_i)\delta u_i + (D_{i,i} - q)\delta\phi \right] dV$$
$$- \int_{t_0}^{t_1} dt \int_{S_T} (T_{ji}n_j - \bar{t}_i)\delta u_i dS - \int_{t_0}^{t_1} dt \int_{S_D} (D_i n_i + \bar{\sigma})\delta\phi \, dS,$$
$$(1.27)$$

where we have denoted

$$\mathbf{T} = \frac{\partial H}{\partial \mathbf{S}}, \qquad \mathbf{D} = -\frac{\partial H}{\partial \mathbf{E}}. \tag{1.28}$$

Therefore, the stationary condition of Π is

$$T_{ji,j} + \rho f_i = \rho\ddot{u}_i \quad \text{in} \quad V, \quad t_0 < t < t_1, \quad D_{i,i} = q \quad \text{in} \quad V, \quad t_0 < t < t_1,$$
$$T_{ji}n_j = \bar{t}_i \quad \text{on} \quad S_T, \quad t_0 < t < t_1, \quad D_i n_i = -\bar{\sigma} \quad \text{on} \quad S_D, \quad t_0 < t < t_1.$$
$$(1.29)$$

Hamilton's principle can be stated as: among all the admissible $\{\mathbf{u}, \phi\}$, the one that also satisfies Equation (1.29) makes Π stationary.

1.5 Poynting's Theorem and Energy Integral

We begin with the rate of change of the total internal energy density, given as

$$\begin{aligned}
\dot{U} &= \frac{\partial U}{\partial S_{ij}}\dot{S}_{ij} + \frac{\partial U}{\partial D_i}\dot{D}_i \\
&= T_{ij}\dot{S}_{ij} + E_i\dot{D}_i = T_{ij}\dot{u}_{i,j} - \phi_{,i}\dot{D}_i \\
&= (T_{ij}\dot{u}_i)_{,j} - T_{ij,j}\dot{u}_i - (\phi\dot{D}_i)_{,i} + \phi\dot{D}_{i,i} \\
&= (T_{ij}\dot{u}_i)_{,j} - (\rho\ddot{u}_i - \rho f_i)\dot{u}_i - (\phi\dot{D}_i)_{,i} + \phi\dot{q} \\
&= (T_{ji}\dot{u}_j)_{,i} - \frac{\partial}{\partial t}\left(\frac{1}{2}\rho\dot{u}_i\dot{u}_i\right) + \rho f_i\dot{u}_i - (\phi\dot{D}_i)_{,i} + \phi\dot{q}.
\end{aligned} \tag{1.30}$$

Therefore,

$$\frac{\partial}{\partial t}(T + U) = \rho f_i\dot{u}_i + \phi\dot{q} - (\phi\dot{D}_i - T_{ji}\dot{u}_j)_{,i}, \tag{1.31}$$

where

$$T = \frac{1}{2}\rho\dot{u}_i\dot{u}_i \tag{1.32}$$

is the kinetic energy density, and $\phi\dot{D}_i$ is the quasi-static Poynting vector. Equation (1.31) is the Poynting theorem of piezoelectricity.

Integration of Equation (1.31) over V gives

$$\frac{\partial}{\partial t}\int_V (T+U)dV = \int_V \rho(f_i\dot{u}_i + \phi\dot{q})dV + \int_{S_u} T_{ji}n_j\dot{\bar{u}}_i dS$$

$$+ \int_{S_T} \bar{t}_i\dot{u}_i dS - \int_{S_\phi} \dot{D}_i n_i\bar{\phi}dS + \int_{S_D} \dot{\bar{\sigma}}\phi dS. \quad (1.33)$$

Integrating Equation (1.33) from t_0 to t, we obtain

$$\int_V (T+U)|_t dV = \int_V (T+U)|_{t_0}dV + \int_{t_0}^t dt \int_V \rho(f_i\dot{u}_i + \phi\dot{q})dV$$

$$+ \int_{t_0}^t dt \int_{S_u} T_{ji}n_j\dot{\bar{u}}_i dS + \int_{t_0}^t dt \int_{S_T} \bar{t}_i\dot{u}_i dS$$

$$- \int_{t_0}^t dt \int_{S_\phi} \dot{D}_i n_i\bar{\phi}dS + \int_{t_0}^t dt \int_{S_D} \dot{\bar{\sigma}}\phi dS. \quad (1.34)$$

Equation (1.34) is called the energy integral which states that the energy at time t is the energy at time t_0 plus the work done to the body from t_0 to t.

1.6 Uniqueness

Consider two solutions to the following initial boundary value problem:

$$\begin{aligned}
T_{ji,j} + \rho f_i &= \rho\ddot{u}_i \quad \text{in} \quad V, \quad t > t_0, \\
D_{i,i} &= q \quad \text{in} \quad V, \quad t > t_0, \\
T_{ij} &= c_{ijkl}S_{kl} - e_{kij}E_k \quad \text{in} \quad V, \quad t > t_0, \\
D_i &= e_{ijk}S_{jk} + \varepsilon_{ij}E_j \quad \text{in} \quad V, \quad t > t_0, \\
S_{ij} &= (u_{i,j} + u_{j,i})/2 \quad \text{in} \quad V, \quad t > t_0,
\end{aligned} \quad (1.35)$$

and

$$\begin{aligned}
u_i &= \bar{u}_i \quad \text{on} \quad S_u, \quad t > t_0, & T_{ji}n_j &= \bar{t}_i \quad \text{on} \quad S_T, \quad t > t_0, \\
\phi &= \bar{\phi} \quad \text{on} \quad S_\phi, \quad t > t_0, & D_i n_i &= -\bar{\sigma} \quad \text{on} \quad S_D, \quad t > t_0, \\
u_i &= u_i^0 \quad \text{in} \quad V, \quad t = t_0, & \dot{u}_i &= v_i^0 \quad \text{in} \quad V, \quad t = t_0, \\
\phi &= \phi^0 \quad \text{in} \quad V, \quad t = t_0.
\end{aligned} \quad (1.36)$$

From the principle of superposition, the difference of the two solutions satisfies the homogeneous version of Equations (1.35) and (1.36). Let $\mathbf{u}^*, \phi^*$, $\mathbf{S}^*, \mathbf{T}^*, \mathbf{E}^*$, and $\mathbf{D}^*$ denote the differences of the corresponding fields and apply Equation (1.34) to them. The initial energy and the external work for the difference fields are zero. Then the energy integral implies that, for the difference fields, at any $t > t_0$,

$$\int_V (T^* + U^*)\big|_t \, dV = 0, \qquad t > t_0. \tag{1.37}$$

Because both T and U are nonnegative,

$$U^* = 0, \qquad T^* = 0 \quad \text{in} \quad V, \quad t > t_0. \tag{1.38}$$

From the positive definiteness of T and U,

$$\mathbf{S}^* = 0, \qquad \mathbf{E}^* = 0, \qquad \dot{\mathbf{u}}^* = 0 \quad \text{in} \quad V, \quad t > t_0. \tag{1.39}$$

Hence the two solutions are identical for $\mathbf{S}, \mathbf{E}, \mathbf{T}, \mathbf{D}$, and the velocity fields but may differ by a static rigid body displacement and a constant potential [2].

1.7 Four-Vector Formulation

Let us define the four-space coordinate system [6]

$$x_p = \{x_i, t\}, \tag{1.40}$$

and the four-vector

$$U_p = \{u_i, \phi\}, \tag{1.41}$$

where subscripts p, q, r, and s are assumed to run from 1 to 4. Also, define the second-rank four-tensor

$$\rho_{pq} = \begin{cases} \rho \delta_{pq}, & p, q = 1, 2, 3, \\ 0, & p, q = 4, \end{cases} = \begin{bmatrix} \rho & 0 & 0 & 0 \\ 0 & \rho & 0 & 0 \\ 0 & 0 & \rho & 0 \\ 0 & 0 & 0 & 0 \end{bmatrix}, \tag{1.42}$$

and the fourth-rank four-tensor M_{pqrs}, where

$$\begin{aligned} M_{ijkl} &= c_{ijkl}, & M_{4jkl} &= e_{jkl}, & M_{ijk4} &= e_{kij}, \\ M_{4jk4} &= -\varepsilon_{jk}, & M_{p44s} &= -\rho_{ps}, \end{aligned} \tag{1.43}$$

and all other components of $M_{pqrs} = 0$. Then

$$\begin{aligned} (U_{p,q} & M_{pqrl})_{,r} \\ &= (U_{i,j} M_{ijrl} + U_{4,j} M_{4jrl} + U_{i,4} M_{i4rl} + U_{4,4} M_{44rl})_{,r} \\ &= (U_{i,j} M_{ijkl} + U_{4,j} M_{4jkl} + U_{i,4} M_{i4kl} + U_{4,4} M_{44kl})_{,k} \\ &\quad + (U_{i,j} M_{ij4l} + U_{4,j} M_{4j4l} + U_{i,4} M_{i44l} + U_{4,4} M_{444l})_{,4} \end{aligned}$$

$$= (u_{i,j}c_{ijkl} + \phi_{,j}e_{jkl})_{,k} + (-\dot{u}_i\rho_{il})_{,4}$$
$$= c_{ijkl}u_{i,jk} + e_{jkl}\phi_{,jk} - \rho\ddot{u}_l, \tag{1.44}$$

and

$$\begin{aligned}
(U_{p,q}M_{pqr4})_{,r} \\
= (U_{i,j}M_{ijr4} + U_{4,j}M_{4jr4} + U_{i,4}M_{i4r4} + U_{4,4}M_{44r4})_{,r} \\
= (U_{i,j}M_{ijk4} + U_{4,j}M_{4jk4} + U_{i,4}M_{i4k4} + U_{4,4}M_{44k4})_{,k} \\
+ (U_{i,j}M_{ij44} + U_{4,j}M_{4j44} + U_{i,4}M_{i444} + U_{4,4}M_{4444})_{,4} \\
= (u_{i,j}e_{kij} - \phi_{,j}\varepsilon_{jk})_{,k} \\
= u_{i,jk}e_{kij} - \phi_{,jk}\varepsilon_{jk}. \tag{1.45}
\end{aligned}$$

Therefore,
$$(U_{p,q}M_{pqrs})_{,r} = 0 \tag{1.46}$$

yields the homogeneous equation of motion and the charge equation.

1.8 Cylindrical Coordinates

The cylindrical coordinates (r, θ, z) are defined by

$$x_1 = r\cos\theta, \qquad x_2 = r\sin\theta, \qquad x_3 = z. \tag{1.47}$$

In cylindrical coordinates, we have the strain-displacement relation

$$S_{rr} = u_{r,r}, \qquad S_{\theta\theta} = \frac{1}{r}u_{\theta,\theta} + \frac{u_r}{r}, \qquad S_{zz} = u_{z,z},$$
$$2S_{r\theta} = u_{\theta,r} + \frac{1}{r}u_{r,\theta} - \frac{u_\theta}{r}, \qquad 2S_{\theta z} = \frac{1}{r}u_{z,\theta} + u_{\theta,z},$$
$$2S_{zr} = u_{r,z} + u_{z,r}. \tag{1.48}$$

The electric field-potential relation is given by

$$E_r = -\phi_{,r}, \qquad E_\theta = -\frac{1}{r}\phi_{,\theta}, \qquad E_z = -\phi_{,z}. \tag{1.49}$$

The equations of motion are

$$\frac{\partial T_{rr}}{\partial r} + \frac{1}{r}\frac{\partial T_{\theta r}}{\partial \theta} + \frac{\partial T_{zr}}{\partial z} + \frac{T_{rr} - T_{\theta\theta}}{r} + \rho f_r = \rho\ddot{u}_r,$$
$$\frac{\partial T_{r\theta}}{\partial r} + \frac{1}{r}\frac{\partial T_{\theta\theta}}{\partial \theta} + \frac{\partial T_{z\theta}}{\partial z} + \frac{2}{r}T_{r\theta} + \rho f_\theta = \rho\ddot{u}_\theta,$$
$$\frac{\partial T_{rz}}{\partial r} + \frac{1}{r}\frac{\partial T_{\theta z}}{\partial \theta} + \frac{\partial T_{zz}}{\partial z} + \frac{1}{r}T_{rz} + \rho f_z = \rho\ddot{u}_z. \tag{1.50}$$

The electrostatic charge equation is

$$\frac{1}{r}(rD_r)_{,r} + \frac{1}{r}D_{\theta,\theta} + D_{z,z} = q. \tag{1.51}$$

1.9 Spherical Coordinates

The spherical coordinates (r, θ, φ) are defined by

$$x_1 = r\sin\theta\cos\phi, \qquad x_2 = r\sin\theta\sin\varphi, \qquad x_3 = r\cos\theta. \tag{1.52}$$

In spherical coordinates we have the strain-displacement relation

$$S_{rr} = \frac{\partial u_r}{\partial r}, \qquad S_{\theta\theta} = \frac{1}{r}\frac{\partial u_\theta}{\partial\theta} + \frac{u_r}{r}, \qquad S_{\varphi\varphi} = \frac{1}{r\sin\theta}\frac{\partial u_\varphi}{\partial\varphi} + \frac{u_r}{r} + \frac{u_\theta}{r}\cot\theta, \tag{1.53}$$

$$2S_{r\theta} = \frac{\partial u_\theta}{\partial r} + \frac{1}{r}\frac{\partial u_r}{\partial\theta} - \frac{u_\theta}{r}, \qquad 2S_{\theta\varphi} = \frac{1}{r}\frac{\partial u_\varphi}{\partial\theta} + \frac{1}{r\sin\theta}\frac{\partial u_\theta}{\partial\varphi} - \frac{u_\varphi}{r}\cot\theta,$$

$$2S_{\varphi r} = \frac{1}{r\sin\theta}\frac{\partial u_r}{\partial\varphi} + \frac{\partial u_\varphi}{\partial r} - \frac{u_\varphi}{r}. \tag{1.54}$$

The electric field-potential relation is

$$E_r = -\frac{\partial\phi}{\partial r}, \qquad E_\theta = -\frac{1}{r}\frac{\partial\phi}{\partial\theta}, \qquad E_\varphi = -\frac{1}{r\sin\theta}\frac{\partial\phi}{\partial\varphi}. \tag{1.55}$$

The equations of motion are

$$\frac{\partial T_{rr}}{\partial r} + \frac{1}{r}\frac{\partial T_{\theta r}}{\partial\theta} + \frac{1}{r\sin\theta}\frac{\partial T_{\varphi r}}{\partial\varphi} + \frac{1}{r}(2T_{rr} - T_{\theta\theta} - T_{\varphi\varphi} + T_{\theta r}\cot\theta) + \rho f_r$$

$$= \rho\ddot{u}_r, \tag{1.56}$$

$$\frac{\partial T_{r\theta}}{\partial r} + \frac{1}{r}\frac{\partial T_{\theta\theta}}{\partial\theta} + \frac{1}{r\sin\theta}\frac{\partial T_{\varphi\theta}}{\partial\varphi} + \frac{1}{r}[3T_{r\theta} + (T_{\theta\theta} - T_{\varphi\varphi})\cot\theta] + \rho f_\theta$$

$$= \rho\ddot{u}_\theta, \tag{1.57}$$

$$\frac{\partial T_{r\varphi}}{\partial r} + \frac{1}{r}\frac{\partial T_{\theta\varphi}}{\partial\theta} + \frac{1}{r\sin\theta}\frac{\partial T_{\varphi\varphi}}{\partial\varphi} + \frac{1}{r}(3T_{r\varphi} + 2T_{\theta\varphi}\cot\theta) + \rho f_z$$

$$= \rho\ddot{u}_\varphi. \tag{1.58}$$

The electrostatic charge equation is

$$r^2\frac{\partial}{\partial r}(r^2 D_r) + \frac{1}{r\sin\theta}\frac{\partial}{\partial\theta}(D_\theta\sin\varphi) + \frac{1}{r\sin\theta}\frac{\partial}{\partial\varphi}D_\varphi = q. \tag{1.59}$$

1.10 Compact Matrix Notation

We now introduce a compact matrix notation [1, 2]. This notation consists of replacing pairs of indices, ij or kl, by single indices, p or q, where i, j, k, and l take the values of 1, 2, and 3; and p and q take the values of 1, 2, 3, 4, 5, and 6 according to

$$
\begin{aligned}
&ij \text{ or } kl\text{: } 11 \; 22 \; 33 \; 23 \text{ or } 32 \; 31 \text{ or } 13 \; 12 \text{ or } 21\\
&p \text{ or } q\text{: } \; 1 \quad 2 \quad 3 \qquad 4 \qquad\quad 5 \qquad\quad 6
\end{aligned}
\tag{1.60}
$$

Thus

$$
c_{ijkl} \rightarrow c_{pq}, \quad e_{ikl} \rightarrow e_{ip}, \quad T_{ij} \rightarrow T_p.
\tag{1.61}
$$

For the strain tensor, we introduce S_p such that

$$
\begin{aligned}
S_1 &= S_{11}, & S_2 &= S_{22}, & S_3 &= S_{33},\\
S_4 &= 2S_{23}, & S_5 &= 2S_{31}, & S_6 &= 2S_{12}.
\end{aligned}
\tag{1.62}
$$

The constitutive relations can then be written as

$$
T_p = c_{pq}^E S_q - e_{kp} E_k, \qquad D_i = e_{iq} S_q + \varepsilon_{ik}^S E_k.
\tag{1.63}
$$

In matrix form, Equation (1.63) becomes

$$
\begin{Bmatrix} T_1 \\ T_2 \\ T_3 \\ T_4 \\ T_5 \\ T_6 \end{Bmatrix}
=
\begin{pmatrix}
c_{11}^E & c_{12}^E & c_{13}^E & c_{14}^E & c_{15}^E & c_{16}^E \\
c_{21}^E & c_{22}^E & c_{23}^E & c_{24}^E & c_{25}^E & c_{26}^E \\
c_{31}^E & c_{32}^E & c_{33}^E & c_{34}^E & c_{35}^E & c_{36}^E \\
c_{41}^E & c_{42}^E & c_{43}^E & c_{44}^E & c_{45}^E & c_{46}^E \\
c_{51}^E & c_{52}^E & c_{53}^E & c_{54}^E & c_{55}^E & c_{56}^E \\
c_{61}^E & c_{62}^E & c_{63}^E & c_{64}^E & c_{65}^E & c_{66}^E
\end{pmatrix}
\begin{Bmatrix} S_1 \\ S_2 \\ S_3 \\ S_4 \\ S_5 \\ S_6 \end{Bmatrix}
-
\begin{pmatrix}
e_{11} & e_{21} & e_{31} \\
e_{12} & e_{22} & e_{32} \\
e_{13} & e_{23} & e_{33} \\
e_{14} & e_{24} & e_{34} \\
e_{15} & e_{25} & e_{35} \\
e_{16} & e_{26} & e_{36}
\end{pmatrix}
\begin{Bmatrix} E_1 \\ E_2 \\ E_3 \end{Bmatrix},
\tag{1.64}
$$

$$
\begin{Bmatrix} D_1 \\ D_2 \\ D_3 \end{Bmatrix}
=
\begin{bmatrix}
e_{11} & e_{12} & e_{13} & e_{14} & e_{15} & e_{16} \\
e_{21} & e_{22} & e_{23} & e_{24} & e_{25} & e_{26} \\
e_{31} & e_{32} & e_{33} & e_{34} & e_{35} & e_{36}
\end{bmatrix}
\begin{Bmatrix} S_1 \\ S_2 \\ S_3 \\ S_4 \\ S_5 \\ S_6 \end{Bmatrix}
+
\begin{pmatrix}
\varepsilon_{11}^S & \varepsilon_{12}^S & \varepsilon_{13}^S \\
\varepsilon_{21}^S & \varepsilon_{22}^S & \varepsilon_{23}^S \\
\varepsilon_{31}^S & \varepsilon_{32}^S & \varepsilon_{33}^S
\end{pmatrix}
\begin{Bmatrix} E_1 \\ E_2 \\ E_3 \end{Bmatrix}.
\tag{1.65}
$$

Similarly, Equations (1.10), (1.12), and (1.13) can also be written in matrix form. The matrices of the material constants in various expressions are related by

$$
\begin{aligned}
c_{pr}^E s_{qr}^E &= \delta_{pq}, & c_{pr}^D s_{qr}^D &= \delta_{pq},\\
\beta_{ik}^S \varepsilon_{jk}^S &= \delta_{ij}, & \beta_{ik}^T \varepsilon_{jk}^T &= \delta_{ij},
\end{aligned}
\tag{1.66}
$$

$$c_{pq}^{D} = c_{pq}^{E} + e_{kp}h_{kq}, \qquad s_{pq}^{D} = s_{pq}^{E} - d_{kp}g_{kq},$$
$$\varepsilon_{ij}^{T} = \varepsilon_{ij}^{S} + d_{iq}e_{jq}, \qquad \beta_{ij}^{T} = \beta_{ij}^{S} - g_{iq}h_{jq}, \tag{1.67}$$
$$e_{ip} = d_{iq}c_{pq}^{E}, \qquad d_{ip} = \varepsilon_{ik}^{T}g_{kp},$$
$$g_{ip} = \beta_{ik}^{T}d_{kp}, \qquad h_{ip} = g_{ig}c_{qp}^{D}. \tag{1.68}$$

References

[1] Meitzler AH, Tiersten HF, Warner AW et al. (1988) *IEEE Standard on Piezoelectricity.* IEEE, New York
[2] Tiersten HF (1969) *Linear Piezoelectric Plate Vibrations.* Plenum, New York
[3] Yang JS (2005) *An Introduction to the Theory of Piezoelectricity.* Springer, New York
[4] Tiersten HF (1971) On the nonlinear equations of thermoelectroelasticity. *Int J Engng Sci* 9: 587–604
[5] Lax M, Nelson DF (1971) Linear and nonlinear electrodynamics in elastic anisotropic dielectrics. *Phys Rev B* 4:3694–3731
[6] Holland R, EerNisse EP (1969) *Design of Resonant Piezoelectric Devices.* MIT Press, Cambridge, MA

Chapter 2
Green's Functions

Ernian Pan

2.1 Introduction

Coupling between mechanical and electric fields has stimulated interesting research related to the microelectromechanical system [1, 2]. The major applications are in sensor and actuator devices by which an electric voltage can induce an elastic deformation and vice versa. Because many novel materials, such as the nitride group semiconductors, are piezoelectric, study on quantum nanostructures is currently a cutting-edge topic with the strain energy band engineering in the center [3, 4]. Novel laminated composites (with adaptive and smart components) are continuously attracting great attention from mechanical, aerospace, and civil engineering branches [5]. In materials property study, the Eshelby-based micromechanics theory has been very popular [6]. In most of these exciting research topics, the fundamental solution of a given system under a unit concentrated force/charge or simply the Green's function solution is required. This motivates the writing of this chapter. In this chapter, however, only the static case with general anisotropic piezoelectricity is considered, even though a couple of closely related references on vibration and/or dynamics (time-harmonic) wave propagation are briefly reviewed. Furthermore, although emphasis is given to the generalized point and line forces, the Green's functions to the corresponding point and line dislocations, as well as point and line eigenstrain are also discussed or presented based on Betti's reciprocal theorem.

Ernian Pan
Department of Civil Engineering, Dept. of Applied Mathematics, The University of Akron, Akron, OH 44325-3905, USA, e-mail: pan2@uakron.edu

J. Yang (ed.), *Special Topics in the Theory of Piezoelectricity*, DOI: 10.1007/978-0-387-89498-0_2, 13
© Springer Science + Business Media, LLC 2009

2.2 Governing Equations

Consider a linear, anisotropic piezoelectric and heterogeneous solid occupying the domain V bounded by the boundary S. In discussing the Green's functions, the problem domain and the corresponding boundary conditions are clearly described later. We also assume that the deformation is static, and thus the field equations for such a solid consist of [7]:

(a) *Equilibrium equations (including Gauss equation):*

$$\sigma_{ji,j} + f_i = 0 \qquad D_{i,i} - q = 0, \tag{2.1}$$

where σ_{ij} and D_i are the stress and electric displacement, respectively; f_i and q are the body force and electric charge, respectively. In this and the following sections, summation from 1 to 3 (1 to 4) over repeated lowercase (uppercase) subscripts is implied. A subscript comma denotes the partial differentiation.

In the Cartesian coordinate system, the equilibrium equations are

$$\frac{\partial \sigma_{xx}}{\partial x} + \frac{\partial \sigma_{xy}}{\partial y} + \frac{\partial \sigma_{xz}}{\partial z} + f_x = 0$$

$$\frac{\partial \sigma_{yx}}{\partial x} + \frac{\partial \sigma_{yy}}{\partial y} + \frac{\partial \sigma_{yz}}{\partial z} + f_y = 0$$

$$\frac{\partial \sigma_{zx}}{\partial x} + \frac{\partial \sigma_{zy}}{\partial y} + \frac{\partial \sigma_{zz}}{\partial z} + f_z = 0 \tag{2.2a}$$

$$\frac{\partial D_x}{\partial x} + \frac{\partial D_y}{\partial y} + \frac{\partial D_z}{\partial z} - q = 0. \tag{2.2b}$$

In the cylindrical coordinate system, the equilibrium equations are

$$\frac{\partial \sigma_{rr}}{\partial r} + \frac{\partial \sigma_{r\theta}}{r\partial \theta} + \frac{\partial \sigma_{rz}}{\partial z} + \frac{\sigma_{rr} - \sigma_{\theta\theta}}{r} + f_r = 0$$

$$\frac{\partial \sigma_{r\theta}}{\partial r} + \frac{\partial \sigma_{\theta\theta}}{r\partial \theta} + \frac{\partial \sigma_{\theta z}}{\partial z} + \frac{2\sigma_{r\theta}}{r} + f_\theta = 0 \tag{2.3a}$$

$$\frac{\partial \sigma_{rz}}{\partial r} + \frac{\partial \sigma_{\theta z}}{r\partial \theta} + \frac{\partial \sigma_{zz}}{\partial z} + \frac{\sigma_{rz}}{r} + f_z = 0$$

$$\frac{\partial D_r}{\partial r} + \frac{\partial D_\theta}{r\partial \theta} + \frac{\partial D_z}{\partial_z} - q = 0. \tag{2.3b}$$

(b) *Constitutive relations:*

$$\sigma_{ij} = C_{ijlm}\gamma_{lm} - e_{kji}E_k \qquad D_i = e_{ijk}\gamma_{jk} + \varepsilon_{ij}E_j, \tag{2.4}$$

where γ_{ij} is the strain and E_i the electric field; C_{ijlm}, e_{ijk}, and ε_{ij} are the elastic moduli, piezoelectric coefficients, and dielectric constants, respectively. The uncoupled state (purely elastic and purely electric deformation) can be obtained by simply setting $e_{ijk} = 0$. For transversely isotropic piezoelectric materials with the z-axis being the material symmetric (or the poling) axis, the constitutive relation in the Cartesian coordinate system is (using the reduced indices for C_{ijkl} and e_{ijk}, with the following correspondence between the one and two indices: $1 = 11, 2 = 22, 3 = 33, 4 = 23, 5 = 13, 6 = 12$)

$$\sigma_{xx} = C_{11}\gamma_{xx} + C_{12}\gamma_{yy} + C_{13}\gamma_{zz} - e_{31}E_z$$
$$\sigma_{yy} = C_{12}\gamma_{xx} + C_{11}\gamma_{yy} + C_{13}\gamma_{zz} - e_{31}E_z$$
$$\sigma_{zz} = C_{13}\gamma_{xx} + C_{13}\gamma_{yy} + C_{33}\gamma_{zz} - e_{33}E_z$$
$$\sigma_{yz} = 2C_{44}\gamma_{yz} - e_{15}E_y$$
$$\sigma_{xz} = 2C_{44}\gamma_{xz} - e_{15}E_x$$
$$\sigma_{xy} = 2C_{66}\gamma_{xy} \tag{2.5a}$$
$$D_x = 2e_{15}\gamma_{xz} + \varepsilon_{11}E_x$$
$$D_y = 2e_{15}\gamma_{yz} + \varepsilon_{11}E_y$$
$$D_z = e_{31}(\gamma_{xx} + \gamma_{yy}) + e_{33}\gamma_{zz} + \varepsilon_{33}E_z , \tag{2.5b}$$

where $C_{66} = (C_{11} - C_{12})/2$.

Similarly, in the cylindrical coordinate system, the constitutive relation is

$$\sigma_{rr} = C_{11}\gamma_{rr} + C_{12}\gamma_{\theta\theta} + C_{13}\gamma_{zz} - e_{31}E_z$$
$$\sigma_{\theta\theta} = C_{12}\gamma_{rr} + C_{11}\gamma_{\theta\theta} + C_{13}\gamma_{zz} - e_{31}E_z$$
$$\sigma_{zz} = C_{13}\gamma_{rr} + C_{13}\gamma_{\theta\theta} + C_{33}\gamma_{zz} - e_{33}E_z$$
$$\sigma_{\theta z} = 2C_{44}\gamma_{\theta z} - e_{15}E_\theta$$
$$\sigma_{rz} = 2C_{44}\gamma_{rz} - e_{15}E_r$$
$$\sigma_{r\theta} = 2C_{66}\gamma_{r\theta} \tag{2.6a}$$
$$D_r = 2e_{15}\gamma_{rz} + \varepsilon_{11}E_r$$
$$D_\theta = 2e_{15}\gamma_{\theta z} + \varepsilon_{11}E_\theta$$
$$D_z = e_{31}(\gamma_{rr} + \gamma_{\theta\theta}) + e_{33}\gamma_{zz} + \varepsilon_{33}E_z . \tag{2.6b}$$

(c) *Elastic strain-displacement and electric field-potential relations:*

$$\gamma_{ij} = \frac{1}{2}(u_{i,j} + u_{j,i}) \qquad E_i = -\phi_{,i} , \tag{2.7}$$

where u_i and ϕ are the elastic displacement and electric potential, respectively. In the Cartesian coordinate system, we have

$$\gamma_{xx} = \frac{\partial u_x}{\partial x}, \qquad \gamma_{yy} = \frac{\partial u_y}{\partial y}, \qquad \gamma_{zz} = \frac{\partial u_z}{\partial z} \qquad \gamma_{yz} = 0.5\left(\frac{\partial u_y}{\partial z} + \frac{\partial u_z}{\partial y}\right)$$

$$\gamma_{xz} = 0.5\left(\frac{\partial u_x}{\partial z} + \frac{\partial u_z}{\partial x}\right) \qquad \gamma_{xy} = 0.5\left(\frac{\partial u_x}{\partial y} + \frac{\partial u_y}{\partial x}\right) \tag{2.8a}$$

$$E_x = -\frac{\partial \phi}{\partial x}, \qquad E_y = -\frac{\partial \phi}{\partial y}, \qquad E_z = -\frac{\partial \phi}{\partial z} , \tag{2.8b}$$

and in cylindrical coordinate system, we obtain

$$\gamma_{rr} = \frac{\partial u_r}{\partial r}, \qquad \gamma_{\theta\theta} = \frac{\partial u_\theta}{r\partial\theta} + \frac{u_r}{r}, \qquad \gamma_{zz} = \frac{\partial u_z}{\partial z}$$

$$\gamma_{\theta z} = 0.5\left(\frac{\partial u_\theta}{\partial z} + \frac{\partial u_z}{r\partial\theta}\right) \qquad \gamma_{rz} = 0.5\left(\frac{\partial u_z}{\partial r} + \frac{\partial u_r}{\partial z}\right)$$

$$\gamma_{r\theta} = 0.5\left(\frac{\partial u_r}{r\partial\theta} + \frac{\partial u_\theta}{\partial r} - \frac{u_\theta}{r}\right) \tag{2.9a}$$

$$E_r = -\frac{\partial\phi}{\partial r}, \qquad E_\theta = -\frac{\partial\phi}{r\partial\theta}, \qquad E_z = -\frac{\partial\phi}{\partial z}. \tag{2.9b}$$

The notation introduced by Barnett and Lothe [8] has been shown to be very convenient for the analysis of piezoelectric problems. With this notation, the elastic displacement and electric potential, the elastic strain and electric field, the stress and electric displacement, and the elastic and electric moduli (or coefficients) can be grouped together as [9]

$$u_I = \begin{cases} u_i & I = i = 1, 2, 3 \\ \phi & I = 4 \end{cases} \tag{2.10}$$

$$\gamma_{Ij} = \begin{cases} \gamma_{ij} & I = i = 1, 2, 3 \\ -E_j & I = 4 \end{cases} \tag{2.11}$$

$$\sigma_{iJ} = \begin{cases} \sigma_{ij} & J = j = 1, 2, 3 \\ D_i & J = 4 \end{cases} \tag{2.12a}$$

$$T_J = \sigma_{iJ} n_i = \begin{cases} \sigma_{ij} n_i & J = j = 1, 2, 3 \\ D_i n_i & J = 4 \end{cases} \tag{2.12b}$$

$$C_{iJKl} = \begin{cases} C_{ijkl} & J, K = j, k = 1, 2, 3 \\ e_{lij} & J = j = 1, 2, 3; K = 4 \\ e_{ikl} & J = 4; K = k = 1, 2, 3 \\ -\varepsilon_{il} & J, K = 4 \end{cases}. \tag{2.13}$$

It is noted that we have kept the original symbols instead of introducing new ones because they can be easily distinguished by the range of their subscript. In terms of this shorthand notation, the constitutive relations can be unified into a single equation as

$$\sigma_{iJ} = C_{iJKl}\gamma_{Kl}, \tag{2.14}$$

where the material property coefficients C_{iJKl} can be location-dependent in the region.

Similarly, the equilibrium equations in terms of the extended stresses can be recast into

$$\sigma_{iJ,i} + f_J = 0 \tag{2.15}$$

with the extended force f_J being defined as

$$f_J = \begin{cases} f_j & J = j = 1,2,3 \\ -q & J = 4 \end{cases}. \tag{2.16}$$

For the Green's function solutions, the body force and electric charge density are replaced by the following concentrated unit sources ($k = 1, 2, 3$),

$$f_I = \begin{cases} \delta_{ik}\delta(\boldsymbol{x} - \boldsymbol{x}_0), & I = i = 1,2,3 \\ \delta(\boldsymbol{x} - \boldsymbol{x}_0), & I = 4. \end{cases}. \tag{2.17}$$

It is observed that Equations (2.14) and (2.15) are exactly the same as their purely elastic counterparts. The only difference is the dimension of the index of the involved quantities. Therefore, the solution method developed for anisotropic elasticity can be directly applied to the piezoelectric case. For ease of reference, in this chapter, we still use *displacement* to stand for the elastic displacement and electric potential as defined in Equation (2.10), use *stress* for the stress and electric displacement as defined in Equation (2.12a), and use *traction* for the elastic traction and normal electric displacement as defined in Equation (2.12b).

2.3 Relations Among Different Sources and Their Responses

Relations among different concentrated sources and their responses can be studied via Betti's reciprocal theorem, which states that for two systems (1) and (2) belonging to the same material space, the following relation holds (i.e., [9])

$$\sigma_{iJ}^{(1)} u_{J,i}^{(2)} = \sigma_{iJ}^{(2)} u_{J,i}^{(1)}. \tag{2.18}$$

From (2.18), one can easily derive the following integral equation for these two systems

$$\int_S \sigma_{iJ}^{(1)} u_J^{(2)} n_i dS - \int_V \sigma_{iJ,i}^{(1)} u_J^{(2)} dV = \int_S \sigma_{iJ}^{(2)} u_J^{(1)} n_i dS - \int_V \sigma_{iJ,i}^{(2)} u_J^{(1)} dV. \tag{2.19}$$

We let system (1) be the real boundary value problem and (2) be the corresponding "point-force" Green's function problem; that is,

$$\sigma_{iJ,i} = -\delta_{JK}\delta(x_p^f - x_p^s), \tag{2.20}$$

where the field point is at x_p^f and the extended point force is applied at x_p^s in the K-direction (with $K = 4$ corresponding to a negative electric charge). Then (2.19) can be reduced to a well-known integral representation of the displacement field:

$$u_K(x_p^s) = \int_S [\sigma_{iJ}(x_p^f)u_J^{fK}(x_p^s; x_p^f) = \sigma_{iJ}^{fK}(x_p^s; x_p^f)u_J(x_p^f)]n_i(x_p^f)dS(x_p^f)$$

$$+ \int_V u_J^{fK}(x_p^s; x_p^f)f_J(x_p^f)dV(x_p^f), \tag{2.21}$$

where in the Green's function expressions, the first superscript f denotes that the Green's function corresponds to an extended point force, and the second superscript K is the direction of the point force.

Now, we wish to find the displacement response due to a prescribed dislocation (displacement discontinuity) across a surface Σ embedded in V (or the dislocation Green's function). Let $n_i (= n_i^- = -n_i^+)$ be the unit normal to Σ, $b_I = u_I^+ - u_I^-$ being the (extended) dislocation. This dislocation along Σ may have any form provided that the following (extended) traction continuity condition holds.

$$\sigma_{iJ}^+ n_i^+ + \sigma_{iJ}^- n_i^- = 0. \tag{2.22}$$

This type of displacement discontinuity is also called a Somigliana dislocation with the Volterra dislocation (or the dislocation of Volterra–Weingarten) being its special case [10]. In the latter case,

$$\triangle u_I \equiv u_I^+ - u_I^- = \begin{cases} U_i + \Omega_{ij}x_j^s; & I \leq 3, \\ U_4; & I = 4 \end{cases} \tag{2.23}$$

where U_I and Ω_{ij} are constants. If, furthermore, $\Omega_{ij} = 0$, the dislocation then reduces to the (extended) Burger's vector. Assume that the displacement and stress fields satisfy the same homogeneous boundary condition on the outer boundary S, and apply Equation (2.21) to the region bounded internally by Σ and externally by S; we then come to (also omit the volumetric integral term associated with the body force)

$$u_K(x_p^s) = \int_\Sigma \sigma_{iJ}^{fK}(x_p^s; x_p^f)b_J(x_p^f)n_i(x_p^f)d\Sigma(x_p^f). \tag{2.24}$$

This is the integral expression of the displacement due to a dislocation along Σ with its density tensor being defined as

$$D(i, J) \equiv n_i(x_p^f)b_J(x_p^f)d\Sigma(x_p^f). \tag{2.25}$$

In (elastic) seismology, they are defined as fault elements $d\Sigma$ with an outward normal n_i (defined with respect to the positive side of the fault) having a displacement discontinuity $\triangle_i = b_i$. It is noted that the kernel function in Equation (2.24) is the Green's stress with component (iJ) at the field point x_p^f due to a point force at x_p^s in the Kth direction. Alternatively, the displacement response due to the dislocation density tensor can also be expressed by the kernel displacement function due to a point dislocation, namely,

$$u_K(x_p^f) = \int_\Sigma u_K^{diJ}(x_p^s; x_p^f) b_J(x_p^s) n_i(x_p^s) d\Sigma(x_p^s), \qquad (2.26)$$

where the superscript d denotes dislocation and (i, J) is the normal direction of the dislocation plane (i) and the Burger's vector direction (J). Comparing Equation (2.26) to (2.24), we immediately obtain the following important equivalence between the stress due to a point force and the displacement due to a point dislocation

$$u_K^{diJ}(x_p^s; x_p^f) = \sigma_{iJ}^{fK}(x_p^f; x_p^s). \qquad (2.27)$$

That is, the position of the source and field points in the point-force Green's stresses need to be exchanged in order to obtain the point-dislocation Green's displacements. This is a most simple and yet very important relation. Similar results for poroelastic media were derived by Pan [11]. Several important observations of (2.27) are listed below:

(a) In general, once the point-force Green's functions are solved, the corresponding point-dislocation Green's functions can be obtained through the relation (2.27). In deriving relation (2.27), we have assumed that the system is linear piezoelectric, but can be of general *anisotropy* and *heterogeneity*. In particular, this relation can be used to derive the point-dislocation Green's functions in horizontally layered systems, including half-space and bi-material domains as special cases. For example, the point-dislocation Green's functions in horizontally layered media can be derived in both the Fourier transform or the physical domains using Equation (2.27) and the point force solutions [12–14].

(b) Direct solution of the point-dislocation Green's functions is also possible but the procedure may be very complicated. The way to achieve this is to derive the equivalent body force of the point dislocation, find the related discontinuity of the physical quantities, and solve for the unknowns, using the method as previously employed by Pan [12] for the transversely isotropic and layered half-space.

(c) For the elastic isotropic or transversely isotropic bimaterial, half-space, or full-space, each term on the right-hand side of Equation (2.27) is proportional to various eigenstrains, such as the misfit lattice strain, the nucleus of strain (or a nucleus of strain multiplied by the elastic constants), and so on. With Equation (2.27), however, it is unnecessary to add all the related nuclei of strain together and enforce the boundary or interface condition to solve the coefficients involved.

(d) In using Equation (2.27), one must be very careful that on the left-hand side, x_p^s and x_p^f are the source and field points, respectively; and on the right-hand side, x_p^f and x_p^s are the field and source points, respectively.

Therefore, the Green's displacements due to a point dislocation can be obtained from the Green's stresses due to a point force by exchanging the position of the field and source points and by assigning the suitable meanings to the associated indexes.

(e) For a homogeneous and infinite domain, expressing the point-force Green's stresses by the strain and substituting the result back to Equation (2.26), we then have the Volterra relation. It is noted that for this specific case, the point-force Green's stresses are functions of the relative vector from the source to field points $\left(\text{i.e., } x_p^f - x_p^s\right)$, and they satisfy the following relation,

$$\sigma_{iJ}^{fK}(x_P^f; x_p^s) = \sigma_{iJ}^{fK}(x_p^s - x_p^f) = -\sigma_{iJ}^{fK}(x_p^f - x_p^s) = -\sigma_{iJ}^{fK}(x_p^s; x_p^f). \quad (2.28)$$

We therefore have

$$u_K^{diJ}(x_p^s; x_p^f) = -\sigma_{iJ}^{fK}(x_p^s; x_p^f). \quad (2.29)$$

It should be emphasized that only for the homogeneous infinite domain, can the Green's displacements due to a point dislocation be obtained directly from the Green's stresses due to a point force, without exchanging the field and source positions! For all other situations, the dislocation-induced Green displacements should be obtained strictly using Equation (2.27).

For a homogeneous and infinite solid of purely elastic isotropy, (2.29) is reduced to

$$u_k^{dij}(x_p^s; x_p^f) = -\sigma_{ij}^{fk}(x_p^s; x_p^f) = -\lambda u_{l,l}^k \delta_{ij} - \mu(u_{i,j}^k + u_{j,i}^k), \quad (2.30)$$

where λ and μ are the two Lame's elastic constants. The derivatives of the Green's elastic displacements due to a point force at x_p^s in the k-direction are taken with respect to the field point x_p^f.

On the other hand, if we take the derivatives of the point-force Green displacements with respect to the source point x_p^s, we then have

$$\varepsilon^{ij}(x_p^s; x_p^f) = \lambda u_{l,l}^k \delta_{ij} + \mu(u_{i,j}^k + u_{j,i}^k), \quad (2.31)$$

which has an opposite sign to that given by (2.30).

The components in (2.31) are called nuclei of strain by Mindlin [15]. The first term corresponds to the center of compression or dilatation and the other terms are double forces. Therefore, from the physical point of view, the point-dislocation Green's functions can be constructed through super position of various nuclei of strain (or the derivatives of the point-force Green's displacements), with their coefficients being solely related to the elastic constants. This is a physical explanation for the mathematical and arbitrary equivalent relation (2.29). It is obvious that if the point-force Green's functions can be derived in an exact closed-form (or explicit form), the corresponding point dislocation solutions will also have the same features because they are obtained by the superposition of various nuclei of strain. Detailed analyses can be found in [10, 16, 17] for the isotropic elastic case and in [18] for the transversely isotropic elastic case. For materials of either isotropy or transverse isotropy, these exact closed-form nuclei of strain can also be employed to derive the point-force Green's functions in a half-space or a bimaterial space, and to derive the solutions corresponding to various inclusions.

It is noted that the Green's function relations between those due to a point dislocation and those due to a point force are applicable to 3D only. For the 2D case, the Green's functions due to the line force and those due to the line dislocations (open all the way to the half-infinite line) have the same singularity order. These are clearly observed in the following section on 2D Green's functions, and one should pay particular attention to this difference.

2.4 Green's Functions in Anisotropic Two-Dimensional Infinite, Half, and Bimaterial Planes

Our two-dimensional (2D) problem is in the x–z plane, and it is under the assumption that all the field and source quantities are independent of the y-variable (i.e., $\partial()/\partial y = 0$). Therefore, the Green's functions presented are rigorously for the generalized 2D plane strain case. Furthermore, presented below are only the displacements and tractions (on the $z = $ constant plane) based on the Stroh formalism in terms of the complex variables. Summation of the repeated subscript R from 1 to 4 is implied.

2.4.1 Green's Functions in Anisotropic 2D Infinite Planes Due to a Line Force and Line Dislocation

The Green's functions for the displacements and tractions at field point $\boldsymbol{x}(x, z)$ due to a line force at $\boldsymbol{X}(X, Z)$ can be expressed as

$$U_{KJ}(\boldsymbol{x}, \boldsymbol{X}) = \frac{1}{\pi} \operatorname{Im}\{A_{JR} \ln(z_R - s_R) A_{KR}\} \tag{2.32a}$$

$$T_{KJ}(\boldsymbol{x}, \boldsymbol{X}) = -\frac{1}{\pi} \operatorname{Im}\{B_{JR} \frac{p_R n_1 - n_3}{z_R - s_R} A_{KR}\}. \tag{2.32b}$$

The first subscript K is the component of the line force of unit value $\boldsymbol{f} = (f_1, f_2, f_3, -Q)$ and the second subscript J is the component of the displacement (2.32a) and the traction (2.32b). Also in these equations, "Im" stands for the imaginary part of the complex value; A_{IJ} and B_{IJ} are two constant matrices related to the piezoelectric material property; n_1 and n_3 (functions of $\boldsymbol{x}$) are the unit outward normal components along the x- and

z-directions; p_R ($R = 1, 2, 3, 4$) are the Stroh eigenvalues; and $z_R = x + p_R z$ and $s_R = X + p_R Z$ are related to the field $\boldsymbol{x}(x, z)$ and source $\boldsymbol{X}(X, Z)$ points, respectively. These displacement and traction Green's functions are required in the conventional boundary integral equation formulation to solve the general boundary value problems in piezoelectric solids. In order to find the Green's strain and electric fields, one only needs to take the derivation of the Green's displacement (2.32a) with respect to the field point $\boldsymbol{x}$ (refer to (2.7)). The corresponding stress and electric displacements can be obtained thorough the piezoelectric constitutive relations (2.14).

The Stroh eigenvalues and eigenmatrices involved in (2.32) are obtained by solving the following eigensystem of equations [19]. First, the eigenvalue p and the corresponding eigenvector $\boldsymbol{a}$ are solved from the eigenrelation:

$$[\boldsymbol{Q} + p(\boldsymbol{R} + \boldsymbol{R}^T) + p^2 \boldsymbol{T}]\boldsymbol{a} = 0, \tag{2.33}$$

where the superscript T denotes matrix transpose, and

$$Q_{IK} = C_{1IK1}, \qquad R_{IK} = C_{1IK3}, \qquad T_{IK} = C_{3IK3}, \tag{2.34}$$

where C_{iJKl} are the elastic and electric moduli (or coefficients) defined in (2.13).

Then, the eigenvector $\boldsymbol{b}$ is obtained from

$$\boldsymbol{b} = (\boldsymbol{R}^T + p\boldsymbol{T})\boldsymbol{a} = -\frac{1}{p}(\boldsymbol{Q} + p\boldsymbol{R})\boldsymbol{a}. \tag{2.35}$$

Denoting by p_m, $\boldsymbol{a}_m$, and $\boldsymbol{b}_m$ ($m = 1, 2, \ldots, 8$) the eigenvalues and the associated eigenvectors, we then order them in such a way so that

$$\text{Im } p_J > 0, \qquad p_{J+4} = \overline{p}_J, \qquad \boldsymbol{a}_{J+4} = \overline{\boldsymbol{a}}_J, \qquad \boldsymbol{b}_{J+4} = \overline{\boldsymbol{b}}_J \qquad (J = 1, 2, 3, 4)$$

$$\boldsymbol{A} = [\boldsymbol{a}_1, \boldsymbol{a}_2, \boldsymbol{a}_3, \boldsymbol{a}_4], \qquad \boldsymbol{B} = [\boldsymbol{b}_1, \boldsymbol{b}_2, \boldsymbol{b}_3, \boldsymbol{b}_4], \tag{2.36}$$

where an overbar denotes the complex conjugate. We have also assumed that p_J are distinct and the eigenvectors $\boldsymbol{a}_J$ and $\boldsymbol{b}_J$ satisfy the normalization relation [8, 19]

$$\boldsymbol{b}_I^T \boldsymbol{a}_J + \boldsymbol{a}_I^T \boldsymbol{b}_J = \delta_{IJ} \tag{2.37}$$

with δ_{IJ} being the 4×4 Kronecker delta (i.e., the 4×4 identity matrix). We also remark that repeated eigenvalues p_J can be avoided by using slightly perturbed material coefficients with negligible errors [20]. In so doing, the simple structure of the Green's function solutions (2.32) can always be employed.

Similarly, the Green's functions at $\boldsymbol{x}$ due to the generalized line dislocations (Burger's vector and electric potential discontinuity) $\boldsymbol{b} = (\triangle u_1, \triangle u_2, \triangle u_3, \triangle \phi)$ of unit value at $\boldsymbol{X}$ can be found as

$$U_{KJ}(\boldsymbol{x}, \boldsymbol{X}) = \frac{1}{\pi} \text{Im} \left\{ A_{JR} \ln(z_R - s_R) B_{KR} \right\} \tag{2.38a}$$

$$T_{KJ}(\boldsymbol{x}, \boldsymbol{X}) = -\frac{1}{\pi}\mathrm{Im}\left\{B_{JR}\frac{p_R n_1 - n_3}{z_R - s_R}B_{KR}\right\}. \qquad (2.38b)$$

Comparing (2.32) and (2.38), we notice that the line force and line dislocation Green's functions are very similar to each other and they both have the same order of singularity.

2.4.2 Green's Functions in Anisotropic 2D Half-Planes Due to a Line Force and Line Dislocation

The half-plane Green's functions for the displacements and tractions (the Jth component) with outward normal n_1 and n_3 (at $\boldsymbol{x}$) due to a line force at $\boldsymbol{X}$ with component K can be expressed as

$$U_{KJ}(\boldsymbol{x}, \boldsymbol{X}) = \frac{1}{\pi}\mathrm{Im}\left\{A_{JR}\ln(z_R - s_R)A_{KR} + \sum_{v=1}^{4}\left[A_{JR}\ln(z_R - \overline{s}_v)Q_{RK}^v\right]\right\}$$
$$(2.39a)$$

$$T_{KJ}(\boldsymbol{x}, \boldsymbol{X}) = -\frac{1}{\pi}\mathrm{Im}\left\{B_{JR}\frac{p_R n_1 - n_3}{z_R - s_R}A_{KR} + \sum_{v=1}^{4}\left[B_{JR}\frac{p_R n_1 - n_3}{z_R - \overline{s}_v}Q_{RK}^v\right]\right\},$$
$$(2.39b)$$

where

$$Q_{RN}^v = B_{RS}^{-1}\overline{B}_{SP}(I_v)_P\overline{A}_{NP} \qquad (2.40)$$

with

$$\begin{aligned}
\boldsymbol{I}_1 &= \mathrm{diag}[1,0,0,0]; & \boldsymbol{I}_2 &= \mathrm{diag}[0,1,0,0] \\
\boldsymbol{I}_3 &= \mathrm{diag}[0,0,1,0]; & \boldsymbol{I}_4 &= \mathrm{diag}[0,0,0,1].
\end{aligned} \qquad (2.41)$$

Similarly, the half-plane Green's functions for the displacements and tractions (the Jth component) with outward normal n_1 and n_3 (at $\boldsymbol{x}$) due to the line dislocations at $\boldsymbol{X}$ with component K can be expressed as

$$U_{KJ}(\boldsymbol{x}, \boldsymbol{X}) = \frac{1}{\pi}\mathrm{Im}\left\{A_{JR}\ln(z_R - s_R)B_{KR} + \sum_{v=1}^{4}\left[A_{JR}\ln(z_R - \overline{s}_v)Q_{RK}^v\right]\right\}$$
$$(2.42a)$$

$$T_{KJ}(\boldsymbol{x}, \boldsymbol{X}) = -\frac{1}{\pi}\mathrm{Im}\left\{B_{JR}\frac{p_R n_1 - n_3}{z_R - s_R}B_{KR} + \sum_{v=1}^{4}\left[B_{JR}\frac{p_R n_1 - n_3}{z_R - \overline{s}_v}Q_{RK}^v\right]\right\},$$
$$(2.42b)$$

where

$$Q^v_{RN} = B^{-1}_{RS}\overline{B}_{SP}(I_v)_P\overline{B}_{NP}. \tag{2.43}$$

Similar Green's function expressions can be obtained for the general boundary conditions on the surface of the anisotropic elastic and anisotropic piezoelectric half-plane. The detailed discussions can be found in [21, 22].

2.4.3 Green's Functions in 2D Anisotropic Bimaterial Plane Due to a Line Force and Line Dislocation

Depending upon the relative locations of the source and field points, there are four sets of Green's functions for the bimaterial case. We assume that materials 1 and 2 occupy the half-planes $z > 0$ and $z < 0$, respectively. Let us again assume that a line force $\boldsymbol{f} = (f_1, f_2, f_3, -Q)$ or line dislocation $\boldsymbol{b} = (\triangle u_1, \triangle u_2, \triangle u_3, \triangle \phi)$ is applied at the source point (X, Z) in one of the half-planes. To derive the Green's functions, it is sufficient to find the displacement vector $\boldsymbol{u}$ and traction vector $\boldsymbol{t}$ due to the line force or dislocation [19], which are presented below for different combinations of the source and field points.

Assume that the source point (X, Z) is in the half-plane of material $\lambda(\lambda = 1 \text{ or } 2)$. Then if the field point $\boldsymbol{x} = (x, z)$ is in the source plane (i.e., the half-plane of material λ), the displacement and traction vectors can be expressed as [23]

$$\boldsymbol{u}^{(\lambda)} = \frac{1}{\pi}\text{Im}\left\{\boldsymbol{A}^{(\lambda)}\langle\ln(z^{(\lambda)}_* - s^{(\lambda)}_*)\rangle\boldsymbol{q}^{\infty,\lambda}\right\} + \frac{1}{\pi}\text{Im}\sum_{J=1}^{4}\left\{\boldsymbol{A}^{(\lambda)}\langle\ln(z^{(\lambda)}_* - \overline{s}^{(\lambda)}_J)\rangle\boldsymbol{q}^{(\lambda)}_J\right\}$$

$$\boldsymbol{t}^{(\lambda)} = -\frac{1}{\pi}\text{Im}\left\{\boldsymbol{B}^{(\lambda)}\left\langle\frac{p^{(\lambda)}_* n_1 - n_3}{z^{(\lambda)}_* - s^{(\lambda)*}}\right\rangle\boldsymbol{q}^{\infty,\lambda}\right\} - \frac{1}{\pi}\text{Im}\sum_{J=1}^{4}\left\{\boldsymbol{B}^{(\lambda)}\left\langle\frac{p^{(\lambda)}_* n_1 - n_3}{z^{(\lambda)}_* - \overline{s}^{(\lambda)}_*}\right\rangle\boldsymbol{q}^{(\lambda)}_J\right\}. \tag{2.44}$$

If the field point (x, z) is in the other half-plane of material $\mu(\mu \neq \lambda)$ $(\lambda, \mu = 1 \text{ or } 2)$, then the displacement and traction vectors can be expressed as

$$\boldsymbol{u}^{(\mu)} = \frac{1}{\pi}\text{Im}\sum_{J=1}^{4}\left\{\boldsymbol{A}^{(\mu)}\langle\ln(z^{(\mu)}_* - s^{(\lambda)}_J)\rangle\boldsymbol{q}^{(\mu)}_J\right\}$$

$$\boldsymbol{t}^{(\mu)} = -\frac{1}{\pi}\text{Im}\sum_{J=1}^{4}\left\{\boldsymbol{B}^{(\mu)}\left\langle\frac{p^{(\mu)}_* n_1 - n_3}{z^{(\mu)}_* - s^{(\lambda)}_J}\right\rangle\boldsymbol{q}^{(\mu)}_J\right\}. \tag{2.45}$$

In (2.44) and (2.45), the superscripts (λ) and (μ) denote the quantities associated with the material domains 1 and 2; $p^{(\lambda)}_J$, $\boldsymbol{A}^\lambda$, and $\boldsymbol{B}^{(\lambda)}$ $(\lambda = 1 \text{ and } 2$ for the two half-planes) are the Stroh eigenvalues and the corresponding eigenmatrices as given before. Also in (2.44) and (2.45), we defined:

$$\langle \ln(z_*^{(\lambda)} - s_*^{(\lambda)}) \rangle = \mathrm{diag}[\ln(z_1^{(\lambda)} - s_1^{(\lambda)}), \ln(z_2^{(\lambda)} - s_2^{(\lambda)}),$$
$$\ln(z_3^{(\lambda)} - s_3^{(\lambda)}), \ln(z_4^{(\lambda)} - s_4^{(\lambda)})]$$
$$\langle \ln(z_*^{(\lambda)} - \overline{s}_*^{(\lambda)}) \rangle = \mathrm{diag}[\ln(z_1^{(\lambda)} - \overline{s}_J^{(\lambda)}), \ln(z_2^{(\lambda)} - \overline{s}_J^{(\lambda)}),$$
$$\ln(z_3^{(\lambda)} - \overline{s}_J^{(\lambda)}), \ln(z_4^{(\lambda)} - \lambda\overline{s}_J^{(\lambda)})], \tag{2.46}$$

where z_J^α and $s_J^{(\alpha)}$ ($\alpha = 1, 2$) are complex variables associated with the field and source points, respectively. They are defined as

$$z_J^{(\alpha)} = x + p_J^{(\alpha)} z,$$
$$s_J^{(\alpha)} = X + p_J^{(\alpha)} Z. \tag{2.47}$$

We further observe that the first term in (2.44) corresponds to the full-plane Green's functions in material λ with:

$$q^{\infty,\lambda} = (A^{(\lambda)})^T f \tag{2.48}$$

for the line force, and

$$q^{\infty, \lambda} = (B^{(\lambda)})^T b \tag{2.49}$$

for the line dislocation.

The second term in (2.44) and the term in (2.45) are the complementary parts of the Green's function solutions. The complex vectors $q_J^{(\lambda)}$ ($\lambda = 1, 2$; $J = 1, 2, 3, 4$) in (2.44) and $q_J^{(\mu)}$ ($\mu = 1, 2$; $J = 1, 2, 3, 4$) in (2.45) are determined using the continuity conditions along the interface of the two half-planes. Assume that the interface is perfect and after certain algebraic calculations, these vectors can be obtained as ($\lambda, \mu = 1$ or 2, but $\mu \neq \lambda$):

$$q_J^{(\lambda)} = (A^{(\lambda)})^{-1}(M^{(\lambda)} + \overline{M}^{(\mu)})^{-1}(\overline{M}^{(\mu)} - \overline{M}^{(\lambda)})\overline{A}^{(\lambda)}I_J\overline{q}^{\infty,\lambda} \tag{2.50}$$

for (2.44), and

$$q_J^{(\mu)} = (A^{(\mu)})^{-1}(\overline{M}^{(\lambda)} + M^{(\mu)})^{-1}(M^{(\lambda)} + \overline{M}^{(\lambda)})A^{(\lambda)}I_J q^{\infty,\lambda} \tag{2.51}$$

for (2.45).

In (2.50) and (2.51), matrix $M^{(\alpha)}$ is the impedance tensor defined as

$$M^{(\alpha)} = -iB^{(\alpha)}(A^{(\alpha)})^{-1} \qquad (\alpha = 1, 2) \tag{2.52}$$

and the diagonal matrix I_J is the same as that defined by (2.41).

We point out that similar Green's function expressions can be derived for the bimaterial with general (or imperfect) interface conditions. Detailed discussion can be found in [24, 25].

2.5 Green's Functions in Three-Dimensional Infinite, Half, and Bimaterial Spaces: Transverse Isotropy

Green's functions in 3D transversely isotropic piezoelectric infinite, half, and bimaterial spaces were derived by Ding's group at Zhejiang University [26–34] and Dunn's group at the University of Colorado at Boulder [35–37]. However, results presented below are based on the works by the Zhejiang University group using the combined potential function and trial-and-error method (i.e., [32]). To facilitate the discussion, we first present the basic equations with special parameters associated with transversely isotropic materials only.

For the transversely isotropic piezoelectric material with its poling direction along the z-axis, the corresponding constitutive relation is the one given by (2.5) or (2.6). For this material, the characteristic equation is separated into two: one corresponds to the antiplane deformation and another to the in-plane deformation. For the antiplane case, the characteristic root is given by

$$s_0 = \sqrt{c_{66}/c_{44}}. \tag{2.53}$$

For the in-plane case, its three characteristic roots, $s_i (i = 1, 2, 3)$, are the solutions of the following characteristic equation

$$as^6 - bs^4 + cs^2 - d = 0, \tag{2.54}$$

where

$$
\begin{aligned}
a &= c_{44}(e_{33}^2 + c_{33}\varepsilon_{33}) \\
b &= c_{33}[c_{44}\varepsilon_{11} + (e_{15} + e_{31})^2] + \varepsilon_{33}[c_{11}c_{33} + (c_{44}^2 - (c_{13} + c_{44})^2)] \\
  &\quad + e_{33}[2c_{44}e_{15} + c_{11}e_{33} - 2(c_{13} + c_{44})(e_{15} + e_{31})] \\
c &= c_{44}[c_{11}\varepsilon_{33} + (e_{15} + e_{31})^2] + \varepsilon_{11}[c_{11}c_{33} + (c_{44}^2 - (c_{13} + c_{44})^2)] \\
  &\quad + e_{15}[2c_{11}e_{33} + c_{44}e_{15} - 2(c_{13} + c_{44})(e_{15} + e_{31})] \\
d &= c_{11}(e_{15}^2 + c_{44}\varepsilon_{11}).
\end{aligned}
\tag{2.55}
$$

Other parameters used in this section are:

$$
\begin{aligned}
m_1 &= \varepsilon_{11}(c_{13} + c_{44}) + e_{15}(e_{15} + e_{31}) \qquad m_2 = \varepsilon_{33}(c_{13} + c_{44}) + e_{33}(e_{15} + e_{31}) \\
m_3 &= c_{11}\varepsilon_{33} + c_{44}\varepsilon_{11} + (e_{15} + e_{31})^2 \quad m_4 = c_{11}e_{33} + c_{44}e_{15} - (c_{13} + c_{44}) \\
&\qquad\qquad\qquad\qquad\qquad\qquad\qquad\quad \times (e_{15} + e_{31})
\end{aligned}
\tag{2.56}
$$

$$\alpha_{i1} = \frac{c_{11}\varepsilon_{11} - m_3 s_i^2 + c_{44}\varepsilon_{33}s_i^4}{(m_1 - m_2 s_i^2)s_i} \qquad \text{(for } i = 1, 2, 3)$$

$$\alpha_{i2} = \frac{c_{11}e_{15} - m_4 s_i^2 + c_{44}e_{33}s_i^4}{(m_1 - m_2 s_i^2)s_i} \tag{2.57}$$

$$\omega_{01} = c_{44}s_0; \qquad \omega_{02} = e_{15}s_0 \qquad \omega_{i1} = c_{44}(s_i + \alpha_{i1}) + e_{15}\alpha_{i2}$$

$$\omega_{i2} = e_{15}(s_i + \alpha_{i1}) - \varepsilon_{11}\alpha_{i2} \qquad \theta_{i1} = (c_{33}\alpha_{i1} + e_{33}\alpha_{i2})s_i - c_{13}$$

$$\theta_{i2} = (e_{33}\alpha_{i1} + \varepsilon_{33}\alpha_{i2})s_i - e_{13}. \tag{2.58}$$

We also define the following position-related parameters (for $i, j = 0, 1, 2, 3$),

$$z_i = s_i z; \qquad h_i = s_i h; \qquad z_i' = s_i' z; \qquad z_{ij} = z_i + h_j;$$

$$R_{ij} = \sqrt{x^2 + y^2 + z_{ij}^2}; \qquad \overline{z}_{ij} = z_i - h_j; \qquad \overline{R}_{ij} = \sqrt{x^2 + y^2 + \overline{z}_{ij}^2};$$

$$z_{ij}' = z_i' - h_j; \qquad R_{ij}' = \sqrt{x^2 + y^2 + (z_{ij}')^2}. \tag{2.59}$$

The prime "$'$" here and afterwards denotes parameters or quantities in the lower half-space $z < 0$.

2.5.1 Green's Functions for Infinite Space

For a point charge Q and a point force P in the z-direction, applied at the source point $(0, 0, h)$, the elastic displacements and electric potential at the field point (x, y, z) are

$$u = \text{sign}(z - h) \sum_{i=1}^{3} \frac{A_i x}{\overline{R}_{ii}(\overline{R}_{ii} + s_i|z - h|)}$$

$$v = \text{sign}(z - h) \sum_{i=1}^{3} \frac{A_i y}{\overline{R}_{ii}(\overline{R}_{ii} + s_i|z - h|)}$$

$$w = \sum_{i=1}^{3} \frac{\alpha_{i1} A_i}{\overline{R}_{ii}};$$

$$\phi = \sum_{i=1}^{3} \frac{\alpha_{i2} A_i}{\overline{R}_{ii}}. \tag{2.60}$$

For a point force T in the x-direction, applied at the source point $(0, 0, h)$, the elastic displacements and electric potential at the field point (x, y, z) are

$$u = -D_0 \left[\frac{1}{\overline{R}_{00} + s_0|z - h|} - \frac{y^2}{\overline{R}_{00}(\overline{R}_{00} + s_0|z - h|)^2} \right]$$

$$+ \sum_{i=1}^{3} D_i \left[\frac{1}{\overline{R}_{ii} + s_i|z - h|} - \frac{x^2}{\overline{R}_{ii}(\overline{R}_{ii} + s_i|z - h|)^2} \right]$$

$$v = - \frac{D_0 xy}{\overline{R}_{00}(\overline{R}_{00} + s_0|z - h|)^2} - xy \sum_{i=1}^{3} \frac{D_i}{\overline{R}_{ii}(\overline{R}_{ii} + s_i|z - h|)^2}$$

$$w = - \operatorname{sign}(z - h)x \sum_{i=1}^{3} \frac{\alpha_{i1} D_i}{\overline{R}_{ii}(\overline{R}_{ii} + s_i|z - h|)}$$

$$\phi = - \operatorname{sign}(z - h)x \sum_{i=1}^{3} \frac{\alpha_{i2} D_i}{\overline{R}_{ii}(\overline{R}_{ii} + s_i|z - h|)}. \tag{2.61}$$

In (2.60) and (2.61),

$$A_1 = \frac{[P(\theta_{22} - \theta_{32}) + Q(\theta_{21} - \theta_{31})]}{b_1}$$

$$A_2 = \frac{[P(\theta_{32} - \theta_{12}) + Q(\theta_{31} - \theta_{11})]}{b_1}$$

$$b_1 = 4\pi[(\theta_{11} - \theta_{31})(\theta_{22} - \theta_{32}) - (\theta_{21} - \theta_{31})(\theta_{12} - \theta_{32})]$$

$$A_3 = - A_1 - A_2 = \frac{[P(\theta_{12} - \theta_{22}) + Q(\theta_{11} - \theta_{21})]}{b_1} \tag{2.62}$$

$$D_0 = \frac{-T}{(4\pi c_{44} s_0)} \qquad D_1 = \frac{(\alpha_{21}\alpha_{32} - \alpha_{31}\alpha_{22})T}{b_2}$$

$$D_2 = \frac{(\alpha_{31}\alpha_{12} - \alpha_{11}\alpha_{32})T}{b_2} \qquad D_3 = \frac{(\alpha_{11}\alpha_{22} - \alpha_{21}\alpha_{12})T}{b_2}$$

$$b_2 = 4\pi c_{44}[s_1(\alpha_{21}\alpha_{32} - \alpha_{31}\alpha_{22}) + s_2(\alpha_{31}\alpha_{12} - \alpha_{11}\alpha_{32})$$
$$+ s_3(\alpha_{11}\alpha_{22} - \alpha_{21}\alpha_{12})]. \tag{2.63}$$

2.5.2 Green's Functions for Half and Bimaterial Spaces

We first present the bimaterial space Green's functions. The Green's functions in the corresponding half-space can be reduced from the former ones. We assume that along the interface $z = 0$ of the two half-places, the elastic traction and the z-component of the electric displacement are continuous across the interface (i.e., perfect interface condition). For a point charge Q and a point force P in the z-direction, applied at the source point $(0, 0, h > 0)$, the elastic displacements and electric potential at the field point $(x, y, z > 0)$ of the upper half-space are

$$u = \sum_{i=1}^{3} \left[\operatorname{sign}(z - h)\frac{A_i x}{\overline{R}_{ii}(\overline{R}_{ii} + s_i|z - h|)} + \sum_{j=1}^{3} \frac{A_{ij} x}{R_{ij}(R_{ij} + z_{ij})} \right]$$

$$v = \sum_{i=1}^{3} \left[\operatorname{sign}(z-h) \frac{A_i y}{\overline{R}_{ii}(\overline{R}_{ii} + s_i|z-h|)} + \sum_{j=1}^{3} \frac{A_{ij} y}{R_{ij}(R_{ij} + z_{ij})} \right]$$

$$w = \sum_{i=1}^{3} \alpha_{i1} \left[\frac{A_i}{\overline{R}_{ii}} + \sum_{j=1}^{3} \frac{A_{ij}}{R_{ij}} \right] \qquad \phi = \sum_{i=1}^{3} \alpha_{i2} \left[\frac{A_i}{\overline{R}_{ii}} + \sum_{j=1}^{3} \frac{A_{ij}}{R_{ij}} \right]. \qquad (2.64)$$

At the field point $(x, y, z < 0)$ of the lower half-space, the elastic displacements and electric potential are (the prime "$'$" is added to quantities in the lower half-space)

$$u' = \sum_{i=1}^{3} \sum_{j=1}^{3} \frac{A'_{ij} x}{R'_{ij}(R'_{ij} - z'_{ij})}; \qquad v' = \sum_{i=1}^{3} \sum_{j=1}^{3} \frac{A'_{ij} y}{R'_{ij}(R'_{ij} - z'_{ij})}$$

$$w' = -\sum_{i=1}^{3} \alpha'_{i1} \sum_{j=1}^{3} \frac{A'_{ij}}{R'_{ij}}; \qquad \phi' = -\sum_{i=1}^{3} \alpha'_{i2} \sum_{j=1}^{3} \frac{A'_{ij}}{R'_{ij}}. \qquad (2.65)$$

The coefficients A_{ij} and A'_{ij} in (2.64) and (2.65) are solved from the following equations (for $i = 1, 2, 3; m = 1, 2$).

$$-A_i + \sum_{j=1}^{3} A_{ji} = \sum_{j=1}^{3} A'_{ji} \qquad \alpha_{im} A_i + \sum_{j=1}^{3} \alpha_{jm} A_{ji} = -\sum_{j=1}^{3} \alpha'_{jm} A'_{ji}$$

$$-\omega_{i1} A_i - \sum_{j=1}^{3} \omega_{j1} A_{ji} = -\sum_{j=1}^{3} \omega'_{j1} A'_{ji} \qquad \theta_{im} A_i - \sum_{j=1}^{3} \theta_{jm} A_{ji} = -\sum_{j=1}^{3} \theta'_{jm} A'_{jm}.$$

$$(2.66)$$

For a point force T in the x-direction, applied at the source point $(0, 0, h > 0)$, the elastic displacements and electric potential at the field point $(x, y, z > 0)$ of the upper half-space are

$$u = -D_0 \left[\frac{1}{\overline{R}_{00} + s_0|z-h|} - \frac{y^2}{\overline{R}_{00}(\overline{R}_{00} + s_0|z-h|)^2} \right] - D_{00} \left[\frac{1}{R_{00} + z_0} \right.$$

$$\left. - \frac{y^2}{R_{00}(R_{00} + z_{00})^2} \right] + \sum_{i=1}^{3} \left\{ \begin{array}{l} D_i \left[\frac{1}{\overline{R}_{ii} + s_i|z-h|} - \frac{x^2}{\overline{R}_{ii}(\overline{R}_{ii} + s_i|z-h|)^2} \right] \\ + \sum_{j=1}^{3} D_{ij} \left[\frac{1}{R_{ij} + z_{ij}} - \frac{x^2}{R_{ij}(R_{ij} + z_{ij})^2} \right] \end{array} \right\}$$

$$v = -\frac{D_0 xy}{\overline{R}_{00}(\overline{R}_{00} + s_0|z-h|)^2} - \frac{D_{00} xy}{R_{00}(R_{00} + z_{00})^2}$$

$$- xy \sum_{i=1}^{3} \left[\frac{D_i}{\overline{R}_{ii}(\overline{R}_{ii} + s_i|z-h|)^2} + \sum_{j=1}^{3} \frac{D_{ij}}{R_{ij}(R_{ij} + z_{ij})^2} \right]$$

$$w = -x \sum_{i=1}^{3} \alpha_{i1} \left[\text{sign}(z-h) \frac{D_i}{\overline{R}_{ii}(\overline{R}_{ii} + s_{ii}|z-h|)} + \sum_{j=1}^{3} \frac{D_{ij}}{R_{ij}(R_{ij} + z_{ij})} \right]$$

$$\phi = -x \sum_{i=1}^{3} \alpha_{i2} \left[\text{sign}(z-h) \frac{D_i}{\overline{R}_{ii}(\overline{R}_{ii} + s_{ii}|z-h|)} + \sum_{j=1}^{3} \frac{D_{ij}}{R_{ij}(R_{ij} + z_{ij})} \right].$$

$$(2.67)$$

The elastic displacements and electric potential at the field point $(x, y, z < 0)$ of the lower half-space are

$$u' = -L'_{00} \left[\frac{1}{R'_{00} - z'_{00}} - \frac{y^2}{R'_{00}(R'_{00} - z'_{00})^2} \right] + \sum_{i=1}^{3} \sum_{j=1}^{3} L'_{ij} \left[\frac{1}{R'_{ij} - z'_{ij}} \right.$$

$$\left. - \frac{x^2}{R'_{ij}(R'_{ij} - z'_{ij})^2} \right]$$

$$v' = -\frac{L'_{00} xy}{R'_{00}(R'_{00} - z'_{00})^2} - xy \sum_{i=1}^{3} \sum_{j=1}^{3} \frac{L'_{ij}}{R'_{ij}(R'_{ij} - z'_{ij})^2}$$

$$w' = x \sum_{i=1}^{3} \alpha'_{i1} \sum_{j=1}^{3} \frac{L'_{ij}}{R'_{ij}(R'_{ij} - z'_{ij})} \qquad \phi' = x \sum_{i=1}^{3} \alpha'_{i2} \sum_{j=1}^{3} \frac{L'_{ij}}{R'_{ij}(R'_{ij} - z'_{ij})}.$$

$$(2.68)$$

The involved coefficients in (2.67) and (2.68), D_{ij} and L'_{ij}, are solved from the following equations (for $i = 1, 2, 3; m = 1, 2$).

$$D_0 + D_{00} = L'_{00} \qquad \omega_{01}(D_{00} - D_0) = -\omega'_{01} L'_{00}$$

$$D_i + \sum_{j=1}^{3} D_{ji} = \sum_{j=1}^{3} L'_{ji} \qquad \alpha_{im} D_i - \sum_{j=1}^{3} \alpha_{jm} D_{ji} = \sum \alpha'_{jm} L'_{ji}$$

$$-\omega_{i1} D_i + \sum_{j=1}^{3} \omega_{j1} D_{ji} = -\sum_{j=1}^{3} \omega'_{j1} L'_{ij} \qquad \theta_{im} D_i + \sum_{j=1}^{3} \theta_{jm} D_{ji} = \sum_{j=1}^{3} \theta'_{jm} L'_{ji}.$$

$$(2.69)$$

The half-space Green's functions with traction-free (i.e., the elastic traction and the z-component of the electric displacement are zero) on the surface $z = 0$ can be directly reduced from the bimaterial space Green's functions presented in this section. Actually, assuming that the half-space is in the $z > 0$ domain and the source is also at $z = h(> 0)$, then the half-space Green's functions will be exactly the same as the bimaterial case in the source half-space, except that the involved coefficients are simply given as

$$A_{11} = A_1(\theta_{21}\theta_{32}\omega_{11} - \theta_{22}\theta_{31}\omega_{11} - \theta_{12}\theta_{31}\omega_{21}$$
$$+ \theta_{11}\theta_{32}\omega_{21} + \theta_{12}\theta_{21}\omega_{31} - \theta_{11}\theta_{22}\omega_{31})/d_a$$

$$A_{21} = \frac{2A_1(\theta_{12}\theta_{31} - \theta_{11}\theta_{32})\omega_{11}}{d_a}$$

$$A_{31} = \frac{2A_1(\theta_{11}\theta_{22} - \theta_{12}\theta_{21})\omega_{11}}{d_a}$$

$$A_{12} = \frac{2A_2(\theta_{21}\theta_{32} - \theta_{22}\theta_{31})\omega_{21}}{d_a}$$

$$A_{22} = A_2(\theta_{22}\theta_{31}\omega_{11} - \theta_{21}\theta_{32}\omega_{11} + \theta_{12}\theta_{31}\omega_{21}$$
$$- \theta_{11}\theta_{32}\omega_{21} + \theta_{12}\theta_{21}\omega_{31} - \theta_{11}\theta_{22}\omega_{31})/d_a$$

$$A_{32} = \frac{2A_2(\theta_{11}\theta_{22} - \theta_{12}\theta_{21})\omega_{21}}{d_a}$$

$$A_{13} = \frac{2A_3(\theta_{21}\theta_{32} - \theta_{22}\theta_{21})\omega_{31}}{d_a}$$

$$A_{23} = \frac{2A_3(\theta_{12}\theta_{31} - \theta_{11}\theta_{32})\omega_{31}}{d_a}$$

$$A_{33} = A_3(\theta_{22}\theta_{31}\omega_{11} - \theta_{21}\theta_{32}\omega_{11} - \theta_{12}\theta_{31}\omega_{21}$$
$$+ \theta_{11}\theta_{32}\omega_{21} - \theta_{12}\theta_{21}\omega_{31} - \theta_{11}\theta_{22}\omega_{31})/d_a$$

$$d_a = \theta_{22}\theta_{31}\omega_{11} - \theta_{21}\theta_{32}\omega_{11} - \theta_{12}\theta_{32}\omega_{21}$$
$$+ \theta_{11}\theta_{32}\omega_{21} + \theta_{12}\theta_{21}\omega_{31} - \theta_{11}\theta_{22}\omega_{31}$$

$$D_{00} = D_0; \qquad D_{ji} = \frac{D_i A_{ij}}{A_i(i,j=1,2,3)}. \tag{2.70}$$

2.6 Green's Functions in Three-Dimensional Infinite, Half, and Bimaterial Spaces: General Anisotropy

When the piezoelectric material is of general anisotropy, the Green's function solution becomes complicated even for the infinite space case. Therefore, various precise and computationally efficient algorithms have been developed for the evaluation of the infinite space piezoelectric Green's functions [38–40]. A more efficient way to evaluate the piezoelectric Green's function is by calculating the corresponding eigenvalues and eigenvectors [41, 42], as, for example, in [43]. In this section, however, the solutions developed by the author and coworkers are presented. The Green's functions in the infinite space were derived by employing the Radon transform [44] and those in the half and bimaterial spaces were obtained by separating the solution into two parts: the infinite-space Green's function and the complementary part. While the infinite-space Green's function is in an exact closed-form, the complementary part is expressed by a finite-line integral after utilizing the double Fourier transform [i.e., 46].

2.6.1 Infinite Space

Following Pan and Tonon [44], the Green's displacement in the Jth direction at $\boldsymbol{x}$ due to a point force in the Kth direction at the origin can be found as

$$U_{JK}(\mathbf{x}) = -\frac{\mathrm{Im}}{2\pi r} \sum_{m=1}^{4} \frac{A_{JK}(\mathbf{p} + \zeta_m \mathbf{q})}{a_9(\zeta_m - \bar{\zeta}_m) \prod_{\substack{k=1 \\ k \neq m}}^{4} (\zeta_m - \zeta_k)(\zeta_m - \bar{\zeta}_k)} \tag{2.71}$$

$$D(\mathbf{p} + \zeta\mathbf{q}) = \sum_{k=0}^{8} a_{k+1}\zeta^k = a_9 \prod_{m=1}^{4} (\zeta - \zeta_m)(\zeta - \bar{\zeta}_m), \tag{2.72}$$

where $A_{JK}(\mathbf{m})$ is the adjoint matrix of $\Gamma_{JK}(\mathbf{m})$, and $D(\mathbf{m})$ is the determinant of $\Gamma_{JK}(\mathbf{m})$, defined as

$$\Gamma_{JK}(\mathbf{m}) = C_{iJKq}m_i m_q. \tag{2.73}$$

The vector $\mathbf{m}$ is given by

$$\boldsymbol{m} = \mathbf{p} + \zeta\mathbf{q}, \tag{2.74}$$

where

$$\mathbf{p} = \frac{\mathbf{e} \times \mathbf{v}}{|\mathbf{e} \times \mathbf{v}|}; \qquad \mathbf{q} = \mathbf{e} \times \mathbf{p}; \qquad \mathbf{e} = \frac{\mathbf{x}}{|\mathbf{x}|}, \tag{2.75}$$

with $\mathbf{v}$ being an arbitrary unit vector different from $\mathbf{e}(\mathbf{v} \neq \mathbf{e})$.

There are a couple of features associated with this Green's function expression (2.71). First of all, (2.71) is an explicit expression. It is therefore very accurate and efficient [45]. For a given pair of field and source points, we need only to solve the eighth-order polynomial equation (2.72) numerically once to obtain all the components of the Green's displacement. Secondly, in obtaining (2.71), we have assumed that all the poles are simple. Should the poles be multiple, a slight change in the material constants will result in single poles, with negligible errors in the computed Green's tensor, as for the purely elastic case [20]. Thirdly, because Γ_{JK} is symmetric, so is its adjoint A_{JK}. Therefore, the Green's displacement G_{JK} is symmetric [39] and one needs to calculate only 10 out of its 16 elements. The symmetric property of the extended Green's tensor can also be considered as a consequence of the Betti-type reciprocity as presented in Section 3 of this chapter. Finally, although one can choose the vector $\mathbf{v}(\neq \mathbf{e})$ arbitrarily, it should be one of the base vectors in the space-fixed Cartesian coordinates, that is, $(1, 0, 0)$ or $(0, 1, 0)$ or $(0, 0, 1)$. The analytical expression for the Green's displacement is much simpler using such a vector $\mathbf{v}$ than using any other vectors.

2.6.2 Half-Space

For the half-space case $(z > 0)$ with traction-free boundary condition at $z = 0$ (i.e., the elastic traction and z-component of the electric displacement are zero), the Green's function solutions at the field point $\boldsymbol{x}$ $(x_1, x_2, z > 0)$ due to a point force at $\boldsymbol{d}$ $(d_1, d_2, d > 0)$ can be obtained first in the Fourier transformed domain, and then invert back to the physical domain. By so doing, the final half-space Green's function in the physical domain, or the generalized Mindlin solution, can be expressed as a sum of an explicit Kelvin-type solution and a complementary part in terms of a line integral over $[0, 2\pi]$. Furthermore, the latter can be reduced to an integral over $[0, \pi]$. For the half-space displacement tensor (4×4), with its row and column indices being the components of the field quantity and the direction of the point source, respectively, it can be expressed as [21]

$$\boldsymbol{U}(\boldsymbol{x};\boldsymbol{d}) = \boldsymbol{U}^\infty(\boldsymbol{x};\boldsymbol{d}) + \frac{1}{2\pi^2} \int_0^\pi \overline{\boldsymbol{A}} \boldsymbol{G}_1 \boldsymbol{A}^T d\theta, \qquad (2.76)$$

where

$$(\boldsymbol{G}_1)_{IJ} = \frac{(\overline{\boldsymbol{B}}^{-1}\boldsymbol{B})_{IJ}}{-\overline{p}_I z + p_J d - [(x_1 - d_1)\cos\theta + (x_2 - d_2)\sin\theta]}. \qquad (2.77)$$

The first term in (2.76) corresponds to the Green's displacement tensor in an anisotropic and piezoelectric full space, which is given by (2.71) [43, 44]. Consequently, the half-space displacement tensor can be expressed as a sum of an explicit Kelvin tensor and a complementary part in terms of a line integral over $[0, \pi]$. It is emphasized that in (2.76) and (2.77), the eigenvalues p_J and the eigenmatrices $\boldsymbol{A}$ and $\boldsymbol{B}$ are functions of θ, with p_J $(J = 1, 2, 3, 4)$ and $\boldsymbol{A} = [\boldsymbol{a}_1, \boldsymbol{a}_2, \boldsymbol{a}_3, \boldsymbol{a}_4]$ being the eigensolutions of (2.33) for a given θ. In other words, p and $\boldsymbol{a}$ satisfy the same eigenequation (2.33) but with

$$Q_{IK} = C_{jIKs}n_j n_s, \qquad R_{IK} = C_{jIKs}n_j m_s,$$
$$T_{IK} = C_{jIKs}m_j m_s \qquad (2.78)$$

$$\boldsymbol{n} = \begin{bmatrix} n_1 \\ n_2 \\ 0 \end{bmatrix} = \begin{bmatrix} \cos\theta \\ \sin\theta \\ 0 \end{bmatrix}, \qquad \boldsymbol{m} = \begin{bmatrix} 0 \\ 0 \\ 1 \end{bmatrix}. \qquad (2.79)$$

Matrix $\boldsymbol{B}$ is defined by (2.36), with its vector $\boldsymbol{b}_i$ related to $\boldsymbol{a}_i$ via (2.35).

Equation (2.76) is the generalized Mindlin solution, or the Green's displacement under the traction-free boundary conditions in an anisotropic and piezoelectric half-space. It is remarked that similar Mindlin solutions can be presented for many other homogeneous boundary conditions on the surface $z = 0$ [21]. We also point out that if the source and field points are not simultaneously on the surface, then the line integral in (2.76) can be carried out by employing regular numerical quadrature. If, however, $z = d = 0$, then the half-space Green's function is reduced to the special surface Green's function, where the involved singular integration needs special numerical treatment.

2.6.3 Bimaterial Space

The procedure for solving the bimaterial Green's functions is as follows. First, we apply the double Fourier transform to the two horizontal variables (x, y); second, we solve the Green's function problem in the transformed domain; third, we apply the inverse Fourier transform to obtain the physical domain Green's function. To handle the double infinite integrals in the inverse space, the polar coordinate transform is applied so that the infinite integral with respect to the radial variable can be carried out exactly. Thus, the final bimaterial Green's functions in the physical domain can be expressed in terms of a regular line integral over $[0, 2\pi]$, which can be further reduced to $[0, \pi]$ using certain properties of the Stroh eigenvalues and Stroh matrices.

We assume that the upper half-space $(z > 0)$ is occupied by material 1 and the lower half $(z < 0)$ by material 2. The interface at $z = 0$ between the two half-spaces is further assumed to be perfect. In other words, the elastic traction and the z-component of the electric displacement are continuous across the interface. We further assume that the point force is in material 1 at $\boldsymbol{d}$ $(d_1, d_2, d > 0)$; then the 4×4 Green's function tensor at $\boldsymbol{x}$ $(x_1, x_2, z > 0)$ in material 1, with its first index for the displacement component and the second for the extended point force direction, is found to be [46]

$$U^{(1)}(\boldsymbol{x}; \boldsymbol{d}) = U^{\infty}(\mathbf{x}; \mathbf{d}) + \frac{1}{2\pi^2}\left[\int_0^{\pi} \overline{\boldsymbol{A}}^{(1)} \boldsymbol{G}_u^{(1)}(\boldsymbol{A}^{(1)})^T d\theta\right] \tag{2.80}$$

$$(\boldsymbol{G}_u^{(1)})_{IJ} = \frac{(\boldsymbol{G}_1)_{IJ}}{-\overline{p}_I^{(1)}z + p_J^{(1)}d - [(x_1 - d_1)\cos\theta + (x_2 - d_2)\sin\theta]}. \tag{2.81}$$

In (2.80), $\boldsymbol{U}^{\infty}(\mathbf{x}; \mathbf{d})$ denotes the Green's function tensor for the displacements in the full space with the material 1 property (i.e., (2.71)).

In material 2, the Green's tensor at $\boldsymbol{x}$ $(x_1, x_2, z < 0)$ is

$$\boldsymbol{U}^{(2)}(\boldsymbol{x};\boldsymbol{d}) = -\frac{1}{2\pi^2}\left[\int_0^\pi \boldsymbol{A}^{(2)}\boldsymbol{G}_u^{(2)}(\boldsymbol{A}^{(1)})^T d\theta\right] \tag{2.82}$$

$$(\boldsymbol{G}_u^{(2)})_{IJ} = \frac{(\boldsymbol{G}_2)_{IJ}}{-p_I^{(2)}z + p_J^{(1)}d - [(x_1 - d_1)\cos\theta + (x_2 - d_2)\sin\theta]}. \tag{2.83}$$

In (2.80) to (2.83), the superscripts "(1)" and "(2)" denote quantities in materials 1 and 2, respectively, and the matrices $\boldsymbol{G}_1$ and $\boldsymbol{G}_2$ are given by

$$\boldsymbol{G}_1 = -(\overline{\boldsymbol{A}}^{(1)})^{-1}(\overline{\boldsymbol{M}}^{(1)} + \boldsymbol{M}^{(2)})^{-1}(\boldsymbol{M}^{(1)} - \boldsymbol{M}^{(2)})\boldsymbol{A}^{(1)}$$

$$\boldsymbol{G}_2 = (\boldsymbol{A}^{(2)})^{-1}(\overline{\boldsymbol{M}}^{(1)} + \boldsymbol{M}^{(2)})^{-1}(\boldsymbol{M}^{(1)} - \overline{\boldsymbol{M}}^{(1)})\boldsymbol{A}^{(1)}, \tag{2.84}$$

where $\boldsymbol{M}^{(\alpha)}$ is the modified impedance tensor defined as

$$\boldsymbol{M}^{(\alpha)} = -i\boldsymbol{B}^{(\alpha)}(\boldsymbol{A}^{(\alpha)})^{-1} \qquad (\alpha = 1, 2). \tag{2.85}$$

In summary, in material 1, the bimaterial Green's function is expressed as a sum of the explicit full-space Green's function and a complementary part in terms of a line integral over $[0, \pi]$; In material 2, the bimaterial Green's function is expressed in terms of a line integral over $[0, \pi]$. With regard to these physical domain bimaterial Green's functions ((2.80) and (2.82)), the following important observations can be made.

(a) For the complementary part of the solution in material 1 and the solution in material 2, the dependence of the solutions on the field point $\boldsymbol{x}$ and source point $\boldsymbol{d}$ appears only through matrices $\boldsymbol{G}_u^{(1)}$ and $\boldsymbol{G}_u^{(2)}$ defined in (2.81) and (2.83).

(b) The integrals in (2.80) and (2.82) are regular if $z \neq 0$ or $d \neq 0$, and thus can be easily carried out by a standard numerical integral method such as Gaussian quadrature.

(c) If $z \neq 0$ and $d = 0$, the bimaterial Green's function is still mathematically regular although some of its components may not have a direct and apparent physical meaning (see Pan, [25], for the purely elastic counterpart).

(d) When the field and source points are both on the interface (i.e., $z = d = 0$), the bimaterial Green's function is then reduced to the interfacial Green's function. For this special case, the line integral involved in the Green's function expression becomes singular and the resulting finite part integral needs to be handled with special care [47].

(e) Bimaterial Green's functions can be solved similarly for other (imperfect) interface models. To do so, one need only find the modified Stroh matrices $\boldsymbol{A}^{(\alpha)}$ and $\boldsymbol{B}^{(\alpha)}$ for the given interface models. For detailed discussion, one should refer to [25].

2.7 Green's Functions in Layered Half-Space

We solve the Green's functions in layered half-space in terms of two systems of vector functions combined with the propagator matrix method. The vector function method is essentially equivalent to the double Fourier transform or Hankel transform in the horizontal layer plane, but possesses certain advantages over the latter ones (i.e., [48]). The propagator matrix method is utilized to propagate the field quantities from one layer to the other.

We first introduce the two systems of vector functions. The Cartesian coordinate system of vector functions is defined as [12, 48, 49]

$$\boldsymbol{L}(x, y; \alpha, \beta) = \boldsymbol{e}_z S(x, y; \alpha, \beta)$$
$$\boldsymbol{M}(x, y; \alpha, \beta) = (\boldsymbol{e}_x \partial_x + \boldsymbol{e}_y \partial_y) S(x, y; \alpha, \beta)$$
$$\boldsymbol{N}(x, y; \alpha, \beta) = (\boldsymbol{e}_x \partial_y - \boldsymbol{e}_y \partial_x) S(x, y; \alpha, \beta) \tag{2.86}$$

with

$$S(x, y; \alpha, \beta) = \frac{e^{-i(ax+\beta y)}}{(2\pi)}, \tag{2.87}$$

where $\boldsymbol{e}_x$, $\boldsymbol{e}_y$, and $\boldsymbol{e}_z$ are the unit vectors along the x-, y-, and z-axes, respectively; x and y are horizontal axes, while z-axis points to the problem domain; α and β are the transformation variables corresponding to the two horizontal physical variables x and y.

The corresponding cylindrical system of vector functions is defined as [12, 48, 49]

$$\boldsymbol{L}(r, \theta; \lambda, m) = e_z S(r, \theta; \lambda, m)$$
$$\boldsymbol{M}(r, \theta; \lambda, m) = (e_r \frac{\partial}{\partial r} + e_\theta \frac{\partial}{r\partial\theta}) S(r, \theta; \lambda, m)$$
$$\boldsymbol{N}(r, \theta; \lambda, m) = (e_r \frac{\partial}{r\partial\theta} - e_\theta \frac{\partial}{\partial r}) S(r, \theta; \lambda, m) \tag{2.88}$$

with $\boldsymbol{e}_r$, $\boldsymbol{e}_\theta$, and $\boldsymbol{e}_z$ as the unit vectors along the r-, θ-, and z-axes, respectively, and

$$S(t, \theta; \lambda, m) = \frac{1}{\sqrt{2\pi}} J_m(\lambda r) e^{im\theta}, \tag{2.89}$$

where $J_m(\lambda r)$ is the Bessel function of order m with $m = 0$ corresponding to the axial symmetric deformation.

There are several important features associated with the two systems of vector functions.

(a) For plane strain deformation in the (x, z)-plane, one needs only to replace 2π by $\sqrt{2\pi}$ and β by 0, respectively.

(b) While the solution in terms of the $\boldsymbol{L}$ & $\boldsymbol{M}$ vectors is contributed to the dilatational deformation, that of the $\boldsymbol{N}$ vector to the rotational part.

Corresponding to the dynamic counterparts, the $\boldsymbol{L}$ & $\boldsymbol{M}$ part is related to the Rayleigh wave and the $\boldsymbol{N}$ part to the Love wave. Here, we name the solution associated with the $\boldsymbol{L}$ & $\boldsymbol{M}$ vectors the *LM*-type solution and that associated with the $\boldsymbol{N}$ vector the *N*-type solution.

(c) We remark that the general solution and propagator matrix in the cylindrical system of vector functions are exactly the same as those in the Cartesian system. This feature gives certain numerical advantages when programming these formulations in the two systems of vector functions.

(d) Another advantage is that both 2D and 3D Green's functions can be studied uniformly under these systems [12, 48, 49] because the general solutions in terms of the two systems are the same for both 2D and 3D deformation.

For the Green's function problem, we first express the elastic displacement, electric potential, traction, electric displacements, body force, and negative electric charge density in terms of the cylindrical system of vector functions,

$$\boldsymbol{u}(r,\theta,z) = \sum_m \int_0^{+\infty} [U_L(z)\boldsymbol{L}(r,\theta) + U_M(z)\boldsymbol{M}(r,\theta) + U_N(z)\boldsymbol{N}(r,\theta)]\lambda d\lambda$$

$$(2.90)$$

$$\phi(r,\theta,z) = \sum_m \int_0^{+\infty} \Phi(z)S(r,\theta)\lambda d\lambda \tag{2.91}$$

$$\boldsymbol{t}(r,\theta,z) \equiv \sigma_{rz}\boldsymbol{e}_r + \sigma_{\theta z}\boldsymbol{e}_\theta + \sigma_{zz}\boldsymbol{e}_z$$

$$= \sum_m \int_0^{+\infty} [T_L(z)\boldsymbol{L}(r,\theta) + T_M(z)\boldsymbol{M}(r,\theta) + T_N(z)\boldsymbol{N}(r,\theta)]\lambda d\lambda$$

$$(2.92)$$

$$\boldsymbol{D}(r,\theta,z) = \sum_m \int_0^{+\infty} [D_L(z)\boldsymbol{L}(r,\theta) + D_M(z)\boldsymbol{M}(r,\theta) + D_N(z)\boldsymbol{N}(r,\theta)]\lambda d\lambda$$

$$(2.93)$$

$$\boldsymbol{f}(r,\theta,z) = \sum_m \int_0^{+\infty} [F_L(z)\boldsymbol{L}(r,\theta) + F_M(z)\boldsymbol{M}(r,\theta) + F_N(z)\boldsymbol{N}(r,\theta)]\lambda d\lambda$$

$$(2.94)$$

$$-q(r,\theta,z) = \sum_m \int_0^{+\infty} Q(z)S(r,\theta)\lambda d\lambda. \tag{2.95}$$

In (2.90)–(2.95), the left-hand side variables in lowercase are the unknown field quantities in the physical domain, and the right-hand side variables in capitals, such as U, Φ, T, ..., are the unknown expansion coefficients in the transformed domain. Making use of corresponding governing equations presented in Section 2, the layered Green's function problems can be converted into an ordinary differential system of equations for each layer in the transformed domain so that the unknowns in the transformed domain can be obtained. Because the Cartesian and cylindrical systems of vector functions are employed, the problem in the transformed domain can be further separated into two independent problems, which are discussed below.

2.7.1 General N- and LM-Type Solutions in the Transformed Domain

(a) N-type solution

Based on either the Cartesian or cylindrical system of vector functions, one can show easily that the N-type solution is independent of the rest, and furthermore, it is independent of the electric quantities. In other words, it is purely elastic and its general solution in each layer can be expressed as

$$[\boldsymbol{E}^N] = [\boldsymbol{Z}^N(z)][\boldsymbol{K}^N], \tag{2.96}$$

where $[\boldsymbol{K}^N]$ is a column coefficient matrix of 2×1 with its elements to be determined by the continuity and/or boundary conditions. Also in (2.96),

$$[\boldsymbol{E}^N(z)] = \left[U_N(z), \frac{T_N(z)}{\lambda}\right]^T, \tag{2.97}$$

and $[\boldsymbol{Z}^N(z)]$ is the solution matrix, the same as that for the purely elastic case [12, 48, 49].

(b) LM-type solution

For this type of deformation, the elastic and piezoelectric fields are coupled together. The ordinary differential equations in each layer for this type can be derived as

$$[\boldsymbol{E}]_z = \lambda[\boldsymbol{W}][\boldsymbol{E}]. \tag{2.98}$$

It is remarked that the diagonal elements of $[\boldsymbol{W}]$ are zero and independent of λ. Also in (2.98),

$$[\boldsymbol{E}] = \left[U_L, \lambda U_M, \frac{T_L}{\lambda}, T_M, \Phi, \frac{D_L}{\lambda}\right]^T. \tag{2.99}$$

To find the homogeneous solution of (2.98), we assume that

$$[\boldsymbol{E}(z)] = [\boldsymbol{b}]e^{\lambda \eta z}. \tag{2.100}$$

Substituting (2.100) into (2.98) and noticing that all the diagonal elements of $[\boldsymbol{W}]$ are zero, we obtain the following six-dimensional eigenequations for (2.98)

$$\{[\boldsymbol{W}] - \eta[\boldsymbol{I}]\}[\boldsymbol{b}] = 0, \tag{2.101}$$

where $[\boldsymbol{I}]$ is the 6×6 identity matrix.

It is observed from (2.101) that the eigenvalues and their corresponding eigenvectors are independent of the integral variable λ! Therefore, these eigenequations need to be solved only once for each layer for the given material properties. Let us, therefore, assume that the six eigenvalues are distinct, and the general solution to (2.98) is then obtained as

$$[\boldsymbol{E}(z)] = [\boldsymbol{Z}(z)][\boldsymbol{K}], \tag{2.102}$$

where $[\boldsymbol{K}]$ is a 6×1 coefficient matrix with its elements to be determined by the interface and/or boundary conditions, and

$$[\boldsymbol{Z}(z)] = [\boldsymbol{B}] \left\langle e^{\lambda \eta^* z} \right\rangle \tag{2.103}$$

with

$$\left\langle e^{\lambda \eta^* z} \right\rangle = \mathrm{diag}[e^{\lambda \eta_1 z}, \quad e^{\lambda \eta_2 z}, \quad e^{\lambda \eta_3 z}, \quad e^{-\lambda \eta_1 z}, \quad e^{-\lambda \eta_2 z}, \quad e^{-\lambda \eta_3 z}] \tag{2.104}$$

being associated with the six eigenvalues, and

$$[\boldsymbol{B}] = [\boldsymbol{b}_1, \boldsymbol{b}_2, \boldsymbol{b}_3, \boldsymbol{b}_4, \boldsymbol{b}_5, \boldsymbol{b}_6] \tag{2.105}$$

associated with the corresponding eigenvectors.

2.7.2 Propagator Matrix Method for Multilayered Structures

The propagator matrix method is most suitable to layered structures. Application of this method can help avoid the complicated calculation of a large matrix and also save significant computation resource. At the core of this method is the propagator matrix, which relates the N and LM expansion coefficients $[\boldsymbol{E}^N]$ and $[\boldsymbol{E}]$ at the top interface to the bottom interface of layer j. In other words, for layer j, we have

$$[\boldsymbol{E}^N(z_{j-1})] = [\boldsymbol{a}^N][\boldsymbol{E}^N(z_j)] \tag{2.106}$$

and

$$[\boldsymbol{E}(z_{j-1})] = [\boldsymbol{a}][\boldsymbol{E}(z_j)] \tag{2.107}$$

where z_{j-1} and z_j ($>z_{j-1}$) are the depths of the top and bottom interfaces of layer j, $[\boldsymbol{a}^N]$ and $[\boldsymbol{a}]$ are the so-called propagator matrix (or layer matrix, or transfer matrix). Propagating the solution from the surface $z = 0$ to half-space bottom $z = H$, we obtained,

$$[\boldsymbol{E}^N(0)] = [\boldsymbol{a}_1^N][\boldsymbol{a}_2^N] - - - [\boldsymbol{a}_{p-1}^N][\boldsymbol{Z}_p^N(h)][\boldsymbol{K}_p]$$
$$[\boldsymbol{E}(0)] = [\boldsymbol{a}_1][\boldsymbol{a}_2] - - - [\boldsymbol{a}_{p-1}][\boldsymbol{Z}_p(h)][\boldsymbol{K}_p] \qquad (2.108\text{a,b})$$

with the undetermined coefficients having the structure as

$$[\boldsymbol{K}_p^N] = [0, *]^t \qquad [\boldsymbol{K}_p] = [0, 0, 0, *, *, *]^t \qquad (2.109\text{a,b})$$

to satisfy the requirement that the solution vanishes when z approaches $+\infty$. The symbol "*" denotes an unknown coefficient. These coefficients can be solved using the traction-free boundary condition on the surface and the discontinuity condition at the source level due to the point force. After the unknown coefficients in $[\boldsymbol{K}_p^N]$ and $[\boldsymbol{K}_p]$ are determined through the propagator matrix method, the expansion coefficients at any depth (e.g., for $z \geq h$ in layer j, i.e., $z_{j-1} \leq z \leq z_j$) can be derived exactly as

$$[\boldsymbol{E}^N(z)] = [\boldsymbol{a}_j^N(z - z_{j-1})][\boldsymbol{a}_{j+1}^N] - - - [\boldsymbol{a}_{p-1}^N][\boldsymbol{Z}_p^N(H)][\boldsymbol{K}_p^N]$$
$$[\boldsymbol{E}(z)] = [\boldsymbol{a}_j(z - z_{j-1})][\boldsymbol{a}_{j+1}] - - - [\boldsymbol{a}_{p-1}][\boldsymbol{Z}_p(H)][\boldsymbol{K}_p]. \qquad (2.110\text{a,b})$$

2.7.3 Physical Domain Solutions

From (2.90)–(2.95), in order to get the field quantities in the physical domain, numerical integration must be carried out. It is noted that the integrands in the infinite integrals for the Green's functions involve Bessel functions that are oscillatory and go to zero slowly when their variable approaches infinity. Thus, the common numerical integral methods, such as the trapezoidal rule or Simpson's rule, are not suitable for such integrations. On the other hand, numerical integration of this type of function via adaptive Gauss quadrature has been found to be very accurate and efficient. In this adaptive quadrature, we express the infinite integral for each Green's function as a summation of partial integration terms:

$$\int_0^{+\infty} f(\lambda, z) J_m(\lambda r) d\lambda = \sum_{n=1}^{N} \int_{\lambda_n}^{\lambda_{n+1}} f(\lambda, z) J_m(\lambda r) d\lambda. \qquad (2.111)$$

In each subinterval, a starting three-point Gauss rule is applied to approximate the integral. A combined relative–absolute error criterion is used to check the results. If the error criterion is not satisfied, new Gauss points are added optimally so that only the new integrand values need to be calculated. This procedure continues until the selected error criterion is satisfied.

The methodology presented in this section can also be applied to find the Green's function solutions in many different layered material structures. These include transversely isotropic layered thermoelastic half-space [13], layered poroelastic half-space [14], transversely isotropic layered piezoelectric half-space [50] and transversely isotropic functionally graded and layered piezoelectric half-space [51, 52]. Similar approaches have been also developed to derive the Green's functions in general anisotropic and layered elastic spaces [53, 54].

2.8 Conclusions

This chapter presents a review of the Green's function solutions in piezoelectric anisotropic solids. It is limited to the static case and with infinite or semi-infinite domains. Even in this limited case, the author may have missed some of the references by other experts. For example, one interesting area that the author intentionally omitted is the circular loading on the surface of the layered piezoelectric half-space as this can be obtained from the corresponding point-source Green's functions by integration over the loading domain [52]. The Green's functions in the functionally graded piezoelectric space are not covered (i.e., [51]). There are also Green's functions associated with the finite domain, for example, on the Green's function-related issues in layered piezoelectric spheres (i.e., [55–57]), and in layered piezoelectric cylinders(i.e., [58, 59]). The dynamic and transient Green's functions are not reviewed. These include dynamic and transient problems in layered cylinders [58, 60–62], in layered spheres [63], and in horizontally layered plates (i.e., [64]). Dynamic Green's functions in anisotropic infinite space are not covered either, and contributions to this difficult area can be found, for example, in [65, 66]. A recent special issue of *Engineering Analysis and Boundary Elements* edited by the author also includes many interesting Green's function solutions [67]. Another interesting area that the author hasn't reviewed but is extremely attractive is related to the multiferroic materials/structures. The coupling between the electric and magnetic fields via the induced strain inside the system has potential applications to many semiconductor devices using electric and magnetic fields. Various Green's functions have already been developed and interested readers should refer to [68–76] for details.

References

[1]   Yang, J. S. and Hu, Y. T. (2004): Mechanics of electroelastic bodies under biasing fields, *Appl. Mech. Rev.*, 57, 173–189.

[2]   Yang, J. S. (2006): A review of a few topics in piezoelectricity. *Applied Mechanics Reviews*, 59, 335–345.

[3] Bimberg, D., Grundmann, M., and Ledentsov, N. N. (1999): *Quantum Dot Heterostructures*. Wiley, New York.

[4] Maranganti, R. and Sharma, P. (2006): A review of strain field calculations in embedded quantum dots and wires. In *Handbook of Theoretical and Computational Nanotechnology*. Edited by M. Reith and W. Schommers.

[5] Saravanos, D. A., Heyliger, P. R., and Hopkins, D. A. (1997): Layerwise mechanics and finite element for the dynamic analysis of piezoelectric composite plates. *Int. J. Solids Struct.*, 34, 359–378.

[6] Kuvshinov, B. N. (2008): Elastic and piezoelectric fields due to polyhedral inclusions. *Int. J. Solids Struct.*, 45, 1352–1384.

[7] Tiersten, H. F. (1969): *Linear Piezoelectric Plate Vibrations*, Plenum, New York.

[8] Barnett, D. M. and Lothe, J. (1975): Dislocations and line charges in anisotropic piezoelectric insulators, *Phys. Stat. Sol. (b)*, 67, 105–111.

[9] Pan, E. (1999): A BEM analysis of fracture mechanics in 2D anisotropic piezoelectric solids, *Eng. Anal. Bound. Elem*, 23, 67–76.

[10] Steketee, J. A. (1958): On Volterra's dislocation on a semi-infinite elastic medium. *Can. J. Phys.*, 36, 192–205.

[11] Pan, E. (1991): Dislocation in an infinite poroelastic medium. *Acta Mech*, 87, 105–115.

[12] Pan, E. (1989): Static response of a transversely isotropic and layered half-space to general dislocation sources. *Phys. Earth Planet. Inter.*, 58, 103–117.

[13] Pan, E. (1990): Thermoelastic deformation of a transversely isotropic and layered half-space by surface loads and internal sources. *Phys. Earth Planet. Inter.*, 60, 254–264.

[14] Pan, E. (1999): Green's functions in layered poroelastic half-spaces. *Int. J. Numer. Anal. Meth. Geomech.*, 23, 1631–1653.

[15] Mindlin, R. D. (1936): Force at a point in the interior of a semi-infinite solid. *Physics*, 7, 195–202.

[16] Maruyama, T. (1964): Statical elastic dislocations in an infinite and semi-infinite medium. *Bull. Earthq. Res. Inst.*, 42, 289–368.

[17] Maruyama, T. (1966): On two-dimensional elastic dislocations in an infinite and semi-infinite medium. *Bull. Earthq. Res. Inst.*, 44, 811–871.

[18] Yu, H.Y., Sanday, S.C., and Chang, C.I. (1994): Elastic inclusions and inhomogeneities in transversely isotropic solids. *Proc. R. Soc. Lond. A*, 444, 239–252.

[19] Ting, T. C. T. (1996): *Anisotropic Elasticity*, Oxford University Press, Oxford.

[20] Pan, E. (1997): A general boundary element analysis of 2-D linear elastic fracture mechanics. *Int. J. Fract.*, 88, 41–59.

[21] Pan, E. (2002): Mindlin's problem for an anisotropic piezoelectric half space with general boundary conditions. *Proc. R. Soc. Lond. A*, 458, 181–208.

[22] Pan, E. (2003): Three-dimensional Green's functions in an anisotropic half space with general boundary conditions. ASME *J. Appl. Mech.*, 70, 101–110.

[23] Jiang, X. and Pan, E. (2004): Exact solution for 2D polygonal inclusion problem in anisotropic magnetoelectroelastic full-, half-, and bimaterial-planes. *Int. J. Solids Struct.*, 41, 4361–4382.

[24] Pan, E. (2003): Three-dimensional Green's functions in anisotropic elastic bimaterials with imperfect interfaces. ASME *J. Appl. Mech.*, 70, 180–190.

[25] Pan, E. (2003): Some new three-dimensional Green's functions in anisotropic piezoelectric bimaterials. *Electron. J. Bound. Elem.*, 1, 236–269.

[26] Ding, H. J., Chen, B., and Liang, J. (1996): On the general solutions for coupled equation for piezoelectric media. *Int. J. Solids Struct.*, 33, 2283-2298.

[27] Ding, H. J., Wang, G. Q., and Liang, J. (1996): General and fundamental solutions of plane piezoelectroelastic problem. *Acta Mech Sinica*, 28, 441-448.

[28] Ding, H. J., Wang, G. Q., and Chen, W. Q. (1997): Fundamental solutions for plane problem of piezoelectric materials. *Sci. China*, E40, 331–336.

[29] Ding, H. J., Chen, B., and Liang, J. (1997): Fundamental solution for transversely isotropic piezoelectric media. *Sci. China*, A39, 766–775. 27

[30] Ding, H. J., Chi, Y., and Guo, F. (1999): Solutions for transversely isotropic piezoelectric infinite body, semi-infinite body and bimaterial infinite body subjected to uniform ring loading and charge. *Int. J. Solids Struct.*, 36, 2613–2631.

[31] Ding, H. J. and Liang, J. (1999): The fundamental solutions for transversely isotropic piezoelectricity and boundary element method.*Comput. Struct.*, 71, 447–455.

[32] Ding, H. J. and Chen, W. Q. (2001): *Three Dimensional Problems of Piezoelasticity.* Nova Science, Huntington, New York.

[33] Ding, H. J., Chen, W. Q., and Jiang, A. M. (2004): Green's functions and boundary element method for transversely isotropic piezoelectric materials. *Eng. Anal. Bound. Elem*, 28, 975–987.

[34] Ding, H. J., Chen, W. Q., and Zhang, L. (2006): *Elasticity of Transversely Isotropic Materials.* Series: Solid Mechanics and Its Applications (Ed. G. M. L. Galdwell), Vol. 126, Kluwer Academic, Boston.

[35] Dunn, M. L. (1994): Electroelastic Green's functions for transversely isotropic piezoelectric media and their application to the solution of inclusion and inhomogeneity problems. *Int. J. Eng. Sci.*, 32, 119–131.

[36] Dunn, M. L. and Wienecke, H. A. (1996): Green's functions for transversely isotropic piezoelectric solids. *Int. J. Solids Struct.*, 33, 4571–4581.

[37] Dunn, M. L. and Wienecke, H. A. (1999): Half-space Green's functions for transversely isotropic piezoelectric solids. *ASME J. Appl. Mech.*, 66, 675–679.

[38] Barnett, D. M. (1972): The precise evaluation of derivatives of the anisotropic elastic Green's functions, *Phys. Stat. Sol. (b)*, 49, 741–748.

[39] Chen, T. (1993): Green's functions and the non-uniform transformation problem in a piezoelectric medium, *Mech. Res. Comm.*, 20, 271–278.

[40] Chen, T. and Lin, F. Z. (1993): Numerical evaluation of derivatives of the anisotropic piezoelectric Green's functions. *Mech. Res. Comm.*, 20, 501–506.

[41] Malen, K. (1971): A unified six-dimensional treatment of elastic Green's functions and dislocations. *Phys. Stat. Sol. (b)*, 44, 661–672.

[42] Wang, C. Y. (1997): Elastostatic fields produced by a point source in solids of general anisotropy. *J. Eng. Math.*, 32, 41–52.

[43] Akamatsu, M. and Tanuma, K. (1997): Green's function of anisotropic piezoelectricity. *Proc. R. Soc. Lond. A*, 453, 473–487.

[44] Pan, E. and Tonon, F. (2000): Three-dimensional Green's functions in anisotropic piezoelectric solids. *Int. J. Solids Struct.*, 37, 943–958.

[45] Tonon, F., Pan, E., and Amadei, B. (2001): Green's functions and boundary element method formulation for 3D anisotropic media. *Comput. Struct.*, 79, 469–482.

[46] Pan, E. and Yuan, F. G. (2000): Three-dimensional Green's functions in anisotropic piezoelectric bimaterials, *Int. J. Eng. Sci.*, 38, 1939–1960.

[47] Pan, E. and Yang, B. (2003): Three-dimensional interfacial Green's functions in anisotropic bimaterials. *Appl. Math. Model.*, 27, 307–326.

[48] Pan, E. (1989): Static response of a transversely isotropic and layered half-space to general surface loads. *Phys. Earth Planet. Inter.*, 54, 353–363.

[49] Pan, E. (1997): Static Green's functions in multilayered half-spaces. *Appl. Math. Model.*, 21, 509–521.

[50] Pan, E. (2003): Three-dimensional fundamental solutions in multilayered piezoelectric solids. *Chinese J. Mech.*, A19, 127–132.

[51] Pan, E. and Han, F. (2005): Green's functions for transversely isotropic piezoelectric functionally graded multilayered half spaces. *Int. J. Solids Struct.*, 42, 3207-3233.

[52] Han, F., Pan, E., Roy, A. K., and Yue, Z. Q. (2006): Responses of piezoelectric, transversely isotropic, functionally graded, and multilayered half spaces to uniform circular surface loadings. *Comput. Model. Eng. Sci.*, 14, 15–29.

[53] Yang, B. and Pan, E. (2002): Three-dimensional Green's functions in anisotropic trimaterials. *Int. J. Solids Struct.*, 39, 2235–2255.

[54] Yang, B. and Pan, E. (2002): Efficient evaluation of three-dimensional Green's functions in anisotropic elastostatic multilayered composites. *Eng. Anal. Bound. Elements*, 26, 355–366.

[55] Heyliger, P. and Wu, Y. C. (1999): Electroelastic fields in layered piezoelectric spheres. *Int. J. Eng. Sci.*, 38, 143–161.

[56] Chen, W. Q., Ding, H. J., and Xu, R. Q. (2001): Three-dimensional static analysis of multi-layered piezoelectric hollow spheres via the state space method. *Int. J. Solids Struct.*, 38, 4921–4936.

[57] Wang, H. M. and Ding, H. J. (2007): Radial vibration of piezoelectric/ magnetostrictive composite hollow sphere. *J. Sound Vib.*, 307, 330–348.

[58] Wang, H. M., Ding, H. J., and Ge, W. (2007): Transient responses in a two-layered elasto-piezoelectric composite hollow cylinder. *Compos. Struct.*, 79, 192–201.

[59] Tanigawa, Y. and Ootao, Y. (2007): Transient piezothermoelasticity of a two-layered composite hollow cylinder constructed of isotropic elastic and piezoelectric layers due to asymmetrical heating. *J. Thermal Stresses*, 30, 1003–1023.

[60] Sun, C. T. and Cheng, N. C. (1974): Piezoelectric waves on a layered cylinder. *J. Appl. Phys.*, 45, 4288–4294.

[61] Bai, H., Taciroglu, E., Dong, S. B., and Shah, A. H. (2004): Elastodynamic Green's functions for a laminated piezoelectric cylinder. *Int. J. Solids Struct.*, 41, 6335–6350.

[62] Bai, H., Shah, A. H., Dong, S. B., and Taciroglu, E. (2006): End reflections in a layered piezoelectric circular cylinder. *Int. J. Solids Struct.*, 43, 6309–6325.

[63] Ding, H. J., Wang, H. M., and Chen, W. Q. (2003): Transient responses in a piezoelectric spherically isotropic hollow sphere for symmetric problems. ASME *J. Appl. Mech.*, 70, 436–445.

[64] Chakraborty, A., Gopalakrishnan, S., and Kausel, E. (2005): Wave propagation analysis in inhomogeneous piezo-composite layer by the thin-layer method. *Int. J. Num. Meth. Eng.*, 64, 567–598.

[65] Norris, A. N. (1994): Dynamic Green's functions in anisotropic piezoelectric, thermoelastic and poroelastic solids. *Proc. R. Soc. Lond.*, A447, 175–188.

[66] Wang, C. Y. and Zhang, Ch. (2005): 3-D and 2-D dynamic Green's functions and time-domain BIEs for piezoelectric solids. *Eng. Anal. Bound. Elem.*, 29, 454–465.

[67] Pan, E. (2005): *Anisotropic Green's Functions and BEMs* (ed.), special issue: *Eng. Analy. Bound. Elem.*, 29, 511–671.

[68] Liu, J.X., Liu, X.L., and Zhao, Y.B. (2001): Green's functions for anisotropic magnetoelectroelastic solids with an elliptical cavity or a crack. *Int. J. Eng. Sci.*, 39, 1405–1418.

[69] Pan, E. (2002): Three-dimensional Green's functions in anisotropic magnetoelectroelastic bimaterials. *J. Appl. Math. Phys. ZAMP*, 53, 815–838.

[70] Li, J.Y. (2002): Magnetoelectric Green's functions and their application to the inclusion and inhomogeneity problems. *Int. J. Solids Struct.*, 39, 4201–4213.

[71] Qin, Q. H. (2005): 2D Green's functions of defective magnetoelectroelastic solids under thermal loading. *Eng. Analy. Bound. Elem.*, 29, 577–585.

[72] Li, L. J. and Li, J. Y. (2006): Electromagnetic fields induced in a uniaxial multiferroic material by a point source or an ellipsoidal inclusion. *Phys. Rev.*, B73, 184416.

[73] Qin, Q. H. (2007): *Green's Function and Boundary Elements of Multified Materials.* Elsevier, Oxford, UK.

[74] Lee, J. M. and Ma, C. C. (2007): Analytical full-field solutions of a magnetoelectroelastic layered half-plane. *J. Appl. Phys.*, 101, 083502.

[75] Wang, X. and Pan, E. (2007): Electromagnetic fields induced by a point source in a uniaxial multiferroic full-space, half-space, and bimaterial-space. *J. Mat. Res.*, 22, 2144–2155.

[76] Wang, X. and Pan, E. (2007): Electromagnetic fields induced by a cuboidal inclusion with uniform spontaneous polarization and magnetization. *J. Appl. Phys.*, 101, 103524.

[77] Wang, X., Pan, E., Feng, W. J., and Albrecht, J. D. (2008): Electromagnetic fields induced by a concentrated heat source in multiferroic materials. *Phys. Stat. Sol. (b)*, 245, 206–212.

Chapter 3
Two-Dimensional Static Problems: Stroh Formalism

Hui Fan

3.1 Introduction and General Equations

The two-dimensional configuration has had its unique position in boundary value problems since people developed various analytical concepts and procedures from it. The focus of the present chapter is to introduce the Stroh formalism in the two-dimensional piezoelectric boundary value problem. Historically, the scheme was originally proposed by Eshelby et al. [1] for linear elasticity in anisotropic solids. The formulation has been considered to be elegant and neat for studies of dislocation [2], wave propagation [3], and interfacial cracks [4]. Then, the scheme was extended to the piezoelectric solids which are intrinsically anisotropic. Barnett and Lothe [5] extended this formalism to the piezoelectricity when a dislocation in an infinite piezoelectric medium was studied. Most important configurations of boundary value problems in piezoelectricity via Stroh formalism were studied in the 1990s after their corresponding problems in anisotropic elasticity had been worked out in the 1980s and early 1990s. In this chapter, we present a few fundamental boundary value problems in piezoelectricity as follows.

1. Elliptical inclusion in piezoelectric medium [6]: This problem is a fundamental configuration in the composite mechanics, known as the Eshelby [7] inclusion problem in the two-dimensional case.
2. Fracture: A widely cited research paper by Suo et al. [8] for an interfacial crack in a piezoelectric composite is followed.
3. Contact in two-dimensional piezoelectric half-space [9].
4. Saint-Venant end effect in a piezoelectric strip [10].

In a piezoelectric boundary value problem, we have field variables σ_{ij}, S_{ij}, D_i, and E_k. In words, they are stress, strain, electric displacement (or electric

Hui Fan
Nanyang Technological University, Republic of Singapore

J. Yang (ed.), *Special Topics in the Theory of Piezoelectricity*, DOI: 10.1007/978-0-387-89498-0_3, 47
© Springer Science + Business Media, LLC 2009

induction), and electric field, respectively. It is realized that $\boldsymbol{S}$ and $\boldsymbol{D}$ together define the physical distortion of the material. They are assumed to be reversible, so that an energy function, $\Phi(S, D)$, exists; that is,

$$d\Phi = \sigma_{ij} dS_{ij} + E_i dD_i. \tag{3.1}$$

Because it is more convenient to formulate the problem in terms of $\boldsymbol{u}$ and φ (displacement and electric potential), we adopt another energy function

$$w = \Phi - E_i D_i. \tag{3.2}$$

A Legendre transformation shows that

$$dw = \sigma_{ij} dS_{ij} - D_i dE_i. \tag{3.3}$$

The stress and electric induction are therefore defined by

$$\sigma_{ij} = \frac{\partial w}{\partial S_{ij}}, \qquad D_i = -\frac{\partial w}{\partial E_i}. \tag{3.4}$$

The linear piezoelectric materials are described by

$$w = \frac{1}{2} C_{ijkl} S_{ij} S_{kl} - \frac{1}{2} \varepsilon_{ij} E_i E_j - e_{ikl} E_i S_{kl}. \tag{3.5}$$

Function w is particularly meaningful for the fracture mechanics in the piezoelectric solid. The energy release rate or J-integral is defined by Suo et al. [8]

$$G = J = \int (w n_1 - n_i \sigma_{ip} u_{p,1} - n_i D_i \varphi_{,1}) ds$$

which is discussed in Section 3.3.

In the following sections, we focus on a linear piezoelectric material. In a rectangular coordinate system, the boundary value problem for a linear piezoelectric solid is described by
Constitutive laws:

$$\sigma_{ij} = C_{ijkl} S_{kl} - e_{kij} E_k, \tag{3.6a}$$

$$D_i = e_{ikl} S_{kl} + \varepsilon_{ik} E_k, \tag{3.6b}$$

Deformation relations:

$$S_{kl} = \frac{1}{2}(u_{k,l} + u_{l,k}), \tag{3.7a}$$

$$E_k = -\varphi_{,k}, \tag{3.7b}$$

where u_k and φ are mechanical displacement and electric potential, respectively.

Equilibrium equations:

$$\sigma_{ij,i} = 0, \tag{3.8a}$$

$$D_{i,i} = 0. \tag{3.8b}$$

In the above equations, $\boldsymbol{C}$ is the elasticity tensor of rank four, ε is the permittivity tensor of rank two, and $\boldsymbol{e}$ is the piezoelectricity tensor of rank three. When the piezoelectricity vanishes, the problem is decoupled into the anisotropic elastic and the dielectric problems.

Besides the governing equations, the boundary and continuity conditions, on the other hand, are the important issues in the boundary value problems. They vary from one problem to another. We cite them accordingly in the following sections for various boundary value problems.

3.1.1 Two-Dimensional Piezoelectric Boundary Value Problems via Stroh Formalism

Starting with the governing equations of the linear piezoelectric solid, we substitute Equations (3.6) and (3.7) into (3.8):

$$(C_{ijkl}u_k + e_{lji}\varphi)_{,li} = 0 \tag{3.9a}$$

$$(e_{ikl}u_k - \varepsilon_{il}\varphi)_{,li} = 0. \tag{3.9b}$$

If all the fields are independent of the third coordinate, say x_3, the special solutions can be sought in the form of

$$\mathrm{U} = \{u_k, \varphi\}^T = f(\varsigma_1 x_1 + \varsigma_2 x_2)\mathbf{a}, \tag{3.10}$$

where, without losing generality,

$$\varsigma_1 = 1, \qquad \varsigma_2 = p, \tag{3.11}$$

and $\mathbf{a} = (a_1, a_2, a_3, a_4)^T$ is independent of the spatial coordinates.

A direct substitution of Equation (3.10) into (3.9) gives

$$(C_{\alpha ik\beta}a_k + e_{\alpha i\beta}a_4)\varsigma_\alpha\varsigma_\beta = 0, \tag{3.12a}$$

$$(e_{\alpha k\beta}a_k - \varepsilon_{\alpha\beta}a_4)\varsigma_\alpha\varsigma_\beta = 0. \tag{3.12b}$$

For nonzero solution for $\mathbf{a}'$s, we must have

$$\det \begin{bmatrix} C_{\alpha jk\beta}\varsigma_\alpha\varsigma_\beta & e_{\alpha j\beta}\varsigma_\alpha\varsigma_\beta \\ e_{\alpha k\beta}\varsigma_\alpha\varsigma_\beta & -\varepsilon_{\alpha\beta}\varsigma_\alpha\varsigma_\beta \end{bmatrix} = 0. \tag{3.13}$$

As in the anisotropic elasticity formulation, it can be proved that the eigenvalue p cannot be purely real due to the positive definiteness of the tensors C_{ijkl} and ε_{ij}. Four pairs of p can be arranged as

$$p_{I+4} = \bar{p}_I, \qquad (I = 1, 2, 3, \text{ and } 4)$$
$$p_I = \alpha_I + i\beta_I, \qquad \beta_I > 0. \tag{3.14}$$

Corresponding to the eigenvalues $p_I = \alpha_I + i\beta_I$, there are four independent eigenvectors, which form a 4×4 matrix, namely,

$$\mathbf{A} = \{\mathbf{a_1}, \mathbf{a_2}, \mathbf{a_3}, \mathbf{a_4}\}. \tag{3.15}$$

The complex conjugates,

$$\overline{\mathbf{A}} = \{\overline{\mathbf{a}}_1, \overline{\mathbf{a}}_2, \overline{\mathbf{a}}_3, \overline{\mathbf{a}}_4\}, \tag{3.16}$$

are the eigenvectors corresponding to $p_{I+4} = \bar{p}_I$.

Using the eigenvalues and corresponding eigenvectors, the general solution can be written as a linear combination of the eight eigenvectors:

$$\mathbf{U} = 2\mathrm{Re}\left((\mathbf{a_1}, \mathbf{a_2}, \mathbf{a_3}, \mathbf{a_4}) \begin{pmatrix} f_1(z_1) \\ f_2(z_2) \\ f_3(z_3) \\ f_4(z_4) \end{pmatrix} \right), \tag{3.17}$$

where

$$z_I = x_1 + p_I x_2, \qquad (I = 1, 2, 3, 4).$$

For the sake of convenience, we write it in a compact form,

$$\mathbf{U} = 2\mathrm{Re}[\mathbf{A}\mathbf{f}(z)]. \tag{3.18}$$

Furthermore, from the constitutive equations, we have

$$\mathbf{t} = \{\sigma_{2i}, D_2\} = 2\mathrm{Re}[\mathbf{B}\mathbf{f}'(z)] \tag{3.19}$$
$$\mathbf{s} = \{\sigma_{1i}, D_1\} = -2\mathrm{Re}[\mathbf{B}\mathbf{P}\mathbf{f}'(z)], \tag{3.20}$$

where

$$\mathbf{B} = \{\mathbf{b_1}, \mathbf{b_2}, \mathbf{b_3}, \mathbf{b_4}\}, \qquad \mathbf{P} = \mathrm{diag}\{p_1, \ p_2, p_3, p_4\} \tag{3.21}$$
$$\mathbf{b}_j = (C_{2jk\beta}a_k + e_{\beta j2}a_4)\varsigma_\beta, \qquad \mathbf{b}_4 = (-\varepsilon_{1\beta}a_4 + e_{2k\beta}a_k)\varsigma_\beta. \tag{3.22}$$

It is noticed that the matrices $\boldsymbol{A}$ and $\boldsymbol{B}$ are nonsingular when the eigenvalues are distinct. However, they may be singular in some special cases in which eigenvalues coincide. Ting [11] has undertaken extensive studies on these degenerate cases which are not discussed in detail here. In the following sections, we take A and B as nonsingular matrices.

3.1.2 Eight-Dimensional Representation of Piezoelectricity

An alternative formulation via an eight-dimensional scheme was introduced by Barnett and Lothe [5], which is briefly described in the present section. In the following formulation, lowercase subscripts take on the range 1, 2, and 3, and the uppercase subscripts take on the range 1, 2, 3, and 4. The eight-dimensional formulation starts with introducing some auxiliary symbols in order to do matrix formulations.

$$Z_{Mn} = \begin{cases} S_{mn} & M = 1,2,3 \\ -E_n & M = 4 \end{cases} \tag{3.23}$$

$$\Sigma_{Mn} = \begin{cases} \sigma_{mn} & M = 1,2,3 \\ D_n & M = 4 \end{cases} \tag{3.24}$$

$$U_M = \begin{cases} u_m & M = 1,2,3 \\ \phi & M = 4 \end{cases} \tag{3.25}$$

$$E_{iJMn} = \begin{cases} C_{ijmn} & J,\ M = 1,2,3 \\ e_{nij} & J = 1,2,3; \quad M = 4 \\ e_{imn} & J = 4; \quad M = 1,2,3 \\ -\varepsilon_{in} & J,\ M = 4. \end{cases} \tag{3.26}$$

It should be pointed out that they are not tensors. One has to be careful when changing coordinate systems.

The constitutive equation (Equation (3.6)) is then written as

$$\Sigma_{iJ} = E_{iJMn} Z_{Mn} = E_{iJMn} U_{M,n}. \tag{3.27}$$

The equilibrium equation (Equation (3.8)) is written as

$$\Sigma_{iJ,i} = 0. \tag{3.28}$$

To satisfy Equation (3.28), a stress function $\Phi = \{\Phi_1, \Phi_2, \Phi_3, \Phi_4\}^T$ is introduced:

$$\Sigma_{1J} = \Phi_{J,2} \qquad \Sigma_{2J} = \Phi_{J,1} \tag{3.29}$$

Substituting Equation (3.29) into (3.27), one obtains:

$$\mathbf{QU}_{,1} + \mathbf{RU}_{,2} = -\boldsymbol{\Phi}_{,2} \qquad \mathbf{R}^T\mathbf{U}_{,1} + \mathbf{TU}_{,2} = \boldsymbol{\Phi}_{,1}, \tag{3.30}$$

where

$$Q_{JM} = E_{1JM1}, \qquad R_{JM} = E_{1JM2}, \qquad T_{JM} = E_{2JM2}. \tag{3.31}$$

It is noted that matrices $\boldsymbol{Q}$ and $\boldsymbol{T}$ are symmetric. Then, Equation (3.30) can be written in an eight-dimensional form as

$$\frac{\partial \mathbf{w}}{\partial y} = \mathbf{N}\frac{\partial \mathbf{w}}{\partial x}, \tag{3.32}$$

where

$$\mathbf{w} = \begin{pmatrix} U \\ \Phi \end{pmatrix}, \qquad \mathbf{N} = \begin{pmatrix} \mathbf{N}_1 & \mathbf{N}_2 \\ \mathbf{N}_3 & \mathbf{N}_1^T \end{pmatrix} \tag{3.33}$$

$$\mathbf{N}_1 = -\mathbf{T}^{-1}\mathbf{R}^T, \qquad \mathbf{N}_2 = \mathbf{T}^{-1} = \mathbf{N}_2^T \qquad \mathbf{N}_3 = \mathbf{RT}^{-1}\mathbf{R}^T - \mathbf{Q} = \mathbf{N}_3^T. \tag{3.34}$$

The inverse of matrix $\mathbf{T}$ is obtained based on the following argument. From Equations (3.26) and (3.31), it follows

$$\mathbf{T} = \begin{pmatrix} \mathbf{T}_e & \mathbf{e} \\ \mathbf{e}^T & -\varepsilon_{22} \end{pmatrix}, \tag{3.35}$$

where the upper-left corner of the matrix $\mathbf{T}_e$ is made by part of the elastic tensor which is 3 by 3 positive definite [4]; the lower-right corner is contributed from the dielectric tensor, a negative scalar; and $\mathbf{e}$ is formed by proper components of the piezoelectric tensor. It is found that

$$\mathbf{T}^{-1} = \begin{pmatrix} \mathbf{T}_e^{-1}(\mathbf{I} + q\mathbf{e}\mathbf{e}^T\mathbf{T}_e^{-1}) & -q\mathbf{T}_e^{-1}\mathbf{e} \\ -q\mathbf{e}^T\mathbf{T}_e^{-1} & q \end{pmatrix}, \tag{3.36}$$

where, by using the fact that $\mathbf{T}_e$ is positive definite [11],

$$q = \frac{1}{-\varepsilon_{22} - \mathbf{e}^T\mathbf{T}_e^{-1}\mathbf{e}} < 0. \tag{3.37}$$

In order to diagonalize Equation (3.32), the following eigenvalue problem is considered,

$$\mathbf{N}\xi = p\xi, \tag{3.38}$$

where

$$\xi = \begin{pmatrix} \mathbf{a} \\ \mathbf{b} \end{pmatrix}. \tag{3.39}$$

Four-dimensional a and b are defined in Equations (3.15) and (3.22). The eigenvalues are obtained by solving

$$\det(\mathbf{N} - p\mathbf{I}) = 0. \tag{3.40}$$

The explicit form of Equation (3.40) is exactly the same as (3.13) which is in a four-dimensional format. In comparison with Equations (3.12) and (3.13), Equation (3.40) represents a standard eigenvalue problem for a nonsymmetric matrix N.

3.1.3 Historical Remarks on Stroh Formalism

Formulation of the two-dimensional piezoelectric boundary value problems is a natural extension of solutions for the anisotropic elasticity in terms of mathematical approaches. Therefore, development of Stroh formalism in piezoelectricity has been following the steps of research progress in anisotropic elasticity. Historically, the so-called Stroh formalism was originally proposed by Eshelby et al. [1], and used by Stroh [2] in dealing with dislocations in the anisotropic elastic solids. Of course, the approach was not called "Stroh formalism" then. The key feature of this formalism is the presentation of gathering the elastic constants into a couple of matrices, A and B. A number of mathematicians have been carrying out the further development of the matrix formalism. Chadwick, Barnett, and Lothe, among others, have made great contributions to the matrix formalism of anisotropic elasticity. In the 1980s, anisotropic composite research created a new wave of interest in anisotropic elasticity. A persistent and hard-working researcher, T. C. T. Ting, who also has a strong mathematical background, has devoted 25 years to anisotropic elasticity since the beginning of the 1980s. His book [11] summarized almost all the important contributions in Stroh formalism until the early 1990s. The "Stroh formalism" became an "official" name for the matrix formalism after the 1980s by the strong recommendation of researchers with mathematical backgrounds. At the same time, they labeled research work not following matrix formalism as the "Lekhnitskii approach" [12].

Actually, the mathematical background researchers always claimed their result as Stroh formalism because it is elegant and compact, whereas the engineering background researchers directly solved the problem without using matrix notations. Most of the time, engineers simplified their problems from the very beginning of their solution with consideration of the symmetry of the material and structural directions. For example, Yang [13], Pak [14], and Sosa and Castro [15] did not mention that they belong to the non-Stroh camp. Although these research results are not considered as "elegant and neat," they are of great engineering interest and are called up for verification as the special cases for the solutions derived via Stroh formalism. It is interesting to notice that some of the researchers, for example, Suo et al. [8], prefer somewhere in between the aforementioned two camps. They use part of the matrix formalism for the sake of elegance in presenting the general theory. When the materials have certain symmetric features, they prefer the Lekhnitskii format.

3.2 Piezoelectric Solid with an Elliptic Inclusion

3.2.1 Importance of the Inclusion Problem

A three-dimensional ellipsoidal inclusion embedded in an infinitely extended matrix is a fundamental configuration for composite mechanics. The three-dimensional isotropic elastic solution was obtained by Eshelby [7] which has been considered as a cornerstone of composite mechanics [16]. The most important and useful result obtained by Eshelby is that under uniform far-field loading, the strain inside the inclusion is uniform. It has been noticed that all the boundary value problems described by an even order of partial differential equations (sets) of elliptical type have a similar feature. For example, the dielectric problem is described by an elliptical-type partial differential equation (Laplace equation) on the order of two; the linear elastic problem is described by a set of partial equations (Navier equation) on the order of six. Extension of the Eshelby solution to the piezoelectric boundary value problem (a set of partial differential equations on the order of eight) was attempted by Wang [17]. However, due to the anisotropy of the material, the Green's function cannot be integrated into a closed form. Therefore, the so-called Eshelby tensor can only be obtained via some numerical procedure. In turn, it is difficult further to employ inclusion solutions for the anisotropic elasticity and piezoelectricity for micromechanics schemes, such as the self-consistent scheme [18], for composite mechanics applications.

Hereby, in this section, we present a solution for an elliptical inclusion embedded in an infinitely extended space as a two-dimensional boundary value problem. By employing the Stroh formalism, we follow the major steps of the derivation contributed by Chung and Ting [6].

3.2.2 Elliptic Inclusion Embedded in an Infinitely Extended Solid

The ellipse is mathematically described by

$$\Gamma : x_1 = a \cos \psi, \qquad x_2 = b \sin \psi. \tag{3.41}$$

The material inside the ellipse is called inclusion, and the material outside the ellipse is called the matrix. Of composite mechanical interest, we only consider the case of uniform far-field loading. The far-field variables are given as follows.

$$\mathbf{U}^\infty = x_1 \varepsilon_1^\infty \quad \text{and} \quad \mathbf{\Phi}^\infty = x_1 \mathbf{s}^\infty - x_2 \mathbf{t}^\infty, \tag{3.42}$$

where

$$\varepsilon_1^\infty = \mathbf{U}_{,1}^\infty = \begin{bmatrix} \varepsilon_{11}^\infty \\ 0 \\ 2\varepsilon_{13}^\infty \\ -E_1^\infty \end{bmatrix} \quad \text{and} \quad \varepsilon_2^\infty = \mathbf{U}_2^\infty = \begin{bmatrix} 2\varepsilon_{21}^\infty \\ \varepsilon_{22}^\infty \\ 2\varepsilon_{23}^\infty \\ -E_2^\infty \end{bmatrix}. \tag{3.43}$$

$\mathbf{s}^\infty$ and $\mathbf{t}^\infty$ are defined according to Equations (3.20) and (3.19). There is a connection between them,

$$\mathbf{N} \begin{bmatrix} \varepsilon_1^\infty \\ \mathbf{t}^\infty \end{bmatrix} = \begin{bmatrix} \varepsilon_2^\infty \\ -\mathbf{s}^\infty \end{bmatrix}, \tag{3.44}$$

where the $\mathbf{N}$ matrix is given by Equation (3.34). The far-field information is considered to be completely known.

For this boundary value problem, the complex function given in Equation (3.18) takes the form of

$$f_\alpha(z_*) = q_\alpha f(z_*), \tag{3.45}$$

which leads Equation (3.18) to

$$\mathbf{U} = 2\text{Re}[\mathbf{A} < f(z_*) > \mathbf{q}]. \tag{3.46}$$

Equation (3.19), with the definition of (3.29), becomes

$$\mathbf{\Phi} = 2\text{Re}[\mathbf{B} < f(z_*) > \mathbf{q}]. \tag{3.47}$$

In Equations (3.46) and (3.47),

$$< f(z_*) > = \text{diag} < f(z_1), f(z_2), f(z_3), f(z_4) > . \tag{3.48}$$

A mapping function is needed as follows.

$$z_\alpha = c_\alpha \xi_\alpha + d_\alpha \xi_\alpha^{-1} \quad (\alpha = 1, 2, 3, 4), \tag{3.49}$$

where c_α and d_α are the complex constants and $z_\alpha = x_1 + p_\alpha x_2$. c_α and d_α are chosen such that when $(x_1, x_2) \subset \Gamma$, $\xi_\alpha(\alpha = 1, 2, 3, 4)$ is on a unit circle. That is,

$$\xi_\alpha\big|_\Gamma = \cos\psi + i\sin\psi \quad (\alpha = 1, 2, 3, 4) \tag{3.50}$$

and

$$z_\alpha = a\cos\psi + p_\alpha b\sin\psi. \tag{3.51}$$

Substituting (3.47) and (3.48) into (3.46), we have

$$c_\alpha = \frac{a - ip_\alpha b}{2} \quad \text{and} \quad d_\alpha = \frac{a + ip_\alpha b}{2}. \tag{3.52}$$

By means of superposition, the field solution in the matrix is given by Chung and Ting [6],

$$\mathbf{U} = \mathbf{U}^\infty + 2\mathrm{Re}[\mathbf{A} < \xi_*^{-1}] > \mathbf{A}^T]\mathbf{g}_1 + 2\mathrm{Re}[\mathbf{A} < \xi_*^{-1}] > \mathbf{B}^T]\mathbf{h}_1 \tag{3.53a}$$

$$\boldsymbol{\Phi} = \boldsymbol{\Phi}^\infty + 2\mathrm{Re}[\mathbf{B} < \xi_*^{-1}] > \mathbf{A}^T]\mathbf{g}_1 + 2\mathrm{Re}[\mathbf{B} < \xi_*^{-1}] > \mathbf{B}^T]\mathbf{h}_1. \tag{3.53b}$$

In Equations (3.53a) and (3.53b), the first terms are the uniform solution given by (3.42), and the second terms are due to the presence of the inclusion. Comparing Equations (3.53a) and (3.53b) with Equations (3.46) and (3.47), we have introduced

$$\mathbf{q} = \mathbf{A}^T\mathbf{g}_1 + \mathbf{B}^T\mathbf{h}_1, \tag{3.54}$$

where $\boldsymbol{g}$ and $\boldsymbol{h}$ are real constants, whereas $\boldsymbol{q}$ is a complex constant. On the other hand, the field solution inside the inclusion is given by a uniform field inspired by the Eshelby [7] solution.

$$\mathbf{U}^0 = x_1 \varepsilon_1^0 + x_2 \varepsilon_2^0, \qquad \boldsymbol{\Phi}^0 = x_1 \mathbf{s}^0 - x_2 \mathbf{t}^0, \tag{3.55}$$

where

$$\varepsilon_1^0 = \mathbf{U}_{,1}^0 = \begin{bmatrix} \varepsilon_{11}^0 \\ \Omega \\ 2\varepsilon_{13}^0 \\ -E_1^0 \end{bmatrix}, \qquad \varepsilon_2^0 = \mathbf{U}_{,2}^0 = \begin{bmatrix} 2\varepsilon_{21}^0 - \Omega \\ \varepsilon_{22}^0 \\ 2\varepsilon_{23}^0 \\ -E_2^0 \end{bmatrix}. \tag{3.56}$$

$\mathbf{s}^0$ and $\mathbf{t}^0$ are defined according to Equations (3.20) and (3.19).

Summarizing the solutions in the matrix (Equations (3.53a) and (3.53b)) and the inclusion (Equation (3.55)), we have six constants (four-dimensional columns) to be determined, namely, $\boldsymbol{g}$, $\boldsymbol{h}$, ε_1^0, ε_2^0, $\mathbf{s}^0$, and $\mathbf{t}^0$. By reinforcing the continuity condition along the interface Γ, that is,

$$\mathbf{U}\big|_\Gamma = \mathbf{U}^0\big|_\Gamma \tag{3.57}$$

and

$$\boldsymbol{\Phi}\big|_\Gamma = \boldsymbol{\Phi}^0\big|_\Gamma, \tag{3.58}$$

we have four sets of equations as the result of collecting terms associated with $\cos\psi$ and $\sin\psi$, respectively. Another two sets of equations come from the constitutive relation between strains and stresses,

$$\mathbf{N}^0 \begin{bmatrix} \varepsilon_1^0 \\ \mathbf{t}_2^0 \end{bmatrix} = \begin{bmatrix} \varepsilon_2^0 \\ -\mathbf{t}_1^0 \end{bmatrix}. \tag{3.59}$$

Therefore, Equations (3.57) through (3.59) provide the answer for g, h, ε_1^0, ε_2^0, s^0, and t^0.

3.2.3 Remarks

It is seen that the solution to a piezoelectric solid for a specific configuration (e.g., the elliptic inclusion embedded in the infinite plane) can be obtained by modifying the corresponding anisotropic elastic solution. The two-dimensional inclusion problem in the anisotropic elasticity was solved by Hwu and Ting [19]. Then, extension of the solution to the piezoelectric solid was carried out by Chung and Ting [6]. Chung and Ting [20] also presented a solution for the "piezoelectric–piezomagnetic–magnetoelectric" solid, which was described by 10 by 10 N matrices. This natural extension from anisotropic elasticity of six-dimensional formalism to the eight-dimensional or even ten-dimensional formalism is considered as one of the advantages of Stroh formalism. We also encounter this type of extension of the anisotropic elasticity solution for other configurations in the following sections, for example, the crack problem and the contact problem.

The Stroh formalism does not, in general, give a closed-form analytical result. Numerical procedure may be needed at the stage of finding eigenvalues from Equation (3.13) or (3.40). Only for some special materials, as shown in the later sections, we are able to have the eigenvalues obtained analytically. Therefore, Stroh formalism provides us an analytical procedure, rather than an analytical closed-form solution.

When the materials possess certain symmetry, the eigenvalues of Equation (3.13) or (3.40) are not distinct; the matrices A and B may be singular. Therefore, the eigenexpansion of Equations (3.18) and (3.19) are not valid. The advocators of Stroh formalism have been working on these singular cases with some special consideration (e.g., Ting [11] and Suo et al. [8]). Stroh formalism treats the isotropic and highly symmetric materials as the degenerate cases, whereas researchers from the non-Stroh camp prefer material symmetry because they start with isotropic elastic and dielectric solids towards the anisotropic solids.

3.3 Cracks in Piezoelectric Solids

Piezoelectric ceramics are brittle in nature. High mechanical and electrical loadings may result in microcracks. The study of cracks in the piezoelectric solid became a very active research topic after Suo et al. [8]. Because there is an independent chapter in this book devoted to the fracture mechanics of piezoelectric solids, we only touch on the aspects involved with Stroh formalism in the present section.

3.3.1 Crack-Tip Solution

Let us consider the crack-tip field, which is also called the asymptotic solution of the crack-tip. A semi-infinite crack, denoted by L, lies on $y = 0$ between two half-spaces (i.e., $y > 0$) and $y < 0$). The crack-tip coincides with origin $x = 0$; $x > 0$ is bounded. The boundary and continuity conditions along $x = 0$ are stated as follows.

$$\mathbf{t}_1(x) = \mathbf{t}_2(x) = 0, \quad \forall x \in L \tag{3.60}$$

$$\mathbf{t}_1(x) = \mathbf{t}_2(x), \quad \forall x \notin L \tag{3.61}$$

$$\mathbf{U}_1(x) = \mathbf{U}_2(x), \quad \forall x \notin L, \tag{3.62}$$

where subscript "1" stands for the upper half-space ($y > 0$); subscript "2" stands for the lower half-space ($y < 0$).

Following the notation of Equation (3.19), we have the above conditions (3.60) and (3.61) rewritten as

$$\mathbf{B}_1\mathbf{f}_1'(x_1) + \overline{\mathbf{B}_1}\overline{\mathbf{f}_1'(x_1)} = \mathbf{B}_2\mathbf{f}_2'(x_1) + \overline{\mathbf{B}_2}\overline{\mathbf{f}_2'(x_1)}, \qquad \forall x_1 \in (-\infty, \infty) \tag{3.63}$$

or

$$\mathbf{B}_1\mathbf{f}_1'(x_1) - \overline{\mathbf{B}_2}\overline{\mathbf{f}_2'(x_1)} = \mathbf{B}_2\mathbf{f}_2'(x_1) - \overline{\mathbf{B}_1}\overline{\mathbf{f}_1'(x_1)}, \qquad \forall x_1 \in (-\infty, \infty). \tag{3.64}$$

The left-hand side of Equation (3.64) is the boundary value of a function analytic in the upper half-plane, and the right-hand side is the boundary value of another function analytic in the lower half-plane. Therefore, both functions can be analytically continued into the entire plane. Because both functions vanish at any infinity to conform to a zero field, they must vanish at $z = x + py$ (Im $p > 0$). Thus,

$$\mathbf{B}_1\mathbf{f}_1'(z) = \overline{\mathbf{B}_2}\overline{\mathbf{f}_2'}(z), \qquad y > 0. \tag{3.65}$$

Define the displacement jump across the interface $y = 0$,

$$\mathbf{d}(x) = \mathbf{U}_1(x) - \mathbf{U}_2(x). \tag{3.66}$$

By using the definition of Equation (3.18) , with help of (3.65), we have

$$i\mathbf{d}'(x) = \mathbf{H}\mathbf{B}_1\mathbf{f}_1'(x) - \overline{\mathbf{H}}\mathbf{B}_2\mathbf{f}_2'(x), \tag{3.67}$$

where

$$\mathbf{H} = \mathbf{Y}_1 + \overline{\mathbf{Y}}_2 \tag{3.68}$$

with

$$\mathbf{Y} = i\mathbf{A}\mathbf{B}^{-1}. \tag{3.69}$$

It is clear that $\boldsymbol{H}$ is a real matrix for our case where upper and lower materials are the same. The present derivation with different subscripts can be easily extended to the crack along a bimaterial interface.

We define a new analytic function

$$\mathbf{h}(z) = \mathbf{B}_1 \mathbf{f}_1'(z) \qquad \forall x_2 > 0 \qquad \mathbf{h}(z) = \mathbf{B}_2 \mathbf{f}_2'(z) \qquad \forall x_2 < 0 \tag{3.70}$$

which is analytic throughout the whole plane except on the crack. The traction-free condition, Equation (3.60), leads to a Hilbert problem

$$\mathbf{h}^+(x_1) + \mathbf{h}^-(x_1) = 0 \qquad \forall x_1 \in L. \tag{3.71}$$

The superscript "+" stands for the value from the upper plane, and "−" stands for the value from the lower plane. The singular solution to Equation (3.12) that keeps the displacement finite is

$$\mathbf{h}(z) = (8\pi z)^{-1/2}\mathbf{k}, \tag{3.72}$$

where the branch cut is along the crack line $(x < 0)$. Some important results used in the following sections are presented here. The traction ahead of the crack-tip is given by

$$\mathbf{t}(r) = (2\pi r)^{-1/2}\mathbf{k}, \tag{3.73}$$

where $\mathbf{k} = (K_{II}, K_I, K_{III}, K_{IV})^T$ is the so-called stress intensity factor. The square-root singularity is inherited. The crack opening displacement behind the crack-tip is given by

$$\mathbf{d}(r) = \left(\frac{2r}{\pi}\right)^{1/2} \mathbf{H}\mathbf{k}. \tag{3.74}$$

Equations (3.73) and (3.74) are needed for calculation of the energy release rate; that is,

$$J = G = \frac{1}{2\Lambda} \int_0^\Lambda \mathbf{t}^T(\Lambda - r)\, \mathbf{d}(r) dr. \tag{3.75}$$

Equation (3.75) has a very clear physical meaning, that is, the energy change with the crack advanced amount of Λ. By substituting Equations (3.73) and (3.74) into (3.75), we obtain

$$J = G = \frac{1}{4}\mathbf{k}^T\mathbf{H}\mathbf{k}. \tag{3.76}$$

In order to visualize the derivation in this section, we consider a simplified case which was studied by Pak [14] via a non-Stroh formalism approach.

3.3.2 Crack Front Coincides with Poling Direction (x_3 -Axis)

Let us consider the materials, such as PZT-5H, poled in the z-direction. The x–y plane is the isotropic plane. The material constants are given by

$$
\begin{bmatrix} \sigma_{11} \\ \sigma_{22} \\ \sigma_{33} \\ \sigma_{23} \\ \sigma_{13} \\ \sigma_{12} \end{bmatrix} = \begin{bmatrix} c_{11} & c_{12} & c_{13} & 0 & 0 & 0 \\ c_{12} & c_{11} & c_{13} & 0 & 0 & 0 \\ c_{13} & c_{13} & c_{33} & 0 & 0 & 0 \\ 0 & 0 & 0 & c_{44} & 0 & 0 \\ 0 & 0 & 0 & 0 & c_{44} & 0 \\ 0 & 0 & 0 & 0 & 0 & (c_{11}-c_{12})/2 \end{bmatrix} \begin{bmatrix} S_{11} \\ S_{22} \\ S_{33} \\ 2S_{32} \\ 2S_{13} \\ 2S_{12} \end{bmatrix} - \begin{bmatrix} 0 & 0 & e_{31} \\ 0 & 0 & e_{31} \\ 0 & 0 & e_{33} \\ 0 & e_{15} & 0 \\ e_{15} & 0 & 0 \\ 0 & 0 & 0 \end{bmatrix} \begin{bmatrix} E_1 \\ E_2 \\ E_3 \end{bmatrix}
$$

$$\tag{3.77}$$

and

$$
\begin{bmatrix} D_1 \\ D_2 \\ D_3 \end{bmatrix} = \begin{bmatrix} 0 & 0 & 0 & 0 & e_{15} & 0 \\ 0 & 0 & 0 & e_{15} & 0 & 0 \\ e_{31} & e_{31} & e_{33} & 0 & 0 & 0 \end{bmatrix} \begin{bmatrix} S_{11} \\ S_{22} \\ S_{33} \\ 2S_{32} \\ 2S_{13} \\ 2S_{12} \end{bmatrix} + \begin{bmatrix} \varepsilon_{11} & 0 & 0 \\ 0 & \varepsilon_{11} & 0 \\ 0 & 0 & \varepsilon_{33} \end{bmatrix} \begin{bmatrix} E_1 \\ E_2 \\ E_3 \end{bmatrix}. \tag{3.78}
$$

This two-dimensional crack problem in the (x, y)-plane is decomposed into an in-plane deformation (u_x, u_y) and an antiplane field (u_z, φ). The former is identical to an isotropic elastic problem. We focus on the latter which shows the mechanical–electrical coupling effect.

The eigenvalues corresponding to this antiplane problem are obtained from Equation (3.13) as

$$
\begin{aligned}
p_3 &= p_4 = i, \\
\bar{p}_7 &= \bar{p}_8 = -i
\end{aligned} \tag{3.79}
$$

and the related matrices are

$$
\mathbf{A} = \begin{bmatrix} \mathbf{A}_{2e} & 0 \\ 0 & \mathbf{A}_p \end{bmatrix}, \qquad \mathbf{B} = \begin{bmatrix} \mathbf{B}_{2e} & 0 \\ 0 & \mathbf{B}_p \end{bmatrix} \tag{3.80}
$$

$$
\mathbf{Y} = \begin{bmatrix} \mathbf{Y}_{2e} & 0 \\ 0 & \mathbf{Y}_p \end{bmatrix}, \tag{3.81}
$$

where the right upper 2 by 2 matrices correspond to in-plane deformation, and the left lower corner matrices correspond to the coupled antiplane deformation and electric field. It is straightforward to derive:

$$
\mathbf{A}_p = \begin{bmatrix} 1 & 0 \\ 0 & 1 \end{bmatrix} \quad \text{and} \quad \mathbf{B}_p = i \begin{bmatrix} c_{44} & e_{15} \\ e_{15} & -\varepsilon_{11} \end{bmatrix} \tag{3.82}
$$

and

$$\mathbf{Y}_p = \frac{1}{1+k} \begin{bmatrix} c_{44}^{-1} & ke_{15}^{-1} \\ ke_{15}^{-1} & -\varepsilon_{11}^{-1} \end{bmatrix}, \tag{3.83}$$

where

$$k = e_{15}^2/(c_{44}\varepsilon_{11}). \tag{3.84}$$

The energy release rate or J-integral is calculated by substituting Equation (3.83) into (3.76), with the help of Equation (3.68),

$$J = \frac{1-\nu}{2\mu}(K_I^2 + K_{II}^2) + \frac{1}{2(1+k)}\left[\frac{K_{III}^2}{c_{44}} - \frac{K_{IV}^2}{\varepsilon_{11}} + 2k\frac{K_{III}K_{IV}}{e_{15}}\right]. \tag{3.85}$$

3.3.3 Full Field Solutions

(a) *Griffith crack.* Let us consider a center crack, $L : (-a, a)$, along the x-axis. The only load acting on the configuration is the traction-charge $\mathbf{T}(x)$ prescribed on the upper and lower crack surfaces

$$\mathbf{t}_1(x) = \mathbf{t}_2(x) = -\mathbf{T}(x), \qquad \forall x \in L. \tag{3.86}$$

The Hilbert problem of Equation (3.71) becomes nonhomogeneous,

$$\mathbf{h}^+(x_1) + \mathbf{h}^-(x_1) = -\mathbf{T}(x), \qquad \forall x_1 \in L. \tag{3.87}$$

This equation has many solutions [21]. The solution for the center Griffith crack satisfies that

$$\mathbf{h}(z) = o(z) \qquad \text{as } z \to \infty, \tag{3.88}$$

and $\mathbf{h}(z)$ is square root singular at the crack-tips. Also, the Burger's vector for the crack vanishes, from Equation (3.67); that is,

$$\int_{-a}^{a} [\mathbf{h}^+(x) - \mathbf{h}^-(x)]dx = 0. \tag{3.89}$$

If $\mathbf{T}(x)$ is uniform along the crack faces, the solution can be reached in analytical closed-form; that is,

$$\int \mathbf{h}(z)dz = \frac{1}{2}\mathbf{T}[(z^2 - a^2)^{1/2} - z], \tag{3.90}$$

where the branch cut for the square root coincides with the crack. The field variables are obtained with the help of Equations (3.70),(3.18), and (3.19). Of great interest for determination of the stress intensity factor $\mathbf{k}$, we obtained

$$\mathbf{t}(x) = [x(x^2 - a^2)^{-1/2} - 1]\mathbf{T}, \quad \forall |x| > a, \tag{3.91}$$

$$\mathbf{d}(x) = (x^2 - a^2)^{-1/2}\mathbf{HT}, \quad \forall |x| < a. \tag{3.92}$$

Suppose that the traction charge is caused by the remote field $\{\sigma_{2i}^{\infty}, D_2^{\infty}\}^T$; the intensity factor column is given by

$$K_{II} = \sqrt{\pi a}\sigma_{21}^{\infty}, \qquad K_I = \sqrt{\pi a}\sigma_{22}^{\infty},$$
$$K_{III} = \sqrt{\pi a}\sigma_{23}^{\infty}, \qquad K_{IV} = \sqrt{\pi a}D_2^{\infty} \tag{3.93a}$$

or

$$\mathbf{k} = \sqrt{\pi a}\mathbf{t}^{\infty}. \tag{3.93b}$$

(b) *Zener–Stroh Crack.* The mechanism of this type of crack was originally proposed by Zener in 1947 ([22]), and mathematically analyzed by Stroh [23, 24]. That is why it has been named after Zener and Stroh. The Zener–Stroh crack is loaded by the net Burger's vector entering the crack, rather than by the traction acting on the crack faces (shown in Figure (3.1). The following derivation is an extension of [25] for the anisotropic elasticity.

The Hilbert problem becomes (implying the traction continuation along the $y = 0$ interface)

$$\mathbf{h}^+(x_1) + \mathbf{h}^-(x_1) = 0, \qquad \forall x_1 \in L. \tag{3.94}$$

The solution for a Zener–Stroh crack satisfies that

$$\mathbf{h}(z) = O(z^{-1}) \quad \text{as} \quad z \to \infty, \tag{3.95}$$

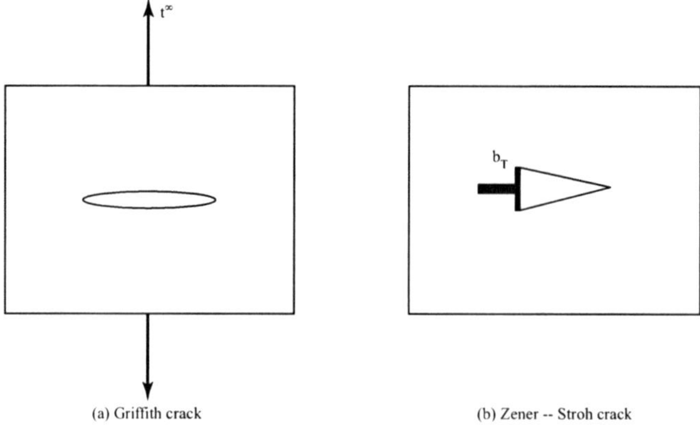

Fig. 3.1 Loading mechanisms of Griffith and Zener-Stroh cracks.

and $\mathbf{h}(z)$ is square root singular at the crack-tips. Also, the Burger's vector for the crack does not vanish, from Equation (3.67); that is,

$$\int_{-a}^{a} \mathbf{d}'(x)dx = \int_{-a}^{a} -i\mathbf{H}[\mathbf{h}^+(x) - \mathbf{h}^-(x)]\, dx = \mathbf{b}_T, \qquad (3.96)$$

where $\mathbf{b}_T$ is the total Burger's vector entering the crack (for the first three components). The fourth component needs some physical description but is beyond the scope of the present section. The solution to the Hilbert problem described above is given by [25]

$$\mathbf{h}(z) = \mathbf{Q}(z^2 - a^2)^{-1/2}. \qquad (3.97)$$

Q can be determined by considering that the crack is equivalent to a super dislocation with Burger's vector of $\mathbf{b}_T$ to the far field; that is,

$$\mathbf{h}(z) = \frac{\mathbf{Q}}{z}, \quad \text{for} \quad z \to \infty. \qquad (3.98)$$

Comparing Equation (3.98) with the dislocation solution ([25] and [26]), we have

$$\mathbf{Q} = \frac{\mathbf{H}^{-1}\mathbf{b}_T^*}{2\pi}, \qquad (3.99)$$

where $\mathbf{b}_T^* = (\mathbf{b}_T, 0)$.

Of our particular interest, we consider the asymptotic behavior of Equation (3.97) by letting $z \to a + r$. The traction ahead of the crack-tip is given by Equation (3.73) in which the stress intensity factor column is given by, with the help of (3.99),

$$\mathbf{k} = \frac{\mathbf{H}^{-1}\mathbf{b}^*T}{\sqrt{\pi a}}. \qquad (3.100)$$

It is noticed that the traction at the crack-tip at $z = -a$ is compression so that there is no crack propagation from this crack-tip.

Comparing Equation (3.100) for the Zener–Stroh crack with (3.93) for the Griffith crack, we reach the conclusion that Zener–Stroh is a stable crack because its stress intensity factor decreases as the crack propagates. On the other hand, the Griffith crack is an unstable crack because its stress intensity factor increases as the crack propagates.

Under a combined loading, the traction, and the net Burger's vector, the crack is called Griffith–Zener–Stroh type. The stress intensity factor column for a center crack embedded in an infinite solid is written as

$$\mathbf{k} = \sqrt{\pi a}\,\mathbf{t}^\infty + \frac{\mathbf{H}^{-1}\mathbf{b}_T}{\sqrt{\pi a}}. \qquad (3.101)$$

Based on the observation of Equation (3.101), in general, we conclude that the Zener–Stroh mechanism is dominant in shorter cracks, and the Griffith mechanism is dominant in longer cracks.

3.4 Contact Problems

Let us consider the configuration that the piezoelectric material occupies the region of $x_2 > 0$, whose boundary, the x_1-axis, is divided into two parts, namely the contact region ($\forall x_1 \in (-a, a)$) and its complement. In the contact region, the traction, electric potential, and induction can be written as

$$\sigma_{2j} = \sigma_{2j}^0 \quad \varphi = \varphi^0 \quad D_2 = D_2^0. \tag{3.102}$$

The right-hand side terms in the above equation are the values in the indentor, which may or may not be known before we solve the problem. The boundary condition on the no-contact part is assumed to be

$$\sigma_{2i}(x_1, 0) = 0 \qquad D_2(x_1, 0) = 0 \qquad x_1 \notin (-a, a). \tag{3.103}$$

Boundary conditions other than Equation (3.103), for instance, a prescribed φ, can also be considered by a formulation provided by Suo [27]. In the present section, we focus on (3.103) to demonstrate the approach.

With Equations (3.102), (3.103) and notation (3.19), we may write

$$\mathbf{Bf}'(x_1) + \overline{\mathbf{Bf}'(x_1)} = \mathbf{t}(x_1) \qquad \forall x_1 \in (-\infty, \infty). \tag{3.104}$$

For later convenience, we introduce a function

$$\mathbf{h}(z) = \mathbf{Bf}'(z) \qquad \forall x_2 > 0 \tag{3.105a}$$

$$\mathbf{h}(z) = -\overline{\mathbf{Bf}'(z)} \qquad \forall x_2 < 0 \tag{3.105b}$$

which is analytic throughout the whole plane except on the contact segment. Using limits of the function $\mathbf{h(z)}$ on $x_2 = 0$ (Muskhelishvili [28]), Equation (3.104) leads to a Hilbert problem

$$\mathbf{h}^+(x_1) - \mathbf{h}^-(x_1) = \mathbf{t}(x_1), \qquad \forall x_1 \in (-a, a), \tag{3.106a}$$

$$\mathbf{h}^+(x_1) - \mathbf{h}^-(x_1) = 0, \qquad \forall x_1 \notin (-a, a). \tag{3.106b}$$

The solution of Equation (3.106) can be obtained as

$$\mathbf{h}(z) = \frac{1}{2\pi i} \int_{-a}^{a} \frac{\mathbf{t}(x)}{x - z} dx \tag{3.107}$$

if the distributions of the traction and electric induction are known over the region $\forall x_1 \in (-a, a)$. As an example, let us take the traction and electric induction to be uniform over this region.

Then

$$\mathbf{h}(z) = \frac{\mathbf{t}}{2\pi i} \ln\left(\frac{z-a}{z+a}\right).$$ (3.108)

If we take:

$$\lim_{a \to 0} 2a\mathbf{t} = \mathbf{T}$$ (3.109)

we have the two-dimensional Green's function for the piezoelectric half-space as

$$\mathbf{h}(z) = -\frac{\mathbf{T}}{2\pi i z}.$$ (3.110)

The one-complex-variable solution Equation (3.110) can be converted to a four-complex-variable format explicitly as follows.

Without losing generality, let us consider a vertical concentrated force acting on the free surface. The column $\mathbf{T}$ in Equation (3.110) is written explicitly as

$$\mathbf{T} = (0, P, 0, 0)^T.$$

Noting (3.105a), we have

$$\mathbf{f}'(z) = \mathbf{B}^{-1} \begin{pmatrix} 0 \\ -P/2\pi i z \\ 0 \\ 0 \end{pmatrix},$$

where

$$\mathbf{B}^{-1} = \begin{bmatrix} b'_{11} & b'_{12} & b'_{13} & b'_{14} \\ b'_{21} & b'_{22} & b'_{23} & b'_{24} \\ b'_{31} & b'_{32} & b'_{33} & b'_{34} \\ b'_{41} & b'_{42} & b'_{43} & b'_{44} \end{bmatrix}$$

is the inverse matrix of $\mathbf{B}$. In terms of the four complex variables, we have

$$\mathbf{f}'(z_1, z_2, z_3, z_4) = -\frac{P}{2\pi i} \begin{pmatrix} b'_{12}/z_1 \\ b'_{22}/z_2 \\ b'_{32}/z_3 \\ b'_{42}/z_4 \end{pmatrix}.$$

Furthermore, the stress induction is calculated from

$$\mathbf{B}\mathbf{f}'(z_1, z_2, z_3, z_4) = -\frac{P}{2\pi i} \begin{pmatrix} \sum_{k=1}^{4} b_{1k}b'_{k2}/z_k \\ \sum_{k=1}^{4} b_{2k}b'_{k2}/z_k \\ \sum_{k=1}^{4} b_{3k}b'_{k2}/z_k \\ \sum_{k=1}^{4} b_{4k}b'_{k2}/z_k \end{pmatrix}.$$

A similar problem has been considered by Sosa and Castro [15] for a simplified constitutive equation.

3.4.1 Nonslip Contact

When the traction induction is unknown in the contact region and all the displacement components and electric potential are prescribed in the contact region, the solution is obtained by considering a Hilbert problem. Using the notation of Equation (3.18), the displacement potential continuity across the interface reads as

$$\mathbf{A}\mathbf{f}'(x_1) + \overline{\mathbf{A}\mathbf{f}'(x_1)} = \mathbf{d}'(x_1) \qquad \forall x_1 \in (-a, a), \tag{3.111}$$

and the traction induction free condition outside the contact zone leads to

$$\mathbf{B}\mathbf{f}'(x_1) + \overline{\mathbf{B}\mathbf{f}'(x_1)} = 0 \qquad \forall x_1 \notin (-a, a), \tag{3.112}$$

where

$$\mathbf{d} = \{u_1, u_2, u_3, \varphi\}^T \tag{3.113}$$

is assumed to be known inside the contact region. By using the function defined in Equations (3.105) and (3.111) is rewritten as

$$\mathbf{h}^+(x_1) + \mathbf{Y}^{-1}\overline{\mathbf{Y}}\mathbf{h}^-(x_1) = i\mathbf{Y}^{-1}\mathbf{d}'(x_1), \qquad \forall x_1 \in (-a, a), \tag{3.114}$$

where

$$\mathbf{Y} = i\mathbf{A}\mathbf{B}^{-1}. \tag{3.115}$$

The matrix $\mathbf{Y}$ has been discussed by Suo et al. [8]. The properties of this matrix are mentioned when it is used later on. If the matrix is real, the solution of (3.114) is given by England [21]

$$\mathbf{h}(z) = \frac{\chi(z)}{2\pi} \int_{-a}^{a} \frac{\mathbf{Y}^{-1}\mathbf{d}'(x)}{\chi^+(x)(x-z)} dx + \chi(z)\mathbf{Q}(z), \tag{3.116}$$

where

$$\chi(z) = ((z-a)(z+a))^{-1/2}, \tag{3.117}$$

and $\mathbf{Q}$ is a polynomial to be determined by considering the resultant force acting on the contact zone (Fan and Keer [29]).

In general $\mathbf{Y}$ is a complex matrix. The Hilbert problem Equation (3.114) is solved only if we can transform the matrix $\mathbf{Y}^{-1}\overline{\mathbf{Y}}$ in (3.114) into a diagonal form. Following this approach was proposed by Ting [4] for anisotropic elasticity, and Suo et al. [8] modified the transformation procedure for piezoelectricity. By considering an eigenvalue problem as

$$\overline{\mathbf{Y}}\mathbf{w} = e^{2\pi\lambda}\mathbf{Y}\mathbf{w}, \tag{3.118}$$

we have four eigenpairs (eigenvalue, eigenvector) as

$$(\varepsilon, \mathbf{w}), \qquad (-\varepsilon, \overline{\mathbf{w}}), \qquad (-ik, \mathbf{w}_3), \qquad (ik, \mathbf{w}_4). \tag{3.119}$$

Any field can be decomposed via this eigensystem; say,

$$\mathbf{h} = h_1 \mathbf{w} + h_2 \overline{\mathbf{w}} + h_3 \mathbf{w}_3 + h_4 \mathbf{w}_4$$
$$\mathbf{d} = d_1 \mathbf{w} + d_2 \overline{\mathbf{w}} + d_3 \mathbf{w}_3 + d_4 \mathbf{w}_4. \tag{3.120}$$

With the eigenexpansion equation (4.19), we can decouple (4.13) as

$$h_1^+ + e^{-2\pi\varepsilon} h_1^- = i\hat{d}'_1 \qquad h_2^+ + e^{-2\pi\varepsilon} h_2^- = i\hat{d}'_2$$
$$h_3^+ + e^{2\pi i k} h_3^- = i\hat{d}'_3 \qquad h_4^+ + e^{-2\pi i k} h_4^- = i\hat{d}'_4, \tag{3.121}$$

where
$$\hat{\mathbf{d}} = \mathbf{Y}^{-1}\mathbf{d}.$$

The solutions are obtained in a form such as Equation (3.116).

It is noticed that there are two kinds of indentor, namely, a sharp-edged and a round-edged indentor. For the sharp-edged one, the contact zone is known. At the two corners, the stress is singular similar to the isotropic elastic contact (e.g., Johnson [30]). However, the singularities for this piezo-electric contact problem are much more complicated than those in the linear elasticity. Fortunately, the detailed structure of the singularities has been dis-cussed by Suo et al. [8] when they studied the interfacial crack in piezoelectric bimaterials. In the case of a round-edged indentor, there is no singularity in the solution. The contact zone size is determined by considering the resultant force acting on the half-plane.

3.4.2 Slip Contact

If the interfacial static friction of the contact is not high enough to prevent the slip between the two bodies, some displacement components in the x-direction may be discontinuous. Thus the boundary condition, Equation (3.111), must be replaced by, for instance,

$$u_2 = \overline{u}_2 \quad \text{and} \quad \sigma_{12} = \mu\sigma_{22} \tag{3.122}$$

together with proper electrical conditions. In Equation (3.122), μ is the sliding friction coefficient.

Let us summarize the displacement potential and traction induction con-ditions inside the contact region:

$$\mathbf{Y}\mathbf{h}^+(x_1) + \overline{\mathbf{Y}}\mathbf{h}^-(x_1) = i\mathbf{d}'(x_1), \qquad \forall x_1 \in (-a, a) \tag{3.123}$$

$$\mathbf{h}^+(x_1) - \mathbf{h}^-(x_1) = \mathbf{t}(x_1), \qquad \forall x_1 \in (-a, a). \tag{3.124}$$

Eliminating $\mathbf{h}^-(x)$ from Equation (3.123), we have

$$(\mathbf{Y} + \overline{\mathbf{Y}})\mathbf{h}^+(x_1) - \overline{\mathbf{Y}}\mathbf{t}(x_1) = i\mathbf{d}'(x_1). \tag{3.125}$$

On the other hand, the Plemelj formula gives:

$$\mathbf{h}^+(x_1) = \frac{1}{2}\mathbf{t}(x_1) + \frac{1}{2\pi i}\int_{-a}^{a}\frac{\mathbf{t}(\xi)}{\xi - x_1}d\xi. \tag{3.126}$$

Substituting Equation (3.126) into (3.125), one has

$$\frac{\mathbf{Y} - \overline{\mathbf{Y}}}{2}\mathbf{t}(x_1) + \frac{\mathbf{Y} + \overline{\mathbf{Y}}}{2\pi i}\int_{-a}^{a}\frac{\mathbf{t}(\xi)}{\xi - x_1}d\xi = i\mathbf{d}'(x_1). \tag{3.127}$$

There are four equations for four unknowns in Equation (3.127). Thus, Equation (3.127) provides the distribution of traction inside the contact region, which in turn allows us to have the whole field solution via Equation (3.107).

3.4.3 Decoupled Elastic and Dielectric Contact

When the piezoelectric tensor vanishes, the problem is decoupled into anisotropic elastic and dielectric ones. The former, the anisotropic elastic contact problem, has been considered by Fan and Keer [29] via Stroh formalism. The latter, a mixed boundary value problem for a general anisotropic dielectric half-space, is presented here. In the case $\boldsymbol{e} = \mathbf{0}$, the eigenvalue problem Equation (3.12) is simplified as

$$C_{\alpha i k\beta}\varsigma_\alpha\varsigma_\beta a_k = 0, \tag{3.128a}$$

and

$$\varepsilon_{\alpha\beta}\varsigma_\alpha\varsigma_\beta a_4 = 0. \tag{3.128b}$$

The corresponding eigenvectors are in the form of

$$\mathbf{A} = \begin{pmatrix} \mathbf{A}_e & 0 \\ 0 & 1 \end{pmatrix},$$

$$\mathbf{B} = \begin{pmatrix} \mathbf{B}_e & 0 \\ 0 & b_4 \end{pmatrix}, \tag{3.129}$$

where $\mathbf{A}_e$ and $\mathbf{B}_e$ are 3 by 3 matrices corresponding to anisotropic elasticity. The dielectric problem and scalars are formed by Equation (3.129).

The eigenvalues for the anisotropic dielectric materials are also obtained from (3.129):

$$\varepsilon_{11} + 2\varepsilon_{12}p + \varepsilon_{22}p^2 = 0. \tag{3.130}$$

They are

$$p_4 = -\frac{\varepsilon_{12}}{\varepsilon_{12}} + i\sqrt{\frac{\varepsilon_{11}}{\varepsilon_{22}} - \left(\frac{\varepsilon_{12}}{\varepsilon_{22}}\right)^2}, \qquad p_8 = \overline{p}_4. \tag{3.131}$$

The last equation in (3.111), corresponding to dielectricity, is decoupled from the first three equations which are associated with anisotropic elasticity. Thus,

$$f_4'(x) + \overline{f}_4'(x) = \varphi_0'(x). \tag{3.132}$$

The solution of Equation (3.132) is easily obtained by applying (3.117).

It is noticed that the degree of electromechanical coupling in a piezoelectric material can be described by a nondimensional parameter formed by the three types of moduli, which is in the range of

$$\frac{e}{\sqrt{\varepsilon C}} = 0.1 \sim 1. \tag{3.133}$$

A weakly coupled material, such as quartz which is widely used in frequency filters and resonators, has moduli on the order of (Salt [31]):

$$C \sim 10^{11} N/m^2, \qquad \varepsilon \sim 10^{-11} F/m^2, \qquad e \sim 10^{-1} C/m^2. \tag{3.134}$$

It is seen that

$$\frac{e}{\sqrt{\varepsilon C}} \sim 0.1. \tag{3.135}$$

The mechanical and electric fields for this weakly coupled piezoelectric material can be approximated by the decoupled elastic and dielectric solutions. On the other hand, a strongly electromechanical coupled material, such as lead–zirconate–titanate (say PZT-5H), has moduli on the order of

$$C \sim 10^{11} N/m^2, \quad \varepsilon \sim 10^{-8} F/m^2, \quad e \sim 10 \ C/m^2. \tag{3.136}$$

The dimensionless coupling parameter is on the order of

$$\frac{e}{\sqrt{\varepsilon C}} \sim 0.3. \tag{3.137}$$

For this kind of material, the mechanical and electric fields have to be calculated based upon the fully coupled formulation.

3.4.4 Piezoelectric Half-Space with Poling Direction Along the x_3-Axis

The piezoelectric ceramic PZT-5H belongs to this material category. Assuming the x_1-x_2 plane is the isotropic plane, the material constants of this material are given in Section 3.3.2.

Let us consider a rigid punch pressed into the piezoelectric half-space. Assuming no slip at the interface, the contact problem is decoupled into elastic in-plane contact and antiplane charge ones, due to the constitutive Equations (3.77) and (3.78). By noting Equation (3.117), the solution of the latter is then given by

$$\mathbf{h}_p(z) = \chi(z)\mathbf{Q}_p. \tag{3.138}$$

The stress and induction show a square root singularity at $z = \pm a$. The constant column $\mathbf{Q}_p$ can be determined by setting

$$|z| \gg a$$

so that

$$\mathbf{h}_p(z) = \frac{\mathbf{Q}_p}{z}. \tag{3.139}$$

Comparing Equations (3.139) and (3.110), the Green's function, we have

$$\mathbf{Q}_p = \frac{i\mathbf{T}_p}{2\pi}. \tag{3.140}$$

If the normal component of flux vanishes everywhere on $x_2 = 0$,

$$\mathbf{T}_p = (T_3, 0)^T,$$

it is easy to obtain that

$$\sigma_{32} = -\left(1 + \frac{1}{k}\right)e_{15}E_2 \qquad \sigma_{31} = -\left(1 + \frac{1}{k}\right)e_{15}E_1, \tag{3.141}$$

where $k = e_{15}^2/(c_{44}\varepsilon_{11})$.

3.5 Decay Analysis for a Piezoelectric Strip

3.5.1 Saint-Venant Principle and Decay Analysis

The Saint-Venant end effect is characterized mathematically by so-called decay analysis. With this decay analysis, the Saint-Venant principle can be extended to modern materials and structures beyond the traditional homogeneous and isotropic solids. In the present section, the decay rate in a piezoelectric strip is considered. Although plate–and beam-shaped piezoelectric devices have been used in communication and sensor industries for many years, the Saint-Venant principle has not been introduced into the piezoelectric material. The difficulties raised by general anisotropy in the piezoelectricity further delayed the decay analysis consideration for the piezoelectric plates and bars. Recent development of anisotropic elasticity via Stroh formalism makes the present two-dimensional decay analysis for piezoelectricity possible.

Decay analysis is a mathematical foundation for the well-known Saint-Venant principle which is considered as one of the basic blocks in linear elasticity. Applicability of beam, plate, and shell theories highly depends on the principle. Since the Saint-Venant principle was proposed over a century ago, his statement was apparently too ambiguous for modern structure and material applications. Understanding the principle from mathematical fundamentals started in the 1960s. Since then establishment of the Saint-Venant principle based upon rigorous mathematical formulation has been an active research topic. With the mathematical form of the Saint-Venant principle, there are no conceptual and descriptive difficulties to define the end effects in modern materials, such as highly anisotropic composites. Thorough reviews about research progress on the Saint-Venant principle were given by Horgan and Knowles [32] and a follow-up paper by Horgan [33]. More recent efforts of extending the concept to piezoelectric materials can be traced in [10], [34] and [35].

The decay rate study makes the qualitative Saint-Venant statement into a quantitative description via the so-called eigenexpansion. Without losing generality, let us consider a nondimensionalized strip bounded in $x_1 \in [0, \infty)$ and $x_2 \in [-1, 1]$ as shown in Figure 3.2. For this configuration, a generic field solution can be expanded as

$$F(x_1, x_2) = F_0(x_1, x_2) + \sum_{k=1}^{\infty} C_k e^{-\lambda_k x_1} F_k(x_2), \qquad (3.142)$$

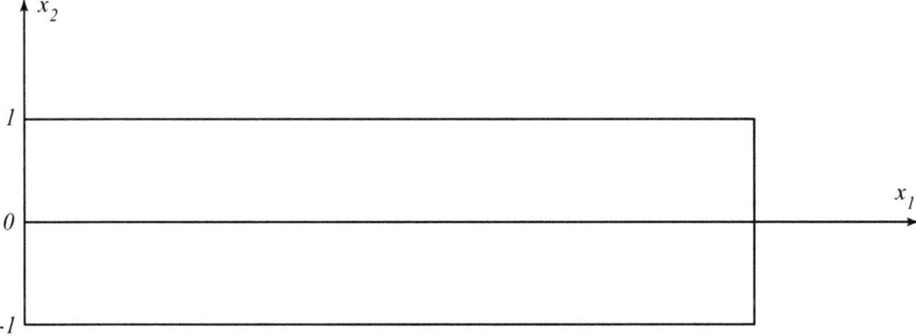

Fig. 3.2 A semi-infinite long strip loaded at the $x = 0$ end.

where $F_0(x_1, x_2)$ is the so-called Saint-Venant solution which is a nondecay term with respect to the coordinate x_1, and all the terms in the summation are exponential decay. In other words, all the $\mathrm{Re}(\lambda) < 0$ are dropped for a finite solution. In all these decay terms, the first term with the smallest value of $\mathrm{Re}(\lambda)$ is the slowest decay term. People refer to this λ as the decay rate (Horgan [33]) which will decide the decay distance of the end effects. Hereby, we introduce the Saint-Venant principle to the piezoelectric material by finding the decay rate in a piezoelectric strip. Because plate-shaped piezoelectric devices have been widely used in resonators, sensors, and actuators, the analysis on a strip configuration is of great theoretical as well as practical importance.

3.5.2 Eigenexpansion for the Piezoelectric Strip

With the concept in Equation (3.142), the decay terms in the displacement and electric potential vector can be expended as

$$\mathbf{U} = \sum_{k=1}^{\infty} C^{(k)} e^{-\lambda_k \chi} \mathbf{U}^{(k)}(y). \tag{3.143}$$

This is called the eigenexpansion for the decay analysis in which λ_k and $\mathbf{U}^{(k)}$ are eigenvalues and eigenfunctions. From the previous sections, we know that the solution can be written in terms of complex variables $x + p_I y$ and their complex conjugates. It is straightforward to assume that a general solution is the superposition of the eight independent solutions corresponding to these eight eigenvalues, p_I and $\bar{p}_I$; that is,

$$\mathbf{U}^{(k)}(y) = \left(\mathbf{A} \langle e^{-\lambda_k p_I y} \rangle \mathbf{q}_k + \overline{\mathbf{A}} \langle e^{-\lambda_k \bar{p}_I y} \rangle \mathbf{h}_k \right), \tag{3.144}$$

where

$$\langle e^{-\lambda p_I y} \rangle = diag\{ e^{-\lambda p_1 y}, \ e^{-\lambda p_2 y}, \ e^{-\lambda p_3 y}, \ e^{-\lambda p_4 y} \}. \tag{3.145}$$

The eigenvalues λ, $\mathbf{q}_k$, and $\mathbf{h}_k$ are determined from the homogeneous boundary conditions along the upper and lower surfaces ($y = \pm 1$) of the strip.

From the eigenexpansion (Equation (3.142)) point of view, the eigenvalue λ with the smallest real part is of interest for the Saint-Venant principle. We focus on this term and drop the subscript of the eigenvalue and eigenfunction if there is no confusion.

Upon the substituting the displacement vector back into the constitutive equations, the stress and induction components can be expressed as

$$\mathbf{t}^{(k)}(y) = \{\sigma_{2i}, D_2\}^T = \lambda_k \left(\mathbf{B}\left\langle e^{-\lambda_k p_I y}\right\rangle \mathbf{q}_k + \overline{\mathbf{B}}\left\langle e^{-\lambda_k \overline{p}_I y}\right\rangle \mathbf{h}_k\right) \tag{3.146}$$

$$\mathbf{s}^{(k)}(y) = \{\sigma_{1i}, D_1\}^T = \lambda_k \left(\mathbf{BP}\left\langle e^{-\lambda_k p_I y}\right\rangle \mathbf{q}_k + \overline{\mathbf{BP}}\left\langle e^{-\lambda_k \overline{p}_I y}\right\rangle \mathbf{h}_k\right), \tag{3.147}$$

where

$$\mathbf{B} = \{\mathbf{b_1}, \mathbf{b_2}, \mathbf{b_3}, \mathbf{b_4}\},$$
$$\mathbf{P} = \mathrm{diag}\{p_1, p_2, p_3, p_4\}. \tag{3.148}$$

With the equations in Section 3.3.1, one has

$$b_j = (C_{2jk\beta}a_k + e_{\beta j2}a_4)\varsigma_\beta,$$
$$b_4 = (-\varepsilon_{1\beta}a_4 + e_{2k\beta}a_k)\varsigma_\beta. \tag{3.149}$$

It is worthwhile to clarify a point that there are two eigenvalue problems in the present derivation. The first one is raised by the anisotropic piezoelectric material. The eigenvalues are determined solely by material constants, and the eigenspace is spanned by the eight eigenvectors. On the other hand, the second eigenvalue problem is introduced by decay analysis (Equation (3.142). There are an infinite number of eigenvalues and eigenfunctions for this eigenvalue problem. Determination of the eigenvalues and eigenfunctions for the second problem depends on the boundary conditions at $y = \pm 1$.

3.5.3 Boundary Conditions and Determination of Eigenvalues λ_k

(a) *Boundary conditions along the upper and lower surfaces of the strip.* For a traditional Saint-Venant principle, the boundary conditions along the upper and lower surfaces of the strip are traction-free. However, there are other alternatives to pose the homogeneous boundary conditions along the surfaces. Wang et al. [34] posed total eight possible boundary conditions with mixed displacement and traction components. Their treatment is extended to the piezoelectric strip in the present section.

The boundary conditions along the upper and lower surfaces can be one of the following 16 combinations. The boundary condition is formed by selecting only one variable from each of the following groups.

$$(\sigma_{21}, u_1), \qquad (\sigma_{22}, u_2), \qquad (\sigma_{23}, u_3), \qquad (D_2, \phi). \tag{3.150}$$

For instance, one of the choices is

$$\sigma_{21} = 0, \qquad \sigma_{22} = 0, \qquad \sigma_{23} = 0, \qquad \varphi = 0, \quad \text{at} \quad y = 1 \text{ and } y = -1. \tag{3.151}$$

A generalized matrix representation of the homogeneous boundary condition is

$$\mathbf{I}_u \mathbf{U}'_x + \mathbf{I}_t \mathbf{t} = \mathbf{0}, \tag{3.152}$$

where $\mathbf{I}_u$ and $\mathbf{I}_t$ are the 4 by 4 diagonal matrices whose diagonal elements are either one or zero. For example, the matrices for Equation (3.151) are

$$\mathbf{I}_u = \mathrm{diag}\{0,0,0,1\},$$
$$\mathbf{I}_t = \mathrm{diag}\{1,1,1,0\}. \tag{3.153}$$

b) *Determination of eigenvalues.* Substituting Equations (3.144) and (3.146) into (3.152) for a given boundary condition, one obtains:

$$\mathbf{K}_+ < e^{-\lambda p} > \mathbf{q} + \overline{\mathbf{K}}_+ < e^{-\lambda \overline{p}} > \mathbf{h} = 0 \quad y = 1, \tag{3.154}$$
$$\mathbf{K}_- < e^{\lambda p} > \mathbf{q} + \overline{\mathbf{K}}_- < e^{\lambda \overline{p}} > \mathbf{h} = 0 \quad y = -1, \tag{3.155}$$

where

$$\mathbf{K} = \mathbf{I}_u \mathbf{A} + \mathbf{I}_t \mathbf{B}. \tag{3.156}$$

It is noted that the lower and upper surfaces may post different boundary conditions. Thus, $\mathbf{K}_+$ and $\mathbf{K}_-$ may not be the same.

It can be proved that matrix $\mathbf{K}$ is nonsingular. The proof is similar to Wang et al. [34] for anisotropic elasticity. Bearing this in mind, we can have the equation of the eigenvalue λ for nonzero $\mathbf{q}$ and $\mathbf{h}$,

$$\det(\mathbf{K}_+ < e^{-2\lambda p} > \mathbf{K}_-^{-1} - \overline{\mathbf{K}}_+ < e^{-2\lambda \overline{p}} > \overline{\mathbf{K}}_-^{-1}) = 0. \tag{3.157}$$

In general, the solution of Equation (3.157) involves a numerical procedure. Some simplified examples are discussed in the next section.

(c) *Boundary conditions at the end. ($x = 0$).* The end conditions are needed to determine the participant factors in eigenexpansion Equation (3.144) with the help of orthogonality among the eigenfunctions. The end condition in a traditional Saint-Venant consideration has to be "self-balanced." For instance,

$$\int_{-1}^{+1} \sigma_{1i}(0, x_2) dx_2 = 0 \tag{3.158}$$

is the self-balanced end loading in the elasticity. As discussed in [35], this condition does not necessarily hold for the cases in which the boundary conditions along the upper and lower surface are the displacement version. Because our primary interest in this chapter is to find the decay factor, this boundary condition is not involved in our formulation. For the sake of completeness, the orthogonality of the eigenfunctions is discussed briefly at the end of this section via the eight-dimensional matrix scheme.

3.5.4 Examples of Decay Factor in the Piezoelectric Strip

(a)*Decay rate when the piezoelectric tensor vanishes.* When the piezoelectric tensor vanishes, the problem is decoupled into anisotropic and dielectric ones. The former, the anisotropic elastic strip, has been considered by Wang et al. [34] for a general anisotropic two-dimensional elastic strip via the Stroh formulation. The latter problem, the dielectric strip which has not been reported in the literature, is presented here.

In order to make the reading fluent, let us repeat a few lines of equations presented in Section 3.4.3. For $e = \mathbf{0}$, the eigenvalue problem Equation (3.12) is simplified as

$$C_{\alpha i k \beta} \varsigma_\alpha \varsigma_\beta a_k = 0, \tag{3.159a}$$

and

$$\varepsilon_{\alpha\beta} \varsigma_\alpha \varsigma_\beta a_4 = 0. \tag{3.159b}$$

The corresponding eigenvectors are in the form of

$$\mathbf{A} = \begin{pmatrix} \mathbf{A}_e & 0 \\ 0 & a_4 \end{pmatrix}, \qquad \mathbf{B} = \begin{pmatrix} \mathbf{B}_e & 0 \\ 0 & b_4 \end{pmatrix}, \tag{3.160}$$

where $\mathbf{A}_e$ and $\mathbf{B}_e$ are 3 by 3 matrices for anisotropic elasticity. The dielectricity problem is governed by Equation (3.159b) and scalars in (3.160). The eigenvalue corresponding to the anisotropic dielectricity is obtained from Equation (3.159b)

$$\varepsilon_{11} + 2\varepsilon_{12}p + \varepsilon_{22}p^2 = 0. \tag{3.161}$$

They are

$$p_4 = \frac{\varepsilon_{12}}{\varepsilon_{22}} + i\sqrt{\frac{\varepsilon_{11}}{\varepsilon_{22}} - \left(\frac{\varepsilon_{12}}{\varepsilon_{22}}\right)^2}, \qquad p_8 = \overline{p}_4. \tag{3.162}$$

For the electric potential version decay analysis, the boundary conditions along the upper and lower surfaces of the strip are read as

$$\varphi = 0 \quad \text{at} \quad y = \pm 1. \tag{3.163}$$

After some straightforward calculations, the equation for the eigenvalue Equation (3.157) corresponding to the dielectricity part is obtained as

$$e^{2\lambda p_4} - e^{-2\lambda \overline{p}_4} = 0. \tag{3.164}$$

Then the solution is

$$2\lambda \mathrm{Im}(p_4) = n\pi. \tag{3.165}$$

The smallest root, normally called the decay factor, is

$$\lambda = \frac{\pi}{2\mathrm{Im}(p_4)} \ .$$
(3.166)

(b) *Decay rate in a piezoelectric material with transverse symmetry.* The materials constants are given in Section 3.3.2. The in-plane deformation (u_x, u_y) is decoupled from the antiplane field (u_z, φ). The former is identical to an elastic problem. We focus on the latter. It is obtained that

$$p_3 = p_4 = i, \qquad \bar{p}_7 = \bar{p}_8 = -i$$
(3.167)

and

$$\mathbf{A} = \begin{bmatrix} \mathbf{A}_{2e} & 0 \\ 0 & \mathbf{A}_p \end{bmatrix},$$
$$\mathbf{B} = \begin{bmatrix} \mathbf{B}_{2e} & 0 \\ 0 & \mathbf{B}_p \end{bmatrix},$$
(3.168)

where the right upper 2 by 2 matrices correspond to in-plane deformation, and the left lower corner matrix corresponds to a coupled antiplane deformation and electric field. Let us concentrate on the latter one. One has:

$$\mathbf{A}_p = \begin{bmatrix} 1 & 0 \\ 0 & 1 \end{bmatrix}, \qquad \mathbf{B}_p = i \begin{bmatrix} c_{44} & e_{15} \\ 0 & -\varepsilon_{11} \end{bmatrix} .$$
(3.169)

Let us consider that the boundary conditions along the upper and lower surfaces are

$$\varphi = 0 \quad \text{and} \quad \sigma_{23} = 0.$$
(3.170)

Thus,

$$\mathbf{K}_p = \mathbf{I}_u \mathbf{A}_p + \mathbf{I}_t \mathbf{B}_p = \begin{bmatrix} 0 & 0 \\ 0 & 1 \end{bmatrix} \begin{bmatrix} 1 & 0 \\ 0 & 1 \end{bmatrix} + i \begin{bmatrix} 1 & 0 \\ 0 & 0 \end{bmatrix} \begin{bmatrix} c_{44} & e_{15} \\ 0 & -\varepsilon_{11} \end{bmatrix} = \begin{bmatrix} ic_{44} & ie_{15} \\ 0 & 1 \end{bmatrix} .$$
(3.171)

After a lengthy calculation, we obtain the equation for the eigenvalues as

$$e^{2\lambda i} - e^{-2\lambda i} = 0.$$
(3.172)

The solution can be obtained as

$$2\lambda = n\pi.$$
(3.173)

The decay factor is

$$\lambda_{\text{smallest}} = \frac{\pi}{2}.$$
(3.174)

3.5.5 Remarks

If we use the eight-dimensional notation introduced in Section 3.1.2, the decay field is expanded as

$$\mathbf{w}(x_1, x_2) = \sum_{k=1}^{\infty} C_k \mathbf{w}^{(k)}(x_2) e^{-\lambda_k x_1}, \qquad (3.175)$$

where the eight-dimensional column $\boldsymbol{w}$ is defined in Equation (3.33). Substituting Equation (3.175) into (3.32), we have:

$$\frac{d\mathbf{w}^{(k)}}{dx_2} = -\lambda_k \mathbf{N} \mathbf{w}^{(k)}. \qquad (3.176)$$

Solution of Equation (3.176) comes out as

$$\mathbf{w}^{(k)}(x_2) = e^{-\lambda_k p x_2} \xi, \qquad (3.177)$$

where ξ is the eigenvector in Equation (3.38). It is noted that the solution, equation (3.177), has been taken intuitively in Equation (3.144).

The orthogonality of the eigenfunctions in Equation (3.175) is sought out by considering:

$$\frac{d}{dx_2}((\mathbf{w}^{(-m)})^T \mathbf{J} \mathbf{w}^{(k)}) = (\lambda_m - \lambda_k)(\mathbf{w}^{(-m)})^T \mathbf{J} \mathbf{N} \mathbf{w}^{(k)}, \qquad (3.178)$$

where

$$\mathbf{J} = \begin{pmatrix} \mathbf{0} & \mathbf{I} \\ \mathbf{I} & \mathbf{0} \end{pmatrix}, \qquad \mathbf{N}^T \mathbf{J} = \mathbf{J} \mathbf{N} \qquad (3.179)$$

have been used. By integrating both sides of Equation (3.178) with respect to x_2 from $(-1, 1)$, it is obtained that

$$\{\mathbf{U}^{(-m)^T} \boldsymbol{\Phi}^{(k)} + \boldsymbol{\Phi}^{(-m)^T} \mathbf{U}^{(k)}\}^1_{-1} = (\lambda_m - \lambda_k) \int_{-1}^{1} (\mathbf{w}^{(-m)})^T \mathbf{J} \mathbf{N} \mathbf{w}^{(k)} dx_2.$$

$$(3.180)$$

The left-hand side vanishes because of the side boundary conditions, Equation (3.152). Thus the orthogonality of the eigenfunctions is proved. The detailed derivation is similar to Wang et al. [34] for the anisotropic elasticity problem. With this orthogonality condition, one can have participation factors C_k in Equation (3.175) determined. In the present analysis, when we are just concerned with the decay rate in the strip, only the first eigenvalue is needed. Therefore, the participation factor is not needed for the decay analysis.

References

[1] Eshelby, J.D., Read, W.T., and Shockley, W.(1953) Anisotropic elasticity with applications to dislocation theory, *Act. Metall.* 1, 251–259.

[2] Stroh, A.N.(1958) Dislocation and crack in anisotropic elasticity, *Philosoph. Mag.*, 3, 625–646.

[3] Barnett, D.M. and Lothe, J.(1985) Free surface (Rayleigh) waves in an anisotropic elastic half space: The surface impedance method, *Proc. Roy. Soc.* A, 402, 135–152.

[4] Ting, T.C.T. (1986) Explicit solution and invariance of the singularities at an interface crack in anisotropic composites, *Int. J. Solids Struct.*, 22(9), 965–983.

[5] Barnett, D.M. and Lothe, J. (1975) Dislocation and line charges in anisotropic piezoelectric insulators, *Physica Status Solidi (b)*, 67,105–111.

[6] Chung, M.Y. and Ting, T.C.T. (1996) Piezoelectric solids with an elliptic inclusion or hole, *Int. J. Solids Struct.*, 33(23), 1343–1361.

[7] Eshelby, J.D.(1957) The determination of elastic field of an ellipsoidal inclusion, and related problems, *Proc. Roy. Soc.* A, 241, 376–396.

[8] Suo, Z., Kuo, C.M., Barnett, D.M., and Willis, J.R. (1992) Fracture mechanics for piezoelectric ceramics, *J. Mech. Phys. Solids*, 40(4), 739–765.

[9] Fan H., Sze, K.Y., and Yang W. (1996) Two dimensional contact on a piezoelctric half space, *Int. J. Solids Struct.*, 33(9), 1305–1315.

[10] Fan, H. (1995) Decay rate in a piezoelectric strip, *Int. J. Eng. Sci.*, 33(8), 1095–1103.

[11] Ting, T.C.T (1996) *Anisotropic Elasticity: Theory and Applications*, Oxford University Press, Oxford, UK.

[12] Lekhnitskii, S.G. (1961) *Theory of Elasticity of an Anisotropic Body*, Holden–Day, San Francisco.

[13] Yang, J.S. (2006) *The Mechanics of Piezoelectric Structures*, World Scientific, NJ.

[14] Pak Y.E. (1990) Crack extension force in a piezoelectric material, *ASME J. Appl. Mech.* 57, 647–653.

[15] Sosa, H.A. and Castro, M.A.(1994) On concentrated loads at the boundary of a piezoelectric half-plane, *J. Phys. Mech. of Solids*, 42(7), 1105–1122.

[16] Mura, T. (1987) *Micromechanics of Defects in Solids*, 2nd edition, Martinus Nijhoof, Netherlands.

[17] Wang, B. (1992) Three dimensional analysis of an ellipsoidal inclusion in a piezoelectric material, *Int. J. Solids Struct.*, 29(3), 293–308.

[18] Budiansky, B. (1965) On the elastic moduli of some heterogeneous materials. *J. Mech. Phys. Solids*, 13, 223–227.

[19] Hwu, C., and Ting, T.C.T. (1989) Two-dimensional problem of the anisotropic elastic solids with an elliptic inclusion, *Q. J. Mech. Appl. Math.*, 42, 553–572.

[20] Chung, M. Y. and Ting, T.C.T. (1995) The Green function for a piezoelectric piezomagnetic magneteletric anisotropic elastic medium with an elliptic hole or rigid inclusion, *Philosophi. Mag. Lett.*, 72(6), 405–410.

[21] England, A.H. (1971) *Complex Variable Methods in Elasticity*, John Wiley & Sons, London, UK.

[22] Weertman, J.R. (1986) Zener–Stroh crack, Zener–Hollomon parameter and other topics, *J. Appl. Phys.*, 60(6), 1877–1887.

[23] Stroh, A.N. (1954) Formation of cracks as a results of plastic flow, *Proc. Roy. Soc.* A, 223, pp404–414.

[24] Stroh, A.N. (1955) Formation of cracks in plastic flow II, *Proc. Roy. Soc.* A, 232, 548–560.

[25] Fan, H. (1994) Interfacial Zener-Stroh crack, *ASME J. Appl. Mech.*, 61(4), 829–834.

[26] Suo, Z. (1990) Singularities, interfaces and cracks in the dissimilar anisotropic medium, *Proc. Roy. Soc.* A, 427, 331–358.

[27] Suo, Z. (1993) Models for breakdown-resistant dielectric and ferroelectric ceramics, *J. Mech. Phys. Solids*, 41(7) 1155–1176.

[28] Muskhelishvili, N.I. (1977) *Some Basic Problems of the Mathematical Theory of Elasticity*, Noordhoff International, Leyden, Netherlands.

[29] Fan, H. and Keer, L.M. (1994) Two-dimensional contact on an anisotropic elastic half space, *ASME J Appl. Mech.*, 61, 250–25.

[30] Johnson, K.L. (1985) *Contact Mechanics*, Cambridge University Press, Cambridge, UK.

[31] Salt, D. (1987) *Hy-Q: Handbook of Quartz Crystal Devices*, Van Nostrand Reinhold, UK.

[32] Horgan, C.O. and Knowles, J. (1983) Recent development concerning Saint-Venant principle, *Adv. Appl. Mech.*, 23, 180–271.

[33] Horgan, C.O. (1989) Recent development concerning Saint-Venant principle: An update, *Appl. Mech. Rev.*, 42(1), 295–303.

[34] Wang, M.Z., Ting, T.C.T., and Yan, G. (1993) The anisotropic elastic semi-infinite strip, *Quart. Appl. Math.*, L1(2), 283–297.

[35] Borrelli, A., C.O. Horgan, and M.C. Patria (2006) Saint-Venant end effect for plane deformation of linear piezoelectric solid, *Int. J. Solids Struct.*, 43, 943–956.

Chapter 4
Fracture and Crack Mechanics

Yasuhide Shindo

4.1 Introduction

In the development of smart material systems and structures, piezoelectric ceramics are extensively used in sensors and actuators. The main weakness of piezoelectric ceramics is their brittleness. Stress and electric field concentrations near the tips of defects or electrodes can also induce crack initiation and propagation, which will lead to the failure of these piezoelectric ceramics. Therefore, the piezoelectric fracture and crack problems have received considerable attention due to practical importance.

In the theoretical studies of the piezoelectric fracture and crack problems, there are two commonly used electrical boundary conditions across the crack face, the permeable crack model [1,2], and the impermeable crack model [3–5]. Recently, the open piezoelectric crack model [6] was used in [7–9], and the effect of electric fields on the fracture mechanics parameters such as energy release rate was discussed. Also an extensive body of theoretical and experimental data on fracture behavior for a wide variety of piezoelectric ceramics over the last two decades was summarized in books and review papers [10–12]. Although the impermeable and open crack models may provide the mathematical solutions of the piezoelectric cracks, it is still questionable to search for fracture design parameters characterizing the electric failure. It has also been found that the permeable crack model can be appropriate for the piezoelectric cracks.

Some experimental results show that the fracture loads are increased or decreased, depending on the mechanical loading conditions (applied load or applied displacement) and direction of electric fields [13,14]. The nonlinear effect caused by the polarization switching may also affect the piezoelectric

Yasuhide Shindo
Department of Materials Processing, Graduate School of Engineering, Tohoku University, Aoba-yama 6-6-02, Sendai 980-8579, Japan, e-mail: shindo@material.tohoku.ac.jp

J. Yang (ed.), *Special Topics in the Theory of Piezoelectricity*, DOI: 10.1007/978-0-387-89498-0_4, 81
© Springer Science + Business Media, LLC 2009

fracture behavior [15, 16]. Hence, it is important for reliability and durability to investigate the effects of applied electromechanical loading and polarization switching on the fracture behavior of piezoelectric ceramics.

In this chapter, a theoretical fracture mechanics for piezoelectric ceramics is first developed, based on the permeable, impermeable, and open crack models. Both analytical and simulation methods are formulated to determine the effects of electric field and polarization switching on the fracture mechanics parameters (e.g., energy release rate). Secondly, the results on the electrical loading dependence of fracture of piezoelectric ceramics are reported. The indentation fracture (IF), bending fracture, and some precracked specimen tests are performed. Finite element analyses are then employed to study the crack behavior in piezoelectric ceramics. Finally, the results on fatigue behavior in piezoelectric specimens under electromechanical loading are presented.

4.2 Piezoelectric Crack Mechanics

The electrical boundary conditions along the crack face remain an issue of debate when studying piezoelectric crack problems. Here, the effect of crack face boundary conditions on the piezoelectric fracture mechanics parameters is discussed. Attention is focused on the electric field dependence of the energy release rate. Nonlinear behavior of the piezoelectric fracture mechanics parameters due to localized polarization switching is also examined.

4.2.1 Fundamental Equations

(a) **Linear behavior**

Consider the rectangular Cartesian coordinate system O-x, y, z. Poling direction is the z-axis. For piezoelectric ceramics that exhibit symmetry of a hexagonal crystal of class 6 mm with respect to principal x, y, and z axes, the constitutive relations can be written in the following form [17, 18],

$$
\begin{Bmatrix} T_{xx} \\ T_{yy} \\ T_{zz} \\ T_{yz} \\ T_{zx} \\ T_{xy} \end{Bmatrix} =
\begin{bmatrix}
c_{11}^E & c_{12}^E & c_{13}^E & 0 & 0 & 0 \\
c_{12}^E & c_{11}^E & c_{13}^E & 0 & 0 & 0 \\
c_{13}^E & c_{13}^E & c_{33}^E & 0 & 0 & 0 \\
0 & 0 & 0 & c_{44}^E & 0 & 0 \\
0 & 0 & 0 & 0 & c_{44}^E & 0 \\
0 & 0 & 0 & 0 & 0 & c_{66}^E
\end{bmatrix}
\begin{Bmatrix} S_{xx} - S_{xx}^r \\ S_{yy} - S_{yy}^r \\ S_{zz} - S_{zz}^r \\ 2(S_{yz} - S_{yz}^r) \\ 2(S_{zx} - S_{zx}^r) \\ 2(S_{xy} - S_{xy}^r) \end{Bmatrix}
-
\begin{bmatrix}
0 & 0 & e_{31} \\
0 & 0 & e_{31} \\
0 & 0 & e_{33} \\
0 & e_{15} & 0 \\
e_{15} & 0 & 0 \\
0 & 0 & 0
\end{bmatrix}
\begin{Bmatrix} E_x \\ E_y \\ E_z \end{Bmatrix}
$$

$$\tag{4.1}$$

$$\begin{Bmatrix} D_x \\ D_y \\ D_z \end{Bmatrix} = \begin{bmatrix} 0 & 0 & 0 & 0 & e_{15} & 0 \\ 0 & 0 & 0 & e_{15} & 0 & 0 \\ e_{31} & e_{31} & e_{33} & 0 & 0 & 0 \end{bmatrix} \begin{Bmatrix} S_{xx} - S^r_{xx} \\ S_{yy} - S^r_{yy} \\ S_{zz} - S^r_{zz} \\ 2(S_{yz} - S^r_{yz}) \\ 2(S_{zx} - S^r_{zx}) \\ 2(S_{xy} - S^r_{xy}) \end{Bmatrix}$$

$$+ \begin{bmatrix} \varepsilon^S_{11} & 0 & 0 \\ 0 & \varepsilon^S_{11} & 0 \\ 0 & 0 & \varepsilon^S_{33} \end{bmatrix} \begin{Bmatrix} E_x \\ E_y \\ E_z \end{Bmatrix} + \begin{Bmatrix} P^r_x \\ P^r_y \\ P^r_z \end{Bmatrix}, \qquad (4.2)$$

where $(T_{xx}, T_{yy}, T_{zz}, T_{yz}, T_{zx}, T_{xy})$ and $(S_{xx}, S_{yy}, S_{zz}, S_{yz}, S_{zx}, S_{xy})$ are the components of stress and strain tensors, (E_x, E_y, E_z) and (D_x, D_y, D_z) are the components of electric field intensity and electric displacement vectors, $(S^r_{xx}, S^r_{yy}, S^r_{zz}, S^r_{yz}, S^r_{zx}, S^r_{xy})$ and (P^r_x, P^r_y, P^r_z) are the components of remanent strains and polarizations, and $(c^E_{11}, c^E_{12}, c^E_{13}, c^E_{33}, c^E_{44}, c^E_{66})$, (e_{31}, e_{33}, e_{15}), and $(\varepsilon^S_{11}, \varepsilon^S_{33})$ are the elastic stiffness constants measured in a constant electric field, piezoelectric constants, and dielectric permittivities measured at constant strain, respectively. The strain components are

$$S_{xx} = u_{x,x}, \qquad S_{yy} = u_{y,y},$$

$$S_{zz} = u_{z,z}, \qquad S_{yz} = S_{zy} = \frac{1}{2}(u_{y,z} + u_{z,y}),$$

$$S_{zx} = S_{xz} = \frac{1}{2}(u_{z,x} + u_{x,z}), \qquad S_{xy} = S_{yx} = \frac{1}{2}(u_{x,y} + u_{y,x}), \qquad (4.3)$$

where (u_x, u_y, u_z) are the components of the displacement vector. The electric field components are related to the electric potential ϕ by

$$E_x = -\phi_{,x}, \qquad E_y = -\phi_{,y}, \qquad E_z = -\phi_{,z}. \qquad (4.4)$$

Using the displacements and electric potential, the governing equations can be written as

$$\left.\begin{aligned} &c^E_{11} u_{x,xx} + c^E_{66} u_{x,yy} + c^E_{44} u_{x,zz} + (c^E_{12} + c^E_{66}) u_{y,xy} \\ &\quad + (c^E_{13} + c^E_{44}) u_{z,xz} + (e_{31} + e_{15})\phi_{,xz} = 0 \\[6pt] &(c^E_{12} + c^E_{66}) u_{x,xy} + c^E_{66} u_{y,xx} + c^E_{11} u_{y,yy} + c^E_{44} u_{y,zz} \\ &\quad + (c^E_{13} + c^E_{44}) u_{z,yz} + (e_{31} + e_{15})\phi_{,yz} = 0 \\[6pt] &(c^E_{13} + c^E_{44})(u_{x,xz} + u_{y,yz}) + c^E_{44}(u_{z,xx} + u_{z,yy}) + c^E_{33} u_{z,zz} \\ &\quad + e_{15}(\phi_{,xx} + \phi_{,yy}) + e_{33}\phi_{,zz} = 0 \end{aligned}\right\} \qquad (4.5)$$

$$(e_{31} + e_{15})(u_{x,xz} + u_{y,yz}) + e_{15}(u_{z,xx} + u_{z,yy}) + e_{33} u_{z,zz}$$
$$- \varepsilon^S_{11}(\phi_{,xx} + \phi_{,yy}) - \varepsilon^S_{33}\phi_{,zz} = 0. \qquad (4.6)$$

In a vacuum, the constitutive equations (4.2) and the governing equation (4.6) become

$$D_x = \varepsilon_0 E_x, \qquad D_y = \varepsilon_0 E_y, \qquad D_z = \varepsilon_0 E_z \tag{4.7}$$

$$\phi_{,xx} + \phi_{,yy} + \phi_{,zz} = 0, \tag{4.8}$$

where $\varepsilon_0 = 8.85 \times 10^{-12}$ C/Vm is the electric permittivity of the vacuum.

(b) Polarization switching behavior

The most important class of ferroelectric materials is the perovskite oxides ABO_3 (e.g., $PbTiO_3$), as shown in Figure 4.1. A central Ti^{4+} ion displaces off-center with respect to surrounding O^{2-} ions, so that the unit cell

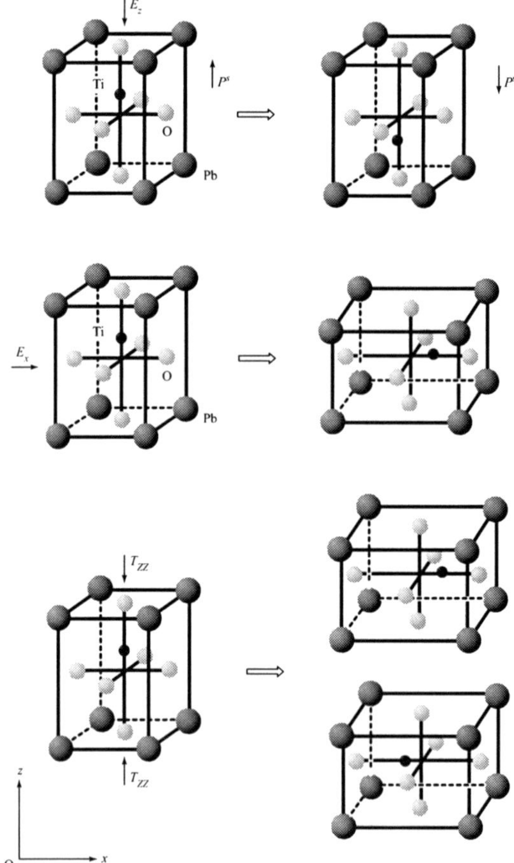

Fig. 4.1 180° or 90° polarization switching for perovskite-type crystal structure of lead titanate.

possesses a spontaneous polarization P^{s} and a spontaneous strain γ^{s} aligned with the dipole moment of the charge distribution. Piezoelectric behavior is induced by the poling process, that is, by applying a high voltage to the material. The electric field aligns dipoles along the field lines. This alignment results in the remanent polarization (P_x^r, P_y^r, P_z^r), and the remanently polarized state has a remanent strain $(S_{xx}^r, S_{yy}^r, S_{zz}^r, S_{yz}^r, S_{zx}^r, S_{xy}^r)$. Polarization switching occurs when an applied electric field exceeds the coercive electric field E_c, and leads to changes in the remanent strain and remanent polarization.

Lead oxide-based ferroelectrics, especially $Pb(Zr,Ti)O_3$ (PZT), exhibit high piezoelectric properties. Due to lead toxicity, however, the introduction of legislation in Europe to limit the usage of lead in automotive and electronic products has led to a worldwide search for lead-free compounds [19, 20].

Some nonlinear constitutive models have been developed for ferroelectric materials. A dynamic macroscopic phenomenological theory for the existence of butterfly and hysteresis loops in ferroelectricity was presented in [21]. The model reproduced the consequences of polarization switching on the mechanical and dielectric responses of the piezoelectric ceramics subjected to a slowly varying cyclic electric field. This is, perhaps, the first quantitative indication of this phenomenon. In [22], a work energy criterion was used to determine the critical loading level at which polarization switching occurs. By using a number of nonlinear functions to represent behavior during switching, a simple smooth switching model (complete phenomenological constitutive model) was developed in [23]. Later, three models were discussed in [24]: (i) a self-consistent polycrystal model; (ii) a crystal plasticity model; and (iii) a rate-independent phenomenological model, and the predictions of each model for the material response in multiaxial electrical loading were compared with the measured responses. In recent years, several existing polarization switching criteria were reviewed, and a new criterion in terms of internal energy density was proposed for combined electromechanical loading [25].

An electric field may rotate the poling direction by either $180°$ or $90°$, but a stress may only rotate it by $90°$ [26]. Figure 4.1 illustrates several possibilities. The criterion in [22] requires that a polarization switches when the combined electrical and mechanical work exceeds a critical value; that is,

$$T_{xx}\Delta S_{xx} + T_{yy}\Delta S_{yy} + T_{zz}\Delta S_{zz} + 2(T_{yz}\Delta S_{yz} + T_{zx}\Delta S_{zx} + T_{xy}\Delta S_{xy})$$
$$+ E_x\Delta P_x + E_y\Delta P_y + E_z\Delta P_z \geq 2P^{\mathrm{s}}E_{\mathrm{c}}, \tag{4.9}$$

where Δ means the changes in the spontaneous strain and polarization. It is assumed that elastic and dielectric constants of the piezoelectric materials remain unchanged after $180°$ or $90°$ polarization switching occurs and only piezoelectric constants vary with switching. It is also assumed, for example, in the zx plane, that for $90°$ switching there are two allowable directions of the poling in the coordinate system: in the positive and negative x-direction (see Figure 4.1). The changes in spontaneous strains and polarizations for $180°$ switching can be expressed as

$$\Delta S_{xx} = 0, \qquad \Delta S_{yy} = 0, \qquad \Delta S_{zz} = 0,$$
$$\Delta S_{yz} = 0, \qquad \Delta S_{zx} = 0, \qquad \Delta S_{xy} = 0 \qquad (4.10)$$
$$\Delta P_x = 0, \qquad \Delta P_y = 0, \qquad \Delta P_z = -2P^{\mathrm{s}}. \qquad (4.11)$$

For $90°$ switching in the zx plane, the changes are

$$S_{xx} = \gamma^{\mathrm{s}}, \qquad \Delta S_{yy} = 0, \qquad \Delta S_{zz} = -\gamma^{\mathrm{s}},$$
$$\Delta S_{yz} = 0, \qquad \Delta S_{zx} = 0, \qquad \Delta S_{xy} = 0 \qquad (4.12)$$
$$\Delta P_x = \pm P^{\mathrm{s}}, \qquad \Delta P_y = 0, \qquad \Delta P_z = -P^{\mathrm{s}}. \qquad (4.13)$$

For $90°$ switching in the yz plane, we have

$$\Delta S_{xx} = 0, \qquad \Delta S_{yy} = \gamma^{\mathrm{s}}, \qquad \Delta S_{zz} = -\gamma^{\mathrm{s}},$$
$$\Delta S_{yz} = 0, \qquad \Delta S_{zx} = 0, \qquad \Delta S_{xy} = 0 \qquad (4.14)$$
$$\Delta P_x = 0, \qquad \Delta P_y = \pm P^{\mathrm{s}}, \qquad \Delta P_z = -P^{\mathrm{s}}. \qquad (4.15)$$

The constitutive equations (4.1) and (4.2) after polarization switching are

$$\begin{Bmatrix} T_{xx} \\ T_{yy} \\ T_{zz} \\ T_{yz} \\ T_{zx} \\ T_{xy} \end{Bmatrix} = \begin{bmatrix} c_{11}^E & c_{12}^E & c_{13}^E & 0 & 0 & 0 \\ c_{12}^E & c_{11}^E & c_{13}^E & 0 & 0 & 0 \\ c_{13}^E & c_{13}^E & c_{33}^E & 0 & 0 & 0 \\ 0 & 0 & 0 & c_{44}^E & 0 & 0 \\ 0 & 0 & 0 & 0 & c_{44}^E & 0 \\ 0 & 0 & 0 & 0 & 0 & c_{66}^E \end{bmatrix} \begin{bmatrix} S_{xx} - S_{xx}^r \\ S_{yy} - S_{yy}^r \\ S_{zz} - S_{zz}^r \\ 2(S_{yz} - S_{yz}^r) \\ 2(S_{zx} - S_{zx}^r) \\ 2(S_{xy} - S_{xy}^r) \end{bmatrix} - \begin{bmatrix} e'_{11} & e'_{21} & e'_{31} \\ e'_{12} & e'_{22} & e'_{32} \\ e'_{13} & e'_{23} & e'_{33} \\ e'_{14} & e'_{24} & e'_{34} \\ e'_{15} & e'_{25} & e'_{35} \\ e'_{16} & e'_{26} & e'_{36} \end{bmatrix} \begin{Bmatrix} E_x \\ E_y \\ E_z \end{Bmatrix}$$

$$(4.16)$$

$$\begin{Bmatrix} D_x \\ D_y \\ D_z \end{Bmatrix} = \begin{bmatrix} e'_{11} & e'_{12} & e'_{13} & e'_{14} & e'_{15} & e'_{16} \\ e'_{21} & e'_{22} & e'_{23} & e'_{24} & e'_{25} & e'_{26} \\ e'_{31} & e'_{32} & e'_{33} & e'_{34} & e'_{35} & e'_{36} \end{bmatrix} \begin{Bmatrix} S_{xx} - S_{xx}^r \\ S_{yy} - S_{yy}^r \\ S_{zz} - S_{zz}^r \\ 2(S_{yz} - S_{yz}^r) \\ 2(S_{zx} - S_{zx}^r) \\ 2(S_{xy} - S_{xy}^r) \end{Bmatrix}$$

$$+ \begin{bmatrix} \varepsilon_{11}^S & 0 & 0 \\ 0 & \varepsilon_{11}^S & 0 \\ 0 & 0 & \varepsilon_{33}^S \end{bmatrix} \begin{Bmatrix} E_x \\ E_y \\ E_z \end{Bmatrix} + \begin{Bmatrix} P_x^r \\ P_y^r \\ P_z^r \end{Bmatrix}. \qquad (4.17)$$

The new piezoelectric constants are related to the elastic and direct piezoelectric constants by

$$e'_{11} = d'_{111}c_{11} + d'_{122}c_{12} + d'_{133}c_{13} \qquad e'_{12} = d'_{111}c_{12} + d'_{122}c_{11} + d'_{133}c_{13}$$
$$e'_{13} = d'_{111}c_{13} + d'_{122}c_{13} + d'_{133}c_{33} \qquad e'_{14} = 2d'_{123}c_{44},$$
$$e'_{15} = 2d'_{131}c_{44}, \qquad e'_{16} = 2d'_{112}c_{66}$$

$$e'_{21} = d'_{211}c_{11} + d'_{222}c_{12} + d'_{233}c_{13} \qquad e'_{22} = d'_{211}c_{12} + d'_{222}c_{11} + d'_{233}c_{13}$$
$$e'_{23} = d'_{211}c_{13} + d'_{222}c_{13} + d'_{233}c_{33} \qquad e'_{24} = 2d'_{223}c_{44},$$
$$e'_{25} = 2d'_{231}c_{44}, \qquad e'_{26} = 2d'_{212}c_{66}$$
$$e'_{31} = d'_{311}c_{11} + d'_{322}c_{12} + d'_{333}c_{13} \qquad e'_{32} = d'_{311}c_{12} + d'_{322}c_{11} + d'_{333}c_{13}$$
$$e'_{33} = d'_{311}c_{13} + d'_{322}c_{13} + d'_{333}c_{33} \qquad e'_{34} = 2d'_{323}c_{44},$$
$$e'_{35} = 2d'_{331}c_{44}, \qquad e'_{36} = 2d'_{312}c_{66}. \tag{4.18}$$

The components of the direct piezoelectric tensor d'_{ikl} are

$$d'_{ikl} = \left\{ d_{33}n_i n_k n_l + d_{31}(n_i\delta_{kl} - n_i n_k n_l) + \frac{1}{2}d_{15}(\delta_{ik}n_l - 2n_i n_k n_l + \delta_{il}n_k) \right\}, \tag{4.19}$$

where n_i is the unit vector in the poling direction, δ_{ij} is the Kroneker delta, and d_{33}, d_{31}, d_{15} are the direct piezoelectric constants.

4.2.2 Crack Face Boundary Conditions

With reference to the theoretical studies of the piezoelectric fracture and crack problems, the investigators differ in opinions on the electrical boundary conditions at the crack faces. As the dielectric constant of the air or the medium between the crack faces is very small compared to that of the piezoelectric material, most of the reported works have assumed the permittivity in the medium interior to the crack to be zero (the so-called condition of impermeability or impermeable condition); that is,

$$D_n^+ = D_n^- = 0, \tag{4.20}$$

where D_n is the normal component of the electric displacement at the crack faces, and superscripts $+$ and $-$ denote upper and lower crack faces, respectively. Across the crack faces, having no charge density, we need to satisfy the relevant boundary conditions [2], which are the continuity of the tangential component of the electric field at the crack faces E_t,

$$E_t^+ = E_t^- \tag{4.21}$$

and the continuity of the normal component of the electric displacement,

$$D_n^+ = D_n^-. \tag{4.22}$$

Here, the effects of the electrical surface conditions on the electroelastic fields are discussed. The geometry of the problem under consideration is shown in Figure 4.2. Two semi-infinite piezoelectric materials are placed a distance $2d$ apart and occupy the region $(-\infty < x < \infty, |z| \geq d)$.

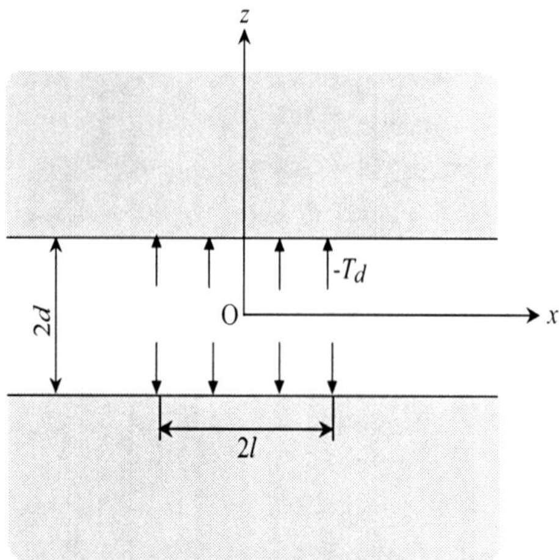

Fig. 4.2 Two semi-infinite piezoelectric materials.

The piezoelectric material is under a state of plane strain in the y-direction. To simulate a crack, the distance d is taken to be very small. This is equivalent to setting $d \to 0$.

The constitutive equations can be written as

$$\left.\begin{aligned}
T_{xx} &= c_{11}^E S_{xx} + c_{13}^E S_{zz} - e_{31} E_z \\
T_{zx} &= 2c_{44}^E S_{zx} - e_{15} E_x \\
T_{zz} &= c_{13}^E S_{xx} + c_{33}^E S_{zz} - e_{33} E_z
\end{aligned}\right\} \tag{4.23}$$

$$\left.\begin{aligned}
D_x &= 2e_{15} S_{zx} + \varepsilon_{11}^S E_x \\
D_z &= e_{31} S_{xx} + e_{33} S_{zz} + \varepsilon_{33}^S E_z
\end{aligned}\right\}. \tag{4.24}$$

The remanent terms are not included in the constitutive equations.

The governing equations become

$$\left.\begin{aligned}
c_{11}^E u_{x,xx} + c_{44}^E u_{x,zz} + (c_{13}^E + c_{44}^E)u_{z,xz} + (e_{31} + e_{15})\phi_{,xz} &= 0 \\
c_{44}^E u_{z,xx} + c_{33}^E u_{z,zz} + (c_{13}^E + c_{44}^E)u_{x,xz} + e_{15}\phi_{,xx} + e_{33}\phi_{,zz} &= 0
\end{aligned}\right\} \tag{4.25}$$

$$(e_{31} + e_{15})u_{x,xz} + e_{15}u_{z,xx} + e_{33}u_{z,zz} - \varepsilon_{11}^S \phi_{,xx} - \varepsilon_{33}^S \phi_{,zz} = 0. \tag{4.26}$$

The piezoelectric half-planes are subjected to a constant normal stress T_d at $z = \pm d$, $-l \le x \le l$. Hence the boundary conditions are given by

$$T_{zx}(x, \pm d) = 0 \tag{4.27}$$

$$T_{zz}(x, \pm d) = \begin{cases} -T_d (0 \le |x| \le l) \\ 0 \ (l < |x| < \infty) \end{cases}. \tag{4.28}$$

We consider next two possible cases of electrical boundary conditions on $z = \pm d$.

Impermeability:

$$D_z(x, \pm d) = 0 \qquad (0 \le |x| < \infty) \tag{4.29}$$

Electrical continuity:

$$E_x(x, \pm d) = E_x^e(x, \pm d) \qquad (0 \le |x| < \infty) \tag{4.30}$$

$$D_z(x, \pm d) = D_z^e(x, \pm d) \qquad (0 \le |x| < \infty), \tag{4.31}$$

where superscript e stands for the component of the electric field quantity outside the solid.

Solutions for the electroelastic fields are obtained using Fourier transform [27]. In the limit $\varepsilon_0 \to 0$, the solutions for the electrical continuity become those for the impermeability. In electrostatics, at a surface separating two dielectric materials with no free charge, the normal component of the electric displacement and the tangential component of the electric field are continuous. When one of the materials is air, these two conditions can be approximated simply by one, namely that the normal component of the electric displacement vanishes at the interface. So the conditions (4.21) and (4.22) are approximated by the single equation (4.20). This assumption is based on the fact that there is a very large difference between the dielectric constants of the material and the air. However, the solutions for the impermeability do not tend to those for the electrical continuity as $d \to 0$. This trend may be more clearly observed in Figure 4.3 [14]. The normalized stress $T_{zz}(l, d)/T_d$, strain $c_{33}^E S_{zz}(l, d)/T_d$, electric displacement $c_{33}^E D_z(l, d)/e_{33}T_d$, and electric field $e_{33}E_z(l, d)/T_d$ are plotted as a function of the normalized distance d/l. Thus the result illustrates how rapidly the applicability of the impermeable assumption deteriorates. Therefore, the electric boundary condition given by Equation (4.29) is not appropriate for a slit crack in piezoelectric materials.

Similarly, the problem of an elliptic hole embedded in a piezoelectric material was addressed in [28]. An infinite piezoelectric material containing an elliptic hole with major and minor axes $2a$ and $2b$ was considered, and the cavity was assumed to be filled with a homogeneous gas of ε_0. The major axis was normal to the poling direction. Expressions for the elastic and electric variables inside and outside the cavity were derived in closed form, and the following were found. (1) If $b/a >> 10^{-4}$, the models for the impermeability and electrical continuity provide virtually the same results;

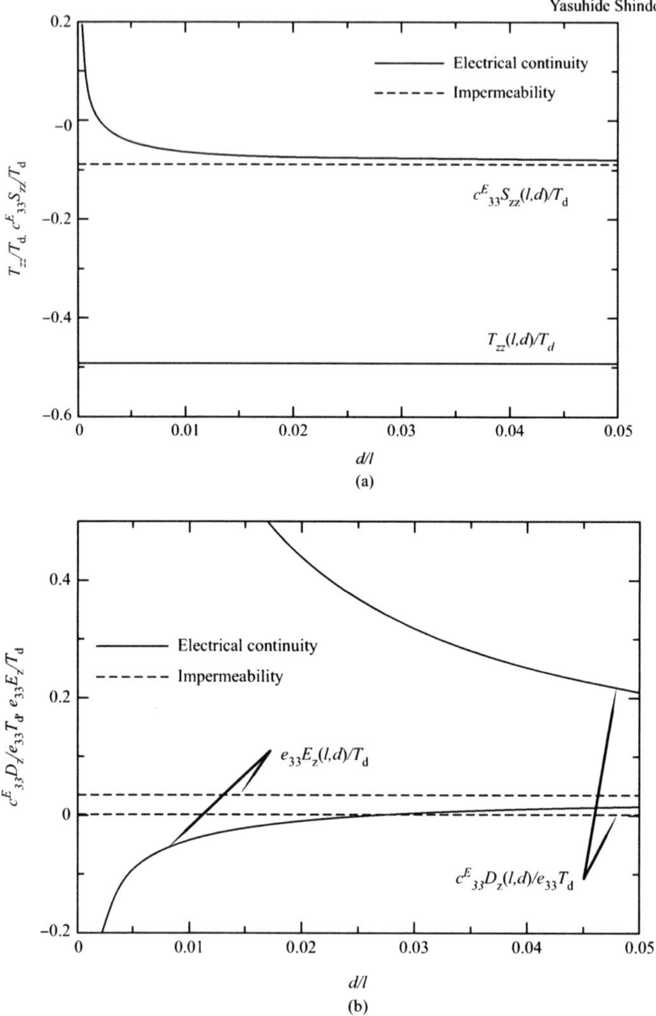

Fig. 4.3 The stress, strain, electric field, and electric displacement versus normalized distance: (a) T_{zz}/T_d and $c_{33}^E S_{zz}/T_d$; (b) $c_{33}^E D_z/e_{33}T_d$ and $e_{33}E_z/T_d$.

(2) if ε_0 is retained, in the limit of $b/a \to 0$, we find the dramatic differences for the fields provided by both models. It was concluded that the condition of electric impermeability at the boundary of the cavity may result in erroneous conclusions; it is particularly relevant for the case of very slender ellipses or sharp cracks. When an internal elliptic cavity problem in piezoelectric materials is analyzed, the impermeable condition is a good approximation. The fracture mechanics parameters for the above two models are presented in the next section.

The electric permeability of the air in a crack gap was considered in [6]. The conditions are

$$D_n^+ = D_n^- \tag{4.32}$$

$$D_n^+(u_n^+ - u_n^-) = -\varepsilon_0(\phi^+ - \phi^-), \tag{4.33}$$

where u_n is the displacement component normal to the crack face. This open crack model is also discussed in the next section.

4.2.3 Fracture Mechanics Parameters

(a) Plane strain crack in infinite material

Consider an infinite piezoelectric material containing a central crack of length $2a$. The crack is assumed with faces normal (Case 1) or parallel (Case 2) to the polarization axis, as shown in Figure 4.4. A set of Cartesian coordinates (x, y, z) is attached to the center of the crack. The piezoelectric material is under a state of plane strain in the y-direction. The material is loaded by mechanical loading and electric field. Only the first quadrant with appropriate boundary conditions needs to be analyzed owing to symmetry.

We first consider the case 1 (see Figure 4.4(a)). The boundary conditions at $z = 0$ can be expressed in the form

$$T_{zx}(x,0) = 0 \qquad (0 \leq x < \infty) \tag{4.34}$$

$$T_{zz}(x,0) = 0 \qquad (0 \leq x < a) \qquad u_z(x,0) = 0 \qquad (a \leq x < \infty) \tag{4.35}$$

$$E_x(x,0) = E_x^c(x,0) \qquad (0 \leq x < a) \qquad \phi(x,0) = 0 \qquad (a \leq x < \infty) \tag{4.36}$$

$$D_z(x,0) = D_z^c(x,0) \qquad (0 \leq x < a), \tag{4.37}$$

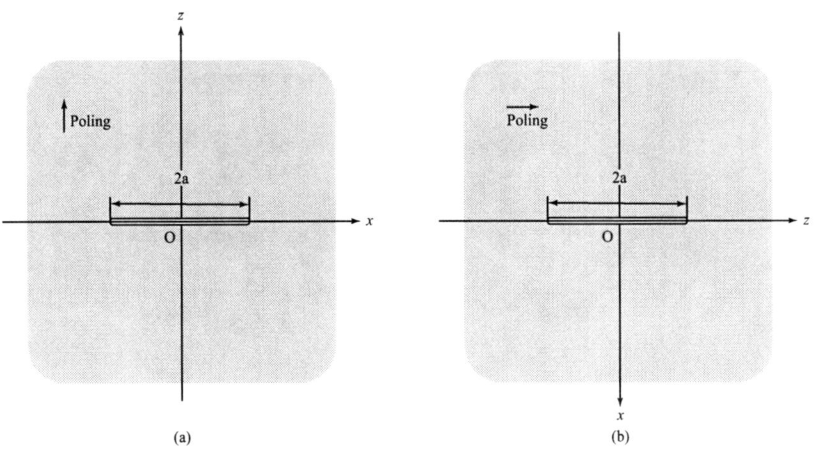

Fig. 4.4 An infinite piezoelectric material with a central crack: (a) Case 1; (b) Case 2.

where superscript c stands for the electric field quantity in the void inside the crack.

The mechanical and electrical loading conditions at infinity are

$$S_{zz}(x, z) = S_0, \qquad E_z(x, z) = E_0 \qquad (0 \le x < \infty, z \to \infty), \qquad (4.38)$$

where S_0 and E_0 are the far-field normal strain and electric field. The electric potential is all zero on the symmetry planes inside the crack and ahead of the crack, so the boundary conditions of Equations (4.36) reduce to $\phi(x, 0) = 0$ $(0 \le x < \infty)$. Equations (4.36) and (4.37) are the permeable boundary conditions.

By applying the loading conditions (4.38), the far-field normal stress T_{E0} is expressed as

$$T_{E0} = T_0 - e_1 E_0, \qquad (4.39)$$

where

$$T_0 = \left\{ c_{33}^E - \frac{(c_{13}^E)^2}{c_{11}^E} \right\} S_0 \qquad (4.40)$$

and

$$e_1 = e_{33} - \left(\frac{c_{13}^E}{c_{11}^E} \right) e_{31}. \qquad (4.41)$$

Note that T_0 is the stress for a closed-circuit condition with the potential forced to remain zero (grounded), and depends only on the strain at infinity. When a uniform strain S_0 is applied and fixed at infinity, the stress T_0 is uniform. On the other hand, when the stress T_{E0} is applied and fixed at infinity, T_{E0} is left unchanged and the strain S_0 depends on E_0.

Fourier transform [27] is applied to Equations (4.25) and (4.26) and the results satisfying the loading conditions of Equation (4.38) are

$$
\left.
\begin{aligned}
u_x(x, z) &= \frac{2}{\pi} \sum_{j=1}^{3} \int_0^\infty a_j A_j(\alpha) \exp(-\gamma_j \alpha z) \sin(\alpha x) d\alpha \\
&\quad + \left\{ \frac{c_{13}^E}{(c_{13}^E)^2 - c_{33}^E c_{11}^E} (T_{E0} + e_1 E_0) + \frac{e_{31}}{c_{11}^E} E_0 \right\} x \\
u_z(x, z) &= \frac{2}{\pi} \sum_{j=1}^{3} \int_0^\infty \frac{1}{\gamma_j} A_j(\alpha) \exp(-\gamma_j \alpha z) \cos(\alpha x) d\alpha \\
&\quad + \frac{c_{11}^E}{c_{33}^E c_{11}^E - (c_{13}^E)^2} (T_{E0} + e_1 E_0) z
\end{aligned}
\right\}
\qquad (4.42)
$$

$$\phi(x, z) = -\frac{2}{\pi} \sum_{j=1}^{3} \int_0^\infty \frac{b_j}{\gamma_j} A_j(\alpha) \exp(-\gamma_j \alpha z) \cos(\alpha x) d\alpha - E_0 z, \qquad (4.43)$$

where $A_j(\alpha)$ $(j = 1, 2, 3)$ are the unknowns to be solved, and a_j and b_j $(j = 1, 2, 3)$ are

$$a_j = \frac{(e_{31} + e_{15})(c_{33}^E \gamma_j^2 - c_{44}^E) - (c_{13}^E + c_{44}^E)(e_{33}\gamma_j^2 - e_{15})}{(c_{44}^E \gamma_j^2 - c_{11}^E)(e_{33}\gamma_j^2 - e_{15}) + (c_{13}^E + c_{44}^E)(e_{31} + e_{15})\gamma_j^2} \tag{4.44}$$

$$b_j = \frac{(c_{44}^E \gamma_j^2 - c_{11}^E)a_j + (c_{13}^E + c_{44}^E)}{e_{31} + e_{15}}. \tag{4.45}$$

γ_j^2 $(j = 1, 2, 3)$ are the roots of the following characteristic equations,

$$a_0 \gamma^6 + b_0 \gamma^4 + c_0 \gamma^2 + d_0 = 0, \tag{4.46}$$

where

$$
\begin{aligned}
a_0 &= c_{44}^E(c_{33}^E \epsilon_{33}^S + e_{33}^2) \\
b_0 &= -2c_{44}^E e_{15} e_{33} - c_{11}^E e_{33}^2 - c_{33}^E(c_{44}^E \varepsilon_{11}^S + c_{11}^E \varepsilon_{33}^S) + \varepsilon_{33}^S(c_{13}^E + c_{44}^E)^2 \\
&\quad + 2e_{33}(c_{13}^E + c_{44}^E)(e_{31} + e_{15}) - (c_{44}^E)^2 \varepsilon_{33}^S - c_{33}^E(e_{31} + e_{15})^2 \\
c_0 &= 2c_{11}^E e_{15} e_{33} + c_{44}^E e_{15}^2 + c_{11}^E(c_{33}^E \varepsilon_{11}^S + c_{44}^E \varepsilon_{33}^S) - \varepsilon_{11}^S(c_{13}^E + c_{44}^E)^2 \\
&\quad - 2e_{15}(c_{13}^E + c_{44}^E)(e_{31} + e_{15}) + (c_{44}^E)^2 \varepsilon_{11}^S + c_{44}^E(e_{31} + e_{15})^2 \\
d_0 &= -c_{11}^E(c_{44}^E \varepsilon_{11}^S + e_{15}^2).
\end{aligned}
\tag{4.47}
$$

Application of the Fourier transform to Equation (4.8) yields

$$\phi^c = \frac{2}{\pi} \int_0^\infty C(\alpha)\sinh(\alpha z)\cos(\alpha x)d\alpha \qquad (0 \le x < a), \tag{4.48}$$

where $C(\alpha)$ is also unknown.

Application of the crack face boundary conditions in Equations (4.34)–(4.36) gives rise to a pair of dual integral equations. The closed-form solutions of these dual integral equations are then obtained. The unknowns $A_j(\alpha)(j = 1, 2, 3)$ are related to the stress T_{E0} by the following equation,

$$A_j(\alpha) = -\frac{\pi}{2}\frac{d_j}{F}\frac{aJ_1(a\alpha)}{\alpha}T_{E0}, \tag{4.49}$$

where $J_1(\ )$ is the first-order Bessel function of the first kind, and

$$F = \sum_{j=1}^{3} g_j d_j \tag{4.50}$$

$$d_1 = \gamma_1(b_2 f_3 - b_3 f_2), \qquad d_2 = \gamma_2(b_3 f_1 - b_1 f_3), \qquad d_3 = \gamma_3(b_1 f_2 - b_2 f_1) \tag{4.51}$$

$$f_j = c_{44}^E(a_j\gamma_j^2 + 1) - e_{15}b_j \qquad (j = 1, 2, 3) \tag{4.52}$$

$$g_j = c_{13}^E a_j - c_{33}^E + e_{33}b_j \qquad (j = 1, 2, 3). \tag{4.53}$$

The displacement u_z and electric potential ϕ on the crack face are given by

$$u_z(x, 0) = -\frac{b}{F}T_{E0}(a^2 - x^2)^{1/2} \qquad \phi(x, 0) = 0 \tag{4.54}$$

in which

$$b = b_1(f_2 - f_3) + b_2(f_3 - f_1) + b_3(f_1 - f_2). \tag{4.55}$$

The tangential component of electric field E_x and the normal component of electric displacement D_z on the crack face are

$$E_x(x, 0) = 0 \qquad D_z(x, 0) = D_{T0} - \frac{1}{F}\left(\sum_{j=1}^{3} h_j d_j\right)T_{E0}, \tag{4.56}$$

where the far-field electric displacement D_{T0} is given by

$$D_{T0} = \frac{e_{31}c_{13}^E - e_{33}c_{11}^E}{(c_{13}^E)^2 - c_{33}^E c_{11}^E}T_0 + \left(\frac{e_{31}^2}{c_{11}^E} + \epsilon_{33}^S\right)E_0 \tag{4.57}$$

and

$$h_j = e_{31}a_j - e_{33} - \epsilon_{33}^S b_j \qquad (i = 1, 2, 3). \tag{4.58}$$

The stress intensity factor K_{I} is obtained as

$$K_{\mathrm{I}} = \lim_{x \to a^+} \{2\pi(x - a)\}^{1/2}T_{zz}(x, 0) = T_{E0}(\pi a)^{1/2}. \tag{4.59}$$

The electric displacement intensity factor K_D is also given by

$$K_D = \lim_{x \to a^+} \{2\pi(x - a)\}^{1/2}D_z(x, 0) = \frac{1}{F}\left(\sum_{j=1}^{3} h_j d_j\right)K_I. \tag{4.60}$$

The displacement component u_z and electric potential ϕ near the crack-tip can be written as

$$\left.\begin{aligned}
u_x &= \frac{K_{\mathrm{I}}}{F}\left(\frac{r}{\pi}\right)^{1/2}\sum_{j=1}^{3} a_j d_j\{(\cos^2\theta + \gamma_j^2\sin^2\theta)^{1/2} + \cos\theta\}^{1/2} \\
u_z &= -\frac{K_{\mathrm{I}}}{F}\left(\frac{r}{\pi}\right)^{1/2}\sum_{j=1}^{3} \frac{d_j}{\gamma_j}\{(\cos^2\theta + \gamma_j^2\sin^2\theta)^{1/2} - \cos\theta\}^{1/2}
\end{aligned}\right\} \tag{4.61}$$

$$\phi = \frac{K_{\mathrm{I}}}{F}\left(\frac{r}{\pi}\right)^{1/2}\sum_{j=1}^{3}\frac{b_j d_j}{\gamma_j}\{(\cos^2\theta + \gamma_j^2\sin^2\theta)^{1/2} - \cos\theta\}^{1/2}, \qquad (4.62)$$

where the polar coordinates r and θ are defined by $r = \{(x-a)^2 + z^2\}^{1/2}$, $\theta = \tan^{-1}(z/(x-a))$. The singular parts of the strains, stresses, electric field intensities, and electric displacements in the neighborhood of the crack-tip are

$$\left.\begin{array}{l} S_{xx} = \dfrac{K_{\mathrm{I}}}{2F(\pi r)^{1/2}}\displaystyle\sum_{j=1}^{3} a_j d_j R_j^c(\theta) \\[2ex] S_{xz} = S_{zx} = -\dfrac{K_{\mathrm{I}}}{4F(\pi r)^{1/2}}\displaystyle\sum_{j=1}^{3}\dfrac{d_j(a_j\gamma_j^2+1)}{\gamma_j}R_j^s(\theta) \\[2ex] S_{zz} = -\dfrac{K_{\mathrm{I}}}{2F(\pi r)^{1/2}}\displaystyle\sum_{j=1}^{3} d_j R_j^c(\theta) \end{array}\right\} \qquad (4.63)$$

$$\left.\begin{array}{l} T_{xx} = \dfrac{K_{\mathrm{I}}}{2F(\pi r)^{1/2}}\displaystyle\sum_{j=1}^{3} m_j d_j R_j^c(\theta) \\[2ex] T_{zz} = \dfrac{K_{\mathrm{I}}}{2F(\pi r)^{1/2}}\displaystyle\sum_{j=1}^{3} g_j d_j R_j^c(\theta) \\[2ex] T_{zx} = -\dfrac{K_{\mathrm{I}}}{2F(\pi r)^{1/2}}\displaystyle\sum_{j=1}^{3}\dfrac{f_j d_j}{\gamma_j} R_j^s(\theta) \end{array}\right\} \qquad (4.64)$$

$$\left.\begin{array}{l} E_x = -\dfrac{K_{\mathrm{I}}}{2F(\pi r)^{1/2}}\displaystyle\sum_{j=1}^{3}\dfrac{b_j d_j}{\gamma_j} R_j^s(\theta) \\[2ex] E_z = -\dfrac{K_{\mathrm{I}}}{2F(\pi r)^{1/2}}\displaystyle\sum_{j=1}^{3} b_j d_j R_j^c(\theta) \end{array}\right\} \qquad (4.65)$$

$$\left.\begin{array}{l} D_x = -\dfrac{K_{\mathrm{I}}}{2F(\pi r)^{1/2}}\displaystyle\sum_{j=1}^{3}\dfrac{n_j d_j}{\gamma_j} R_j^s(\theta) \\[2ex] D_z = \dfrac{K_{\mathrm{I}}}{2F(\pi r)^{1/2}}\displaystyle\sum_{j=1}^{3} h_j d_j R_j^c(\theta) \end{array}\right\}, \qquad (4.66)$$

where

$$R_j^c(\theta) = \left\{\frac{(\cos^2\theta + \gamma_j^2\sin^2\theta)^{1/2} + \cos\theta}{\cos^2\theta + \gamma_j^2\sin^2\theta}\right\}^{1/2}$$

$$R_j^s(\theta) = -\left\{\frac{(\cos^2\theta + \gamma_j^2\sin^2\theta)^{1/2} - \cos\theta}{\cos^2\theta + \gamma_j^2\sin^2\theta}\right\}^{1/2} \qquad (4.67)$$

and

$$m_j = c_{11}^E a_j - c_{13}^E + e_{31} b_j \qquad (i = 1, 2, 3) \tag{4.68}$$

$$n_j = e_{15}(a_j \gamma_j^2 + 1) + \epsilon_{11}^S b_j \qquad (i = 1, 2, 3). \tag{4.69}$$

By using the concept of crack closure energy, the energy release rate G [4] can be obtained as

$$G = \lim_{\Delta a \to 0} \frac{1}{\Delta a} \int_0^{\Delta a} \{T_{zz}(x)u_z(\Delta a - x) + T_{zx}(x)u_x(\Delta a - x)$$
$$+ D_z(x)\phi(\Delta a - x)\}dx, \tag{4.70}$$

where Δa is the assumed crack extension. The energy release rate G is also obtained from the following path-independent integral J [4]:

$$J = \int_{\Gamma_0} \{Hn_x - (T_{xx}u_{x,x} + T_{zx}u_{z,x})n_x - (T_{zx}u_{x,x} + T_{zz}u_{z,x})n_z$$
$$+ D_x E_x n_x + D_z E_x n_z\}d\Gamma, \tag{4.71}$$

where Γ_0 is a small contour closing a crack-tip (see Figure 4.5) and n_x, n_z are the components of the outer unit normal vector. The electrical enthalpy density H is [17]

$$H(S_{xx}, S_{zz}, S_{zx}, E_x, E_z) = \frac{1}{2}(c_{11}^E S_{xx}^2 + c_{33}^E S_{zz}^2 + 2c_{13}^E S_{xx}S_{zz}$$
$$+ 4c_{44}^E S_{zx}^2) - \frac{1}{2}(\epsilon_{11}^S E_x^2 + \epsilon_{33}^S E_z^2) - \{2e_{15}S_{zx}E_x + (e_{31}S_{xx} + e_{33}S_{zz})E_z\}. \tag{4.72}$$

Writing the energy release rate expression for the permeable crack model in terms of the stress intensity factor [29], it results in

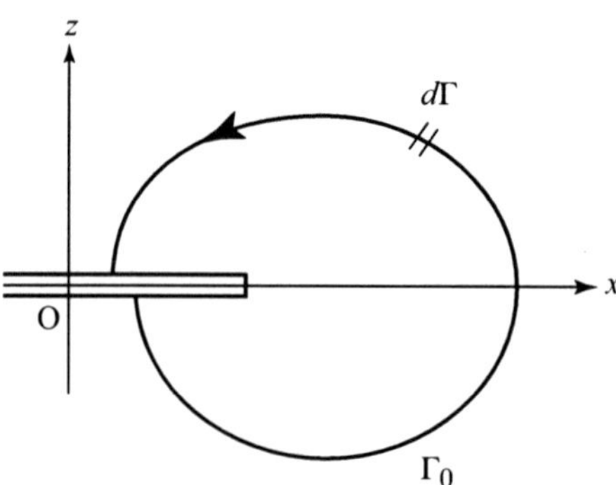

Fig. 4.5 Contour for evaluation of path-independent integral.

$$G = J = \frac{1}{2F^2}\left(-F\sum_{j=1}^{3}\frac{d_j}{\gamma_j} + \sum_{k=1}^{3}h_k d_k \sum_{j=1}^{3}\frac{b_j d_j}{\gamma_j}\right)K_I^2. \tag{4.73}$$

The energy density fracture criterion can be developed by referring to the amount of energy stored in a volume element ahead of the crack [30]. For the piezoelectric material, the energy stored in the volume element dV [31] is

$$dW = \left\{\frac{1}{2}(T_{xx}S_{xx} + T_{xz}S_{xz} + T_{zx}S_{zx} + T_{zz}S_{zz}) \right.$$
$$\left. + \frac{1}{2}(D_x E_x + D_z E_z)\right\}dV. \tag{4.74}$$

The energy density S is expressible in the form

$$S = r\frac{dW}{dV}. \tag{4.75}$$

The energy density criterion can be used to predict the path of mixed mode crack initiation and propagation. The energy density for the permeable crack model [32] is given as

$$S = (a_M + a_E)K_I^2 \tag{4.76}$$

in which the coefficients a_M and a_E depend on the angle θ and they are

$$a_M = \frac{1}{8\pi F^2}\left\{\sum_{j=1}^{3}m_j d_j R_j^c(\theta)\sum_{j=1}^{3}a_j d_j R_j^c(\theta)\right.$$
$$\left. + \sum_{j=1}^{3}\frac{f_j d_j}{\gamma_j}R_j^s(\theta)\sum_{j=1}^{3}\frac{d_j(a_j\gamma_j^2+1)}{\gamma_j}R_j^s(\theta) - \sum_{j=1}^{3}g_j d_j R_j^c(\theta)\sum_{j=1}^{3}d_j R_j^c(\theta)\right\} \tag{4.77}$$

$$a_E = \frac{1}{8\pi F^2}\left\{\sum_{j=1}^{3}\frac{n_j}{\gamma_j}d_j R_j^s(\theta)\sum_{j=1}^{3}\frac{b_j d_j}{\gamma_j}R_j^s(\theta) - \sum_{j=1}^{3}h_j d_j R_j^c(\theta)\sum_{j=1}^{3}b_j d_j R_j^c(\theta)\right\}. \tag{4.78}$$

The stress and electric displacement intensity factors, energy release rate, and energy density for the permeable crack model depend only on the stress T_{E0}, and the electric loading dependence on these parameters is different for the two mechanical loading conditions (applied strain and applied stress). For many practical applications it seems that the piezoelectric devices are operated with the strain fixed at some values. For a general remote boundary condition prescribed by fixed strain, the stress T_{E0} is determined by Equation (4.39) with T_0 and E_0. The stress and electric displacement intensity factors, energy release rate, and energy density for the permeable crack model depend on the electric field E_0, because the stress T_{E0} increases or decreases depending on the magnitude and direction of the electric field E_0. However, if the stress is applied and fixed, the value of the stress T_{E0} is left unchanged. Consequently, the stress and electric displacement intensity factors, energy release rate, and energy density for the permeable crack model are independent of the electric field E_0.

The above phenomena are not observed in the impermeable and open cracks of piezoelectric materials. A solution procedure for the impermeable and open crack models in the infinite piezoelectric materials is outlined in the next two paragraphs.

The crack face electric boundary condition for the impermeable crack model is

$$D_z(x,0) = 0 \qquad (0 \le x < a)$$
$$\phi(x,0) = 0 \qquad (a \le x < \infty). \tag{4.79}$$

The unknowns $A_j(\alpha)$ $(j = 1, 2, 3)$ in Equations (4.42) and (4.43) can be found by the same method of approach as in the permeable case. By applying the crack face boundary conditions of Equations (4.34), (4.35), and (4.79), the unknowns $A_j(\alpha)(j = 1, 2, 3)$ are related to T_{E0} and D_{T0} as follows,

$$\frac{f_1}{\gamma_1} A_1(\alpha) + \frac{f_2}{\gamma_2} A_2(\alpha) + \frac{f_3}{\gamma_3} A_3(\alpha) = 0$$

$$\frac{1}{\gamma_1} A_1(\alpha) + \frac{1}{\gamma_2} A_2(\alpha) + \frac{1}{\gamma_3} A_3(\alpha) = -\frac{\pi}{2F'} \frac{a}{\alpha} J_1(a\alpha)(F_{22}T_{E0} - F_{12}D_{T0})$$

$$\frac{b_1}{\gamma_1} A_1(\alpha) + \frac{b_2}{\gamma_2} A_2(\alpha) + \frac{b_3}{\gamma_3} A_3(\alpha) = \frac{\pi}{2F'} \frac{a}{\alpha} J_1(a\alpha)(F_{21}T_{E0} - F_{11}D_{T0}), \tag{4.80}$$

where

$$F_{11} = \frac{1}{b} \sum_{j=1}^{3} g_j d_j, \qquad F_{12} = \frac{1}{b} \sum_{j=1}^{3} g_j l_j,$$

$$F_{21} = \frac{1}{b} \sum_{j=1}^{3} h_j d_j, \qquad F_{22} = \frac{1}{b} \sum_{j=1}^{3} h_j l_j \tag{4.81}$$

$$F' = F_{11}F_{22} - F_{12}F_{21} \tag{4.82}$$

$$l_1 = \gamma_1(f_2 - f_3), \qquad l_2 = \gamma_2(f_3 - f_1), \qquad l_3 = \gamma_3(f_1 - f_2). \tag{4.83}$$

The displacement u_z, electric potential ϕ, tangential component of electric field E_x, and normal component of electric displacement D_z on the crack face are given by

$$u_z(x,0) = -\frac{F_{22}T_{E0} - F_{12}D_{T0}}{F'}(a^2 - x^2)^{1/2}$$

$$\phi(x,0) = -\frac{F_{21}T_{E0} - F_{11}D_{T0}}{F'}(a^2 - x^2)^{1/2}$$

$$E_x(x,0) = -\frac{F_{21}T_{E0} - F_{11}D_{T0}}{F'}\frac{x}{(a^2 - x^2)^{1/2}} \qquad D_z(x,0) = 0. \qquad (4.84)$$

The energy release rate G^{I} [33] and energy density S^{I} [31] for the impermeable crack model are

$$
\begin{aligned}
G^{\mathrm{I}} = -\frac{1}{2F'^2}\Bigg\{ &\left(F'\sum_{j=1}^{3}\frac{s_j}{\gamma_j} - \sum_{k=1}^{3}h_k s_k \sum_{j=1}^{3}\frac{b_j s_j}{\gamma_j}\right) K_{\mathrm{I}}^2 \\
&+ \left(\sum_{k=1}^{3}h_k t_k \sum_{j=1}^{3}\frac{b_j s_j}{\gamma_j} + \sum_{k=1}^{3}h_k s_k \sum_{j=1}^{3}\frac{b_j t_j}{\gamma_j} - F'\sum_{j=1}^{3}\frac{t_j}{\gamma_j}\right) K_{\mathrm{I}}K_{\mathrm{D}} \\
&- \left(\sum_{k=1}^{3}h_k t_k \sum_{j=1}^{3}\frac{b_j t_j}{\gamma_j}\right) K_{\mathrm{D}}^2 \Bigg\}
\end{aligned} \qquad (4.85)
$$

$$
S^{\mathrm{I}} = \frac{1}{8\pi(F_{11}F_{22} - F_{12}F_{21})^2}\{(\beta_1 K_{\mathrm{I}}^2 + \beta_2 K_{\mathrm{I}}K_{\mathrm{D}} + \beta_3 K_{\mathrm{D}}^2) \\
+ (\beta_4 K_{\mathrm{I}}^2 + \beta_5 K_{\mathrm{I}}K_{\mathrm{D}} + \beta_6 K_{\mathrm{D}}^2)\}, \qquad (4.86)
$$

where

$$s_j = d_j F_{22} - l_j F_{21} \qquad (j=1,2,3) \qquad t_j = d_j F_{12} - l_j F_{11} \qquad (j=1,2,3) \qquad (4.87)$$

and

$$
\begin{aligned}
\beta_1 = &\sum_{k=1}^{3}m_k s_k R_k^c(\theta)\sum_{j=1}^{3}a_j s_j R_j^c(\theta) + 2\sum_{k=1}^{3}\frac{f_k s_k}{\gamma_k}R_k^s(\theta)\sum_{j=1}^{3}\frac{s_j(a_j\gamma_j^2 + 1)}{\gamma_j}R_j^s(\theta) \\
&- \sum_{k=1}^{3}g_k s_k R_k^c(\theta)\sum_{j=1}^{3}s_j R_j^c(\theta)
\end{aligned} \qquad (4.88)
$$

$$
\begin{aligned}
\beta_2 = &-\sum_{k=1}^{3}m_k t_k R_k^c(\theta)\sum_{j=1}^{3}a_j s_j R_j^c(\theta) - 2\sum_{k=1}^{3}\frac{f_k t_k}{\gamma_k}R_k^s(\theta)\sum_{j=1}^{3}\frac{s_j(a_j\gamma_j^2 + 1)}{\gamma_j}R_j^s(\theta) \\
&+ \sum_{k=1}^{3}g_k t_k R_k^c(\theta)\sum_{j=1}^{3}s_j R_j^c(\theta) - \sum_{k=1}^{3}m_k s_k R_k^c(\theta)\sum_{j=1}^{3}a_j t_j R_j^c(\theta) \\
&- 2\sum_{k=1}^{3}\frac{f_k s_k}{\gamma_k}R_k^s(\theta)\sum_{j=1}^{3}\frac{t_j(a_j\gamma_j^2 + 1)}{\gamma_j}R_j^s(\theta) + \sum_{k=1}^{3}g_k s_k R_k^c(\theta)\sum_{j=1}^{3}t_j R_j^c(\theta)
\end{aligned} \qquad (4.89)
$$

$$\beta_3 = \sum_{k=1}^{3} m_k t_k R_k^c(\theta) \sum_{j=1}^{3} a_j t_j R_j^c(\theta) + 2 \sum_{k=1}^{3} \frac{f_k t_k}{\gamma_k} R_k^s(\theta) \sum_{j=1}^{3} \frac{t_j(a_j \gamma_j^2 + 1)}{\gamma_j} R_j^s(\theta)$$

$$- \sum_{k=1}^{3} g_k t_k R_k^c(\theta) \sum_{j=1}^{3} t_j R_j^c(\theta) \tag{4.90}$$

$$\beta_4 = \sum_{k=1}^{3} \frac{n_k s_k}{\gamma_k} R_k^s(\theta) \sum_{j=1}^{3} \frac{b_j s_j}{\gamma_j} R_j^s(\theta) - \sum_{k=1}^{3} h_k s_k R_k^c(\theta) \sum_{j=1}^{3} b_j s_j R_j^c(\theta) \tag{4.91}$$

$$\beta_5 = - \sum_{k=1}^{3} \frac{n_k t_k}{\gamma_k} R_k^s(\theta) \sum_{j=1}^{3} \frac{b_j s_j}{\gamma_j} R_j^s(\theta) + \sum_{k=1}^{3} h_k t_k R_k^c(\theta) \sum_{j=1}^{3} b_j s_j R_j^c(\theta)$$

$$- \sum_{k=1}^{3} \frac{n_k s_k}{\gamma_k} R_k^s(\theta) \sum_{j=1}^{3} \frac{b_j t_j}{\gamma_j} R_j^s(\theta) + \sum_{k=1}^{3} h_k s_k R_k^c(\theta) \sum_{j=1}^{3} b_j t_j R_j^c(\theta) \tag{4.92}$$

$$\beta_6 = \sum_{k=1}^{3} \frac{n_k t_k}{\gamma_k} R_k^s(\theta) \sum_{j=1}^{3} \frac{b_j t_j}{\gamma_j} R_j^s(\theta) - \sum_{k=1}^{3} h_k t_k R_k^c(\theta) \sum_{j=1}^{3} b_j t_j R_j^c(\theta). \tag{4.93}$$

In Equations (4.85) and (4.86), K_I is given by Equation (4.59) and K_D is

$$K_\mathrm{D} = D_{T0}(\pi a)^{\frac{1}{2}}. \tag{4.94}$$

The crack face electric boundary condition for the open crack model becomes

$$
\begin{aligned}
&D_z^+ = D_z^- \qquad (0 \le x < a) \\
&D_z^+(u_z^+ - u_z^-) = \varepsilon_0(\phi^- - \phi^+) \qquad (0 \le x < a) \\
&\phi(x,0) = 0 \qquad (a \le x < \infty).
\end{aligned}
\tag{4.95}
$$

By applying the crack face boundary conditions of Equations (4.34), (4.35), and (4.95), the unknowns $A_j(\alpha)(j = 1, 2, 3)$ in Equations (4.42) and (4.43) are related to T_{E0} and D_{T0} as follows.

$$\frac{f_1}{\gamma_1} A_1(\alpha) + \frac{f_2}{\gamma_2} A_2(\alpha) + \frac{f_3}{\gamma_3} A_3(\alpha) = 0$$

$$\frac{1}{\gamma_1} A_1(\alpha) + \frac{1}{\gamma_2} A_2(\alpha) + \frac{1}{\gamma_3} A_3(\alpha)$$

$$= -\frac{\pi}{2F'} \frac{a}{\alpha} J_1(a\alpha)\{F_{22} T_{E0} + F_{12}(D_0 - D_{T0})\}$$

$$\frac{b_1}{\gamma_1} A_1(\alpha) + \frac{b_2}{\gamma_2} A_2(\alpha) + \frac{b_3}{\gamma_3} A_3(\alpha)$$

$$= \frac{\pi}{2F'} \frac{a}{\alpha} J_1(a\alpha)\{F_{21} T_{E0} + F_{11}(D_0 - D_{T0})\}, \tag{4.96}$$

where

$$D_0 = -\varepsilon_0 \frac{F_{21}T_{E0} + F_{11}(D_0 - D_{T0})}{F_{22}T_{E0} + F_{12}(D_0 - D_{T0})}.$$
(4.97)

If $\varepsilon_0 = 0$, D_0 is equal to zero. When ε_0 becomes very large, the expression for D_0 above shows that $D_0 \to D_{T0} - (F_{21}/F_{11})T_{E0}$. The displacement, electric potential, tangential component of electric field, and normal component of electric displacement on the crack face are

$$u_z(x,0) = -\frac{F_{22}T_{E0} + F_{12}(D_0 - D_{T0})}{F'}(a^2 - x^2)^{1/2}$$

$$\phi(x,0) = -\frac{F_{21}T_{E0} + F_{11}(D_0 - D_{T0})}{F'}(a^2 - x^2)^{1/2}$$

$$E_x(x,0) = -\frac{F_{21}T_{E0} + F_{11}(D_0 - D_{T0})}{F'}\frac{x}{(a^2 - x^2)^{1/2}}$$

$$D_z(x,0) = D_0.$$
(4.98)

The energy release rate G° [34] and energy density S° for the open crack model are given by Equations (4.85) and (4.86), respectively, with

$$K_D = (D_{T0} - D_0)(\pi a)^{1/2}.$$
(4.99)

To examine the electroelastic fields and fracture mechanics parameters of piezoelectric materials, numerical calculations have been carried out for commercially available soft piezoelectric ceramics C-91 (Fuji Ceramics, Japan) [35]. The material properties of C-91 are listed in Table 4.1.

Figure 4.6 shows the crack opening displacement $u_z(x,0^+)$ for an infinite C-91 with a crack under uniform strain $S_0 = 5 \times 10^{-5}$ and electric field $E_0 = 0$ V/m. The results for the permeable, open, and impermeable crack models are shown for comparison purposes. Little difference among three piezoelectric crack models is observed. Similar results are shown in Figures 4.7, through 4.9 for the electric potential $\phi(x,0^+)$, normal component of electric displacement $D_z(x,0^+)$, and tangential component of electric field $E_x(x,0^+)$ along the upper crack face, respectively. There are differences among the crack models. It is noted that the open and impermeable crack models reduce the continuity of the tangential components of the electric field across the crack face.

Table 4.1 Material properties of C-91

	Elastic Stiffnesses $(\times 10^{10}\,\mathrm{N/m^2})$					Piezoelectric Coefficients $(\mathrm{C/m^2})$			Dielectric Constants $(\times 10^{-10}\,\mathrm{C/Vm})$	
	c_{11}^E	c_{12}^E	c_{13}^E	c_{33}^E	c_{44}^E	e_{31}	e_{33}	e_{15}	ε_{11}^S	ε_{33}^S
C-91	12.0	7.7	7.7	11.4	2.4	-17.3	21.2	20.2	226	235

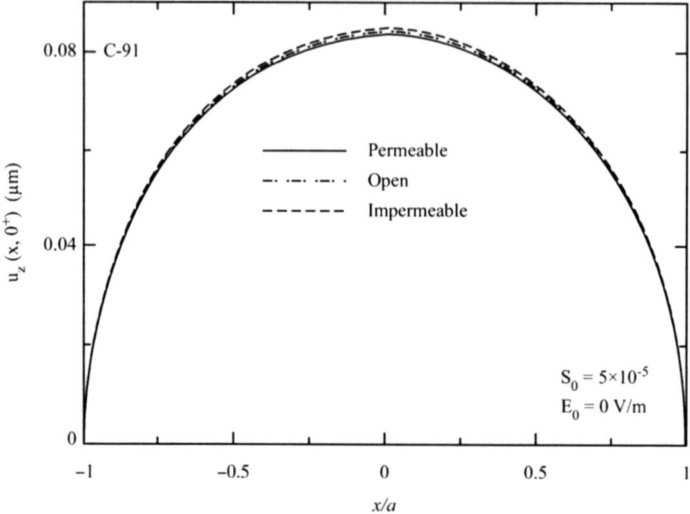

Fig. 4.6 Displacement along the upper crack face for infinite piezoelectric material under uniform strain.

Figure 4.10 shows the dependence of the energy release rate G for the permeable, open, and impermeable crack models under $S_0 = 10^{-4}$ on E_0 in an infinite C-91. The energy release rates are lower for positive electric fields and higher for negative electric fields under applied strain. The energy release rate criteria for the open and impermeable crack models led to

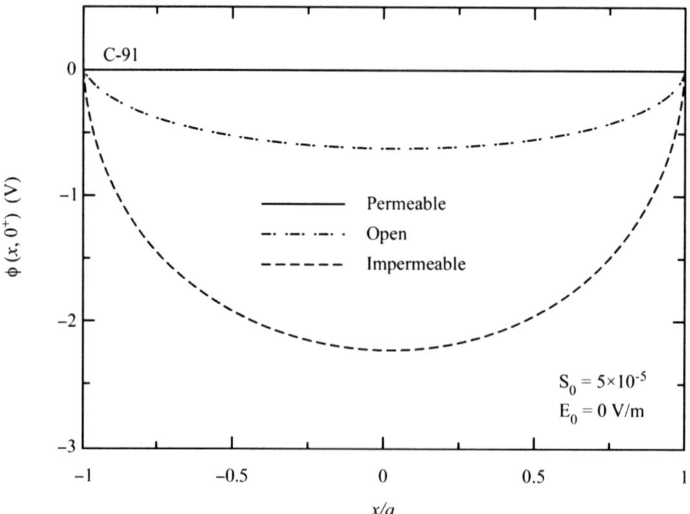

Fig. 4.7 Electric potential along the upper crack face for infinite piezoelectric material under uniform strain.

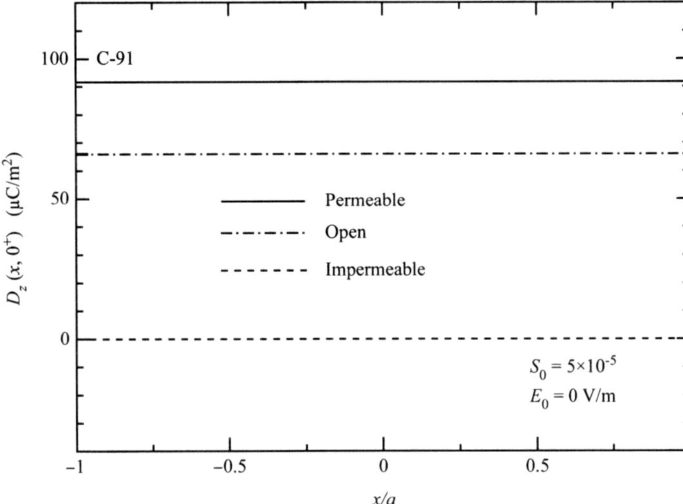

Fig. 4.8 Normal component of electric displacement along the upper crack face for infinite piezoelectric material under uniform strain.

negative values which are unphysical. Note that the energy release rate for the open crack model neglects the energy entering the crack gap. The energy release rate computed near the crack-tip is not equivalent to the energy release rate, including the energy entering the crack gap. If breakdown field

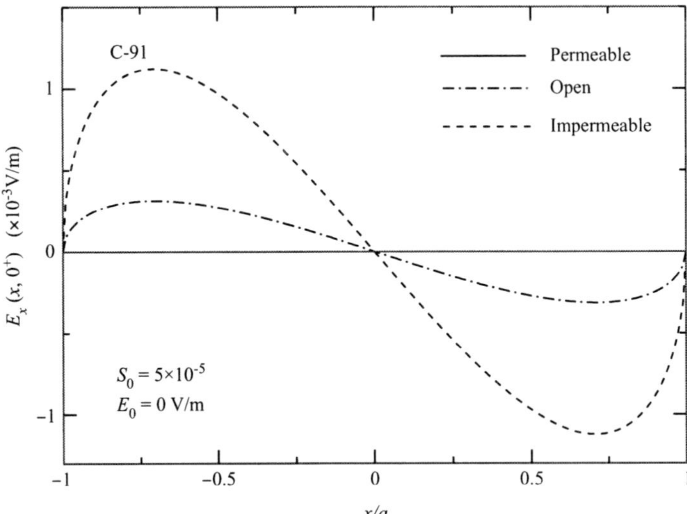

Fig. 4.9 Tangential component of electric field along the upper crack face for infinite piezoelectric material under uniform strain.

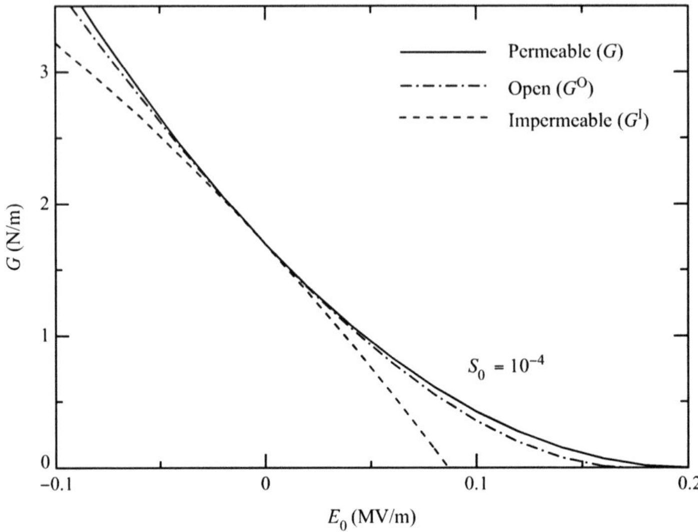

Fig. 4.10 Energy release rate versus electric field for infinite piezoelectric material under uniform strain.

E_d in the crack gap is assumed to be finite (the so-called nonlinear electrically discharging crack model) [8], the energy release rate for the discharging crack model tends to that for the permeable crack model. These are discussed later in detail (see Sections (4.2.3b) and (4.3.3a)). The effect of electric field on the stress intensity factor and energy density for the permeable and impermeable crack models under applied strain was also discussed in [32]. For the permeable crack model, no difference is found in the effect of electric field on the piezoelectric crack behavior for the criteria (the stress intensity factor, energy release rate, and energy density). However, if the open and impermeable crack models are used, different criteria give different results for the piezoelectric crack behavior.

The stress intensity factor, energy release rate, and energy density for the permeable crack model under applied stress are independent of the electric field [32]. It was shown in [33] that positive and negative electric fields decrease the energy release rate for the impermeable crack model under applied stress. Similar phenomena were observed for the open crack model [34]. However, preliminary experimental investigation did not confirm this crack-arresting behavior [36]. Moreover, no consensus is reached on the fracture criteria for the impermeable and open crack models under applied stress.

Next, we consider the case 2 (see Figure 4.4b). The crack face boundary and loading conditions can be written as

$$T_{xz}(0, z) = 0 \qquad (0 \leq z < \infty) \qquad (4.100)$$

$$T_{xx}(0, z) = 0 \qquad (0 \le z < a)$$
$$u_x(0, z) = 0 \qquad (a \le z < \infty) \tag{4.101}$$

$$E_z(0, z) = E_z^c(0, z) \qquad (0 \le z < a)$$
$$\phi(0, z) = 0 \qquad (a \le z < \infty) \tag{4.102}$$

$$D_x(0, z) = D_x^c(0, z) \qquad (0 \le z < a) \tag{4.103}$$
$$S_{xx}(x, z) = S_0, \qquad E_x(x, z) = E_0 \qquad (0 \le z < \infty, x \to \infty). \tag{4.104}$$

The stress intensity factor K_{I} is given by

$$K_{\mathrm{I}} = \lim_{z \to a^+} \{2\pi(z - a)\}^{1/2} T_{xx}(0, z). \tag{4.105}$$

Also, the energy release rate and energy density can easily be obtained.

A comparison of the stress intensity factor was made in [2] between case1 and case 2. Because the values of the stress intensity factor for case 1 were always larger than those for case 2, the crack propagated easily perpendicular to the poling direction. When the piezoelectric materials were poled prior to indentation, increased crack growth normal to the poling direction was observed [37]. The theoretical results are in agreement with the experimental results.

This section has been focused on the static fracture problems. There are, however, some publications concerning dynamic behavior of permeable cracks in piezoelectric ceramics. For case 1, the scattering of normally incident plane harmonic waves by a crack in piezoelectric ceramics was investigated in [38], and the dynamic stress and electric field intensity factors were discussed. In [39], the transient dynamic fracture mechanics parameters were determined for cracked piezoelectric ceramics under normal impact. Although a number of papers on the dynamic behavior of impermeable cracks have also appeared, questions remain regarding the crack face boundary and loading conditions.

(b) Plane strain crack in rectangular material

Consider a rectangular piezoelectric material of width $2h$ and length $2l$ containing a central crack of length $2a$, as shown in Figure 4.11a. A set of Cartesian coordinates (x, y, z) is attached to the center of the crack normal to the z-axis. We assume plane strain perpendicular to the y-axis. The material is loaded by mechanical displacement u_0 with the electric field in the z-direction. Because of the assumed symmetry in geometry and loading, it is sufficient to consider the problem for $0 \le x \le h, 0 \le z \le l$ only.

We consider the permeable, impermeable, and open crack models, with stress-free crack faces. Mechanical boundary conditions at $z = 0$ are given

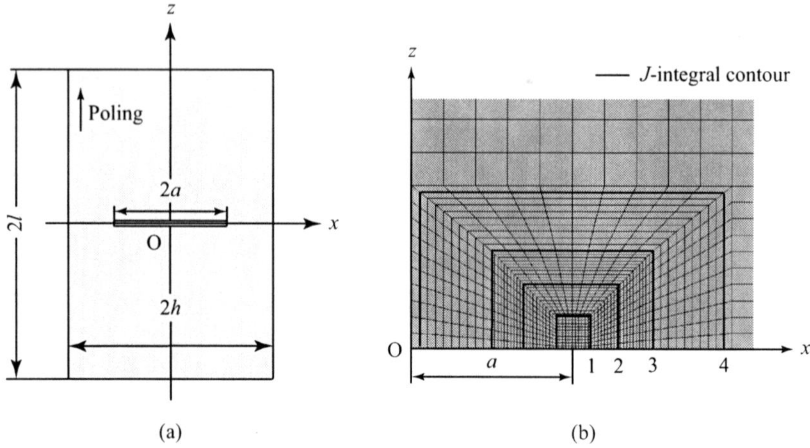

Fig. 4.11 A rectangular piezoelectric material with a central crack: (a) geometry; (b) finite element mesh and paths of J-integral.

by Equations (4.34) and (4.35), where the regions $(0 \leq x < \infty)$ and $(a \leq x < \infty)$ are replaced by $(0 \leq x \leq h)$ and $(a \leq x \leq h)$, respectively. Also, the boundary conditions for the permeable, impermeable, and open crack models are Equations (4.36), (4.37), (4.79), and (4.95), respectively, where $(a \leq x < \infty)$ is replaced by $(a \leq x \leq h)$. The loading conditions may be stated as follows.

$$u_z(x, l) = u_0, \qquad \phi(x, l) = \phi_0 \qquad (0 \leq x \leq h) \tag{4.106}$$

$$T_{xx}(h, z) = 0, \qquad T_{xz}(h, z) = 0 \qquad D_x(h, z) = 0, \tag{4.107}$$

where ϕ_0 is an applied electric potential. The third of Equations (4.107) is valid (see Figure 4.3b). The strain and electric field are

$$S_0 = \frac{u_0}{l}, \qquad E_0 = -\frac{\phi_0}{l}. \tag{4.108}$$

By applying the loading conditions (4.106) and (4.107), the normal stress T_0 at $z = l$ for the uncracked material is given by Equation (4.39).

The stress intensity factor K_{I} is defined by the second side of Equation (4.59). The energy release rate G and energy density S for the permeable crack model are given by Equations (4.73) and (4.76), respectively. These parameters $G^{\mathrm{I}}, S^{\mathrm{I}}, G^{\mathrm{O}}, S^{\mathrm{O}}$ for the impermeable and open crack models are also expressed as Equations (4.85) and (4.86).

For rectangular materials, the finite element method is effective for fracture mechanics parameter calculations. The method can include nonlinear effects such as polarization switching. Due to polarization switching, piezoelectric materials are often nonhomogeneous. The piezoelectric properties vary from one location to the other, and the variations are either continuous or discontinuous. An extension of the J-integral to multiphase materials was proposed in [40]. The energy release rate G can be obtained from the following J-integral [41],

$$
\begin{aligned}
G = J = \int_{\Gamma_0} &\{Hn_x - (T_{xx}u_{x,x} + T_{zx}u_{z,x})n_x - (T_{zx}u_{x,x} + T_{zz}u_{z,x})n_z \\
&+ D_x E_x n_x + D_z E_x n_z\}d\Gamma - \int_{\Gamma_p} \{Hn_x - (T_{xx}u_{x,x} + T_{zx}u_{z,x})n_x \\
&- (T_{zx}u_{x,x} + T_{zz}u_{z,x})n_z + D_x E_x n_x + D_z E_x n_z\}d\Gamma,
\end{aligned}
\tag{4.109}
$$

where Γ_p is a path embracing that part of the phase boundary enclosed by Γ_0.

Finite element analysis was performed in [35] for the permeable, open, and impermeable crack models, and the effects of electric field and polarization switching on the fracture mechanics parameters were discussed in detail for the rectangular piezoelectric material. Each element consists of many grains, and each grain is modeled as a uniformly polarized cell that contains a single domain. The model neglects the domain wall effects and interaction among different domains. In reality, these effects matter, but the assumption does not affect the general conclusions drawn. The polarization of each grain initially aligns as closely as possible to the z-direction. Polarization switching is defined for each element in the material. The displacement u_0 and electric potential ϕ_0 are applied at the edge $0 \leq x \leq h, z = l$, and the electromechanical fields of each element are computed from the finite element analysis. The switching criterion of Equation (4.9) is checked for every element to see if switching will occur. After all possible polarization switches have occurred, the piezoelectric tensor of each element is rotated to the new polarization direction. The electromechanical fields are recalculated, and the process is repeated until the solution converges. The macroscopic response of the material is determined by the finite element model, which is an aggregate of elements. The spontaneous polarization P^s and strain γ^s are assigned representative values of 0.3 C/m^2 and 0.004, respectively. Previous experiments in [42–44] verified the accuracy of the above scheme, and showed that the results obtained are of general applicability. After polarization switching is predicted, J-integral paths are selected, which do not pass exactly through the singular point.

The calculations of the fracture mechanics parameters for the open crack model are more complicated than for the permeable and impermeable crack models. The open crack model calculations start with $\phi = 0$ on the crack face [45]. The crack opening displacement and electric displacement on the crack face are estimated, and the resulting potential difference is applied to the crack face. The fields are again solved, leading to new crack opening displacement and electric displacement on the crack face. If this is accomplished, then the potential difference is applied once more to the crack face. Such a procedure is repeated until the evolution of the objective solutions shows no improvements. If the crack gap can store energy [46], nonzero contributions to G arise from the contour segments along the crack face. The values of G for the open crack model can be obtained by computing contour integrations and then subtracting the electrical enthalpy density of the crack gap $H^c = -\varepsilon_0\{(E_x^c)^2 + (E_z^c)^2\}/2$ times the crack opening displacement $2u_z^+$ evaluated at the intersection x^Γ of the contour with the crack faces [8]. So, for example, Equation (4.109) becomes

$$
\begin{aligned}
G = \int_{\Gamma_0} &\{Hn_x - (T_{xx}u_{x,x} + T_{zx}u_{z,x})n_x - (T_{zx}u_{x,x} + T_{zz}u_{z,x})n_z \\
&+ D_xE_xn_x + D_zE_xn_z\}d\Gamma - \int_{\Gamma_p} \{Hn_x - (T_{xx}u_{x,x} + T_{zx}u_{z,x})n_x \\
&- (T_{zx}u_{x,x} + T_{zz}u_{z,x})n_z + D_xE_xn_x + D_zE_xn_z\}d\Gamma \\
&- 2H^c(x^\Gamma)u_z^+(x^\Gamma).
\end{aligned}
\tag{4.110}
$$

It is required that the contour Γ_0 intersects the upper and lower crack faces at the same position x^Γ. There exists a discrepancy between the energy release rate G in Equation (4.110) and G° in Equation (4.85) with Equation (4.99) for the open crack model.

We now show numerical examples using the piezoelectric ceramics, C-91. The coercive electric field E_c is approximately 0.35 MV/m. Figure 4.12 presents the crack opening displacement $u_z(0, 0^+)$ at the center of the crack versus electric field E_0 from the finite element analysis without polarization switching effect. The results for the permeable, open, and impermeable crack models are shown. The rectangular piezoelectric material with a crack of length $2a = 2$ mm has a length $2l = 20$ mm and width $2h = 20$ mm, and is under applied displacement $u_0 = 0.5$ μm corresponding to the uni-

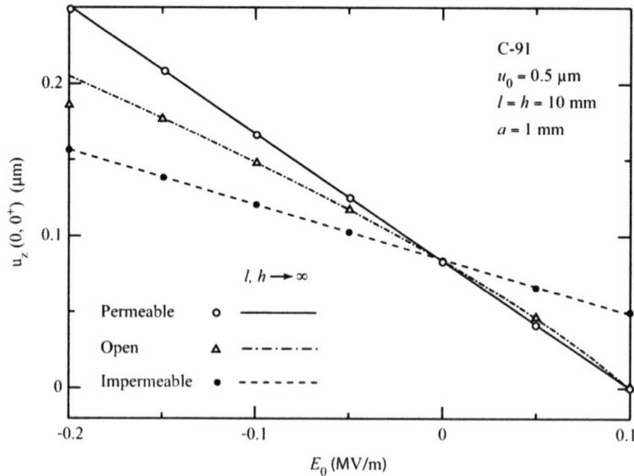

Fig. 4.12 Crack center displacement versus electric field for rectangular piezoelectric material under applied displacement.

Table 4.2 Energy release rate for rectangular piezoelectric material under $u_0 = 0.5$ μm and $E_0 = -0.1$ MV/m^a

			G (N/m^2)	
	Permeable	Impermeable	Open	Discharging
Contour 1	1.67	0.95	1.47 (-3.3×10^{-2})	1.67 (-4.4×10^{-6})
Contour 2	1.63	0.97	1.46 (-4.3×10^{-2})	1.63 (-5.7×10^{-6})
Contour 3	1.58	1.06	1.40 (-7.1×10^{-2})	1.58 (-9.4×10^{-6})
Contour 4	1.53	1.22	1.38 (-9.9×10^{-2})	1.53 (-1.3×10^{-5})
Avg.	1.60	1.05	1.43 (-6.2×10^{-2})	1.60 (-8.1×10^{-6})
$l, h \to \infty$	1.69	1.07	1.59	$-$

^{a}Values in parentheses are the results of $-2H^c(x^\Gamma)u_z^+(x^\Gamma)$ in Equation (4.110).

form strain 5×10^{-5} for the uncracked material. For comparison, the results for the infinite piezoelectric material ($l, h \to \infty$, $S_0 = 5 \times 10^{-5}$) obtained from the theoretical analysis are included. The results for the finite element analysis agree with the theoretical analysis data. Figure 4.13 shows the similar results for the normal component of electric displacement $D_z(0, 0^+)$. Table 4.2 lists the energy release rates for the permeable, impermeable, and open crack models under $u_0 = 0.5$ μm and $E_0 = -0.1$ MV/m for the rectangular piezoelectric material ($2l = 20$ mm, $2h = 20$ mm, $2a = 2$ mm). The J-integrals for the permeable and impermeable crack models are path-independent, and for the calculation of G, four contours are defined in the finite element mesh (see Figure 11b). The average values (Avg.) are also shown, and the values in parentheses are the contribution from the crack interior (see Equation (4.110)). Furthermore, the results for the

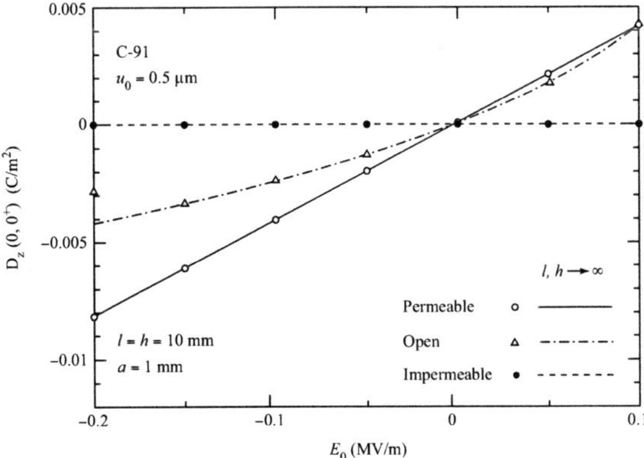

Fig. 4.13 Normal component of electric displacement at the center of the crack versus electric field for rectangular piezoelectric material under applied displacement.

infinite piezoelectric material $(l, h \rightarrow \infty,\ S_0 = 5 \times 10^{-5})$ are listed. It is noted that for the open crack model, little difference between the energy release rates, including and not including the energy entering the crack gap, is observed. Under practical loading conditions, the contribution $-2H^c(x^\Gamma)u_z^+(x^\Gamma)$ from the crack interior is negligibly small. We wish to add that the effect of electrical discharge within the crack is not accounted for in calculations of the energy release rate for the open crack model. In recent years, a nonlinear electrically discharging crack model was proposed in [8]. It was assumed that the crack gap behaves in a linear dielectric manner when the electric field within the crack gap is below a critical discharge level E_d. The solution for the discharging crack (standard air breakdown field $E_d = 3$ MV/m) is also listed in this table. It is found that the energy release rates predicted by the permeable and discharging crack models are not significantly different. Figure 4.14 shows the dependence of the energy release rate G for the permeable, open, and impermeable crack models under $u_0 = 0.5\ \mu$m on E_0 in the same material $(2l = 20$ mm, $2h = 20$ mm, $2a = 2$ mm). Average values of four contours are presented, and the permeable and discharging (not shown) crack models give almost the same results. The results for the infinite piezoelectric material $(l, h \rightarrow \infty,\ S_0 = 5 \times 10^{-5})$ obtained from the theoretical analysis are also shown.

Figure 4.15 displays the variation of G for the permeable crack model with electric field E_0 from the finite element analysis with and without the polarization switching effect. The rectangular piezoelectric material $(2l = 5$ mm, $2h = 5$ mm) with a crack $(2a = 2$ mm) is under applied displacement $u_0 = 0.125\ \mu$m corresponding to the uniform strain 5×10^{-5} for the uncracked material. Positive electric fields decrease the values of G, whereas negative

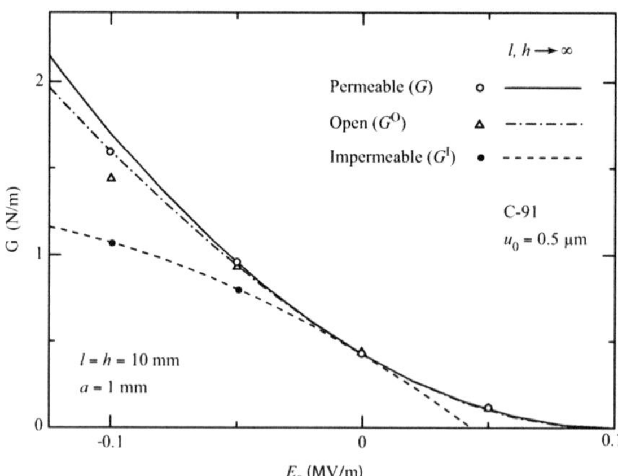

Fig. 4.14 Energy release rate versus electric field for rectangular piezoelectric material under applied displacement without polarization switching effect.

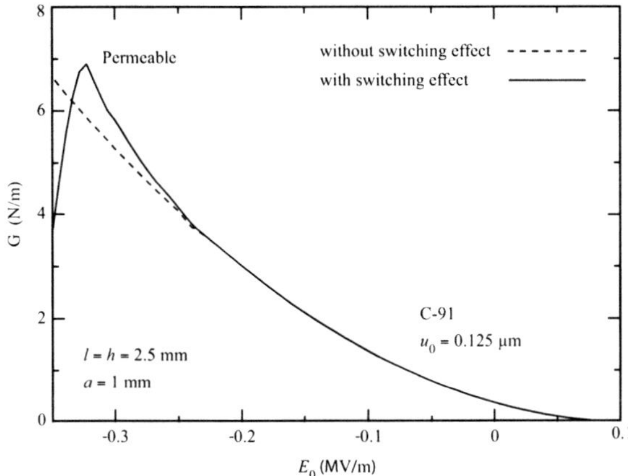

Fig. 4.15 Energy release rate versus electric field for rectangular piezoelectric material under applied displacement with polarization switching effect.

electric fields have an opposite effect. A monotonically increasing negative E_0 causes polarization switching. The value of the electric field associated with the switching is about -0.25 MV/m. When the negative E_0 increases further, G with the polarization switching effect becomes larger than that without the switching effect. After E_0 reaches about -0.32 MV/m, polarization switching in a local region leads to an unexpected decrease in G. The nonlinear effect caused by polarization switching may affect the piezoelectric crack behavior.

Figure 4.16 shows the 180° and 90° switching zones near the permeable crack-tip in the rectangular piezoelectric material ($2l = 5$ mm, $2h = 5$ mm, $2a = 2$ mm) under $u_0 = 0.125$ µm and various values of E_0. The size of the 180° (90°) switching zone behind (ahead of) the crack-tip increases at first when the negative E_0 is increased, and the difference between energy release rate results with and without switching effect becomes larger at a higher negative E_0. As the negative E_0 continues to be increased, the area of the 180° switching zone grows ahead of the crack-tip. Unexpected decrease in G is attributed to 180° switching ahead of the crack-tip. In the impermeable case, the region ahead of the crack-tip is found to undergo 180° switching due to the large negative electric field, and the region behind the crack-tip has 90° switching because of the large intensified electric field E_x [47].

Figure 4.17 shows the energy release rate G versus E_0 under applied stress. The rectangular piezoelectric material ($2l = 5$ mm, $2h = 5$ mm) with a permeable crack ($2a = 2$ mm) is subjected to the stress $T_{E0} = 5.70$ MPa, corresponding to the uniform strain 5×10^{-5} for the uncracked material without the electric field. Also shown are data for $T_{E0} = 22.8$ MPa, corresponding to the uniform strain 2×10^{-4}. The results for the positive E_0 under applied stress are different from those under applied displacement, and the energy release rate for the permeable crack is independent of the positive E_0.

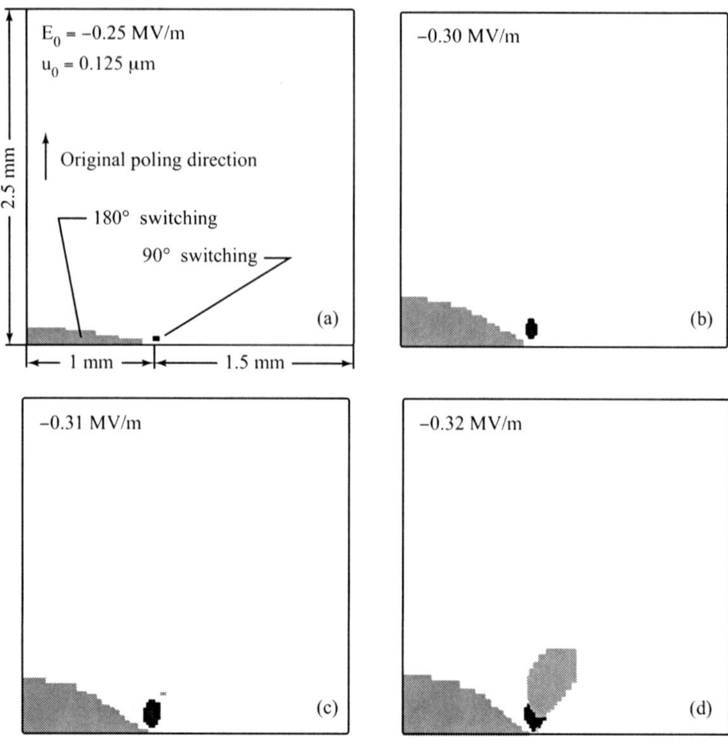

Fig. 4.16 Polarization switching zone induced by displacement $u_0 = 0.125\ \mu$m and electric field E_0.

The behavior of the energy release rate in the negative E_0 is complicated because of the polarization switching phenomena. The critical value of the electric field associated with the polarization switching decreases (being compared to $T_{E0} = 5.70$ MPa) when $T_{E0} = 22.8$ MPa is applied. After E_0 reaches about -0.21 MV/m, the G with the switching effect deviates from the curve without the switching effect. This is due to the 180° switching behind the crack-tip. As the E_0 reaches about -0.32 MV/m, the G falls. In the experimental data [48], crack length deviated from the linear function of the electric field for the case of larger load, especially for negative electric fields. By including the polarization switching effect of the energy release rate, the observed nonlinear dependence of piezoelectric crack behavior on the electric field is explained.

(c) Penny-shaped crack in infinite material

Consider a penny-shaped crack of radius a embedded in an infinite piezoelectric material. The formulation of this class of boundary value problem can be expressed most conveniently in terms of the cylindrical coordinates (r, θ, z). Referring to Figure 4.18, the crack will be located in the plane $z = 0$ of a

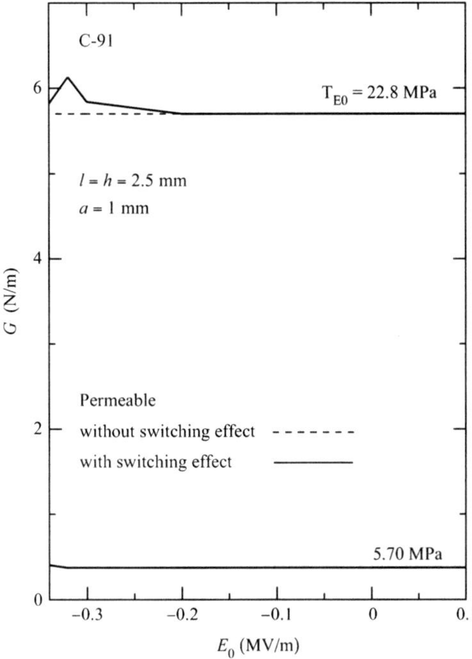

Fig. 4.17 Energy release rate versus electric field for rectangular piezoelectric material under applied stress with polarization switching effect.

piezoelectric material which is centered at the origin $(0, 0, 0)$. The material is transversely isotropic with hexagonal symmetry; it is subjected to far-field normal strain $S_{zz} = S_0$ and electric field $E_z = E_0$.

The equations of force and charge equilibriums without body force and internal charge are

$$\left.\begin{array}{l} T_{rr,r} + T_{zr,z} + \frac{T_{rr}-T_{\theta\theta}}{r} = 0 \\[2ex] T_{zr,r} + T_{zz,z} + \frac{T_{zr}}{r} = 0 \end{array}\right\} \tag{4.111}$$

$$D_{r,r} + \frac{D_r}{r} + D_{z,z} = 0. \tag{4.112}$$

The constitutive equations can be written as

$$\left.\begin{array}{l} T_{rr} = c_{11}^E u_{r,r} + c_{12}^E \frac{u_r}{r} + c_{13}^E u_{z,z} - e_{31}E_z \\[2ex] T_{\theta\theta} = c_{12}^E u_{r,r} + c_{11}^E \frac{u_r}{r} + c_{13}^E u_{z,z} - e_{31}E_z \\[2ex] T_{zz} = c_{13}^E u_{r,r} + c_{13}^E \frac{u_r}{r} + c_{33}^E u_{z,z} - e_{33}E_z \\[2ex] T_{zr} = c_{44}^E (u_{r,z} + u_{z,r}) - e_{15}E_r \end{array}\right\} \tag{4.113}$$

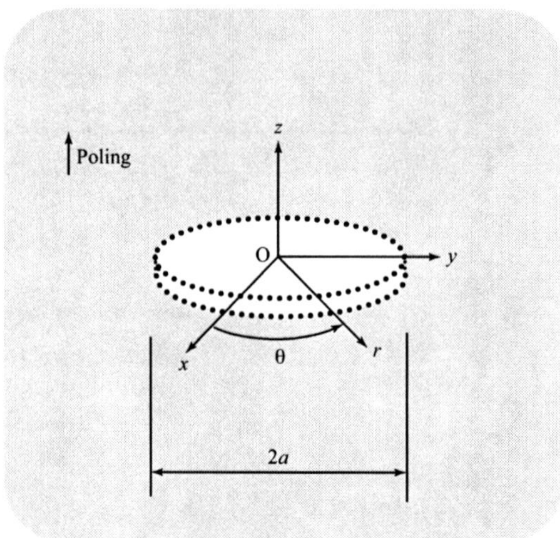

Fig. 4.18 An infinite piezoelectric material with a penny-shaped crack.

$$
\left.
\begin{aligned}
D_r &= e_{15}(u_{r,z} + u_{z,r}) + \varepsilon_{11}^S E_r \\[2mm]
D_z &= e_{31}\left(u_{r,r} + \tfrac{u_r}{r}\right) + e_{33}u_{z,z} + \varepsilon_{33}^S E_z
\end{aligned}
\right\}.
\tag{4.114}
$$

The electric field components are related to $\phi(r,z)$ by

$$
E_r = -\phi_{,r}, \qquad E_z = -\phi_{,z}.
\tag{4.115}
$$

By substituting from Equations (4.113) and (4.114) into Equations (4.111) and (4.112) together with (4.115), the governing equations are obtained as

$$
\left.
\begin{aligned}
&c_{11}^E\left(u_{r,rr} + \tfrac{u_{r,r}}{r} - \tfrac{u_r}{r^2}\right) + c_{44}^E u_{r,zz} + (c_{13}^E + c_{44}^E)u_{z,rz} \\[2mm]
&\qquad\qquad + (e_{31} + e_{15})\phi_{,rz} = 0 \\[3mm]
&(c_{13}^E + c_{44}^E)\left(u_{r,rz} + \tfrac{u_{r,z}}{r}\right) + c_{33}^E u_{z,zz} + c_{44}^E\left(u_{z,rr} + \tfrac{u_{z,r}}{r}\right) \\[2mm]
&\qquad\qquad + e_{15}\left(\phi_{,rr} + \tfrac{\phi_{,r}}{r}\right) + e_{33}\phi_{,zz} = 0
\end{aligned}
\right\}
\tag{4.116}
$$

$$
\begin{aligned}
&(e_{31} + e_{15})\left(u_{r,rz} + \frac{u_{r,z}}{r}\right) + e_{15}\left(u_{z,rr} + \frac{u_{z,r}}{r}\right) + e_{33}u_{z,zz} \\[2mm]
&\qquad - \varepsilon_{11}^S\left(\phi_{,rr} + \frac{\phi_{,r}}{r}\right) - \varepsilon_{33}^S\phi_{,zz} = 0.
\end{aligned}
\tag{4.117}
$$

In a vacuum, the constitutive equations (4.114) and the governing equation (4.117) become

$$D_r = \varepsilon_0 E_r, \qquad D_z = \varepsilon_0 E_z \tag{4.118}$$

$$\phi_{,rr} + \frac{\phi_{,r}}{r} + \phi_{,zz} = 0. \tag{4.119}$$

Referring to the semi-infinite region $z \geq 0$, $0 \leq r < \infty$, $0 \leq \theta \leq 2\pi$, the boundary conditions can be expressed in the form

$$T_{zr}(r,0) = 0 \qquad (0 \leq r < \infty) \tag{4.120}$$

$$\begin{aligned} T_{zz}(r,0) &= 0 & (0 \leq r < a) \\ u_z(r,0) &= 0 & (a \leq r < \infty) \end{aligned} \tag{4.121}$$

$$\begin{aligned} E_r(r,0) &= E_r^c(r,0) & (0 \leq r < a) \\ \phi(r,0) &= 0 & (a \leq r < \infty) \end{aligned} \tag{4.122}$$

$$D_z(r,0) = D_z^c(r,0) \qquad (0 \leq r < a) \tag{4.123}$$

$$S_{zz}(r,z) = S_0, \qquad E_z(r,z) = E_0 \qquad (0 \leq r < \infty, z \to \infty). \tag{4.124}$$

The far-field normal stress T_{E0} is expressed as

$$T_{E0} = T_0 - e_2 E_0, \tag{4.125}$$

where T_0 is given by Equation (4.40), and

$$e_2 = \frac{(c_{11}^E + c_{12}^E)e_{33} - 2c_{13}^E e_{31}}{c_{11}^E + c_{12}^E}. \tag{4.126}$$

The problem is formulated by means of the Hankel transform and the solution is solved exactly [49]. The stress intensity factor k_1 is obtained as

$$k_1 = \lim_{r \to a^+} \{2(r-a)\}^{1/2} T_{zz}(r,0) = \frac{2}{\pi} T_{E0} \sqrt{a}. \tag{4.127}$$

The energy release rate and energy density for the permeable crack model are

$$G = \frac{\pi}{2F^2}\left(-F \sum_{j=1}^{3} \frac{d_j}{\gamma_j} + \sum_{j=1}^{3} h_j d_j \sum_{j=1}^{3} \frac{b_j d_j}{\gamma_j}\right) k_1^2 \tag{4.128}$$

$$S = \pi(a_M + a_E)k_1^2. \tag{4.129}$$

The stress intensity factor, energy release rate, and energy density for the penny-shaped piezoelectric crack depend on the electric field E_0, similar to the plane strain piezoelectric crack. The energy release rate and energy density for the impermeable crack model can also be found in [49].

4.3 Fracture Test and Analysis

In this section, we report the results on the electric field dependence of fracture behavior of piezoelectric ceramics. Some fracture tests are performed, and finite element analyses are then employed to discuss the piezoelectric fracture behavior.

4.3.1 Indentation Fracture

Here, we describe IF test and analysis made on piezoelectric ceramics under combined mechanical and electrical loads; see [37]. Commercially available soft piezoelectric ceramic P-7 (Murata Manufacturing, Japan) samples $5 \times 5 \times 15$ mm are cut and Vickers indentations are made with $P = 9.8$ N load for 15 s under electric fields E_0 parallel to the poling direction. Table 4.3 lists the material properties. Figure 4.19a depicts the specimen and setup for the experiment, and a_n and a_p are the characteristic lengths of indentation cracks normal and parallel to the poling direction, respectively. To generate electric fields, a power supplier (up to 1.25 kV/dc) is used.

Analysis is carried out using the 3-D finite element analysis program, with the two half-penny-shaped cracks model as shown in Figure 4.19b. A rectangular Cartesian coordinate system (x, y, z) is used with the z-axis coinciding with the poling direction. Two point forces $P_\mathrm{En}, P_\mathrm{Ep}$ are applied on the center of the cracks of radius $a = a_\mathrm{n} = a_\mathrm{p}$ in a large piezoelectric material $(A/a = 5.0)$, and electric potentials ϕ_0 and $-\phi_0$ are also added on the edges $z = A$ and $z = -A$, respectively. Because of the double symmetry of the body and loading only one quarter of the body is modeled.

The IF test results show that average values of seven indentation crack lengths normal and parallel to the poling for open (closed) circuit condition

Table 4.3 Material properties of P-7

	Elastic Stiffnesses $(\times 10^{10}$ N/m$^2)$					Piezoelectric Coefficients (C/m^2)			Dielectric Constants $(\times 10^{-10}$ C/Vm$)$	
	c_{11}^E	c_{12}^E	c_{13}^E	c_{33}^E	c_{44}^E	e_{31}	e_{33}	e_{15}	ε_{11}^S	ε_{33}^S
P-7	13.0	8.3	8.3	11.9	2.5	-10.3	14.7	13.5	171	186

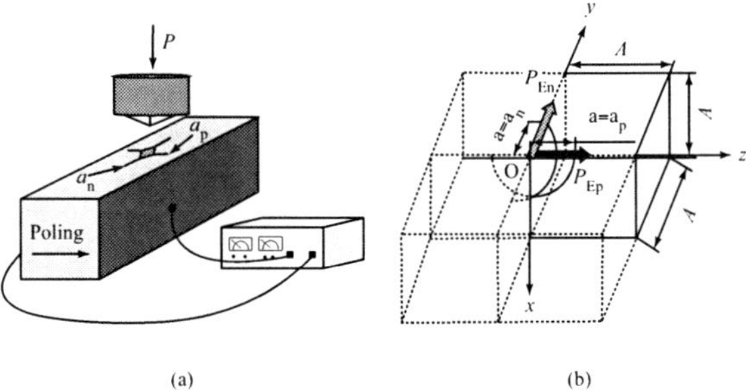

Fig. 4.19 Schematic representation of the IF tests: (a) testing setup; (b) finite element model.

are $a_n = 119.3(106.0)$ and $a_p = 71.3(64.4)$ μm, respectively. The indentation cracks are shorter parallel to the poling direction and longer normal to the poling direction, and longer for the open circuit condition. Figure 4.20 shows the length of indentation crack a_n under various electric fields $E_0 = -\phi_0/A$ divided by an average length a_{n0} with $E_0 = 0$ V/m. The error bars indicate maximum and minimum crack lengths at each electric field, and open circles are average values of seven data. The positive electric field assists crack growth. Results have also been computed giving the values of the energy release rate G at maximum depth point of the crack normal to the poling direction for the residual force $P_E = P_{En} = P_{Ep} = 0.1P = 0.98$ N derived from the indentation plastic zone. The ratio G/G_0 for the permeable crack model and the corresponding ratio for the impermeable crack model are plotted versus E_0 in Figure 4.21, where G_0 is the energy release rate under no

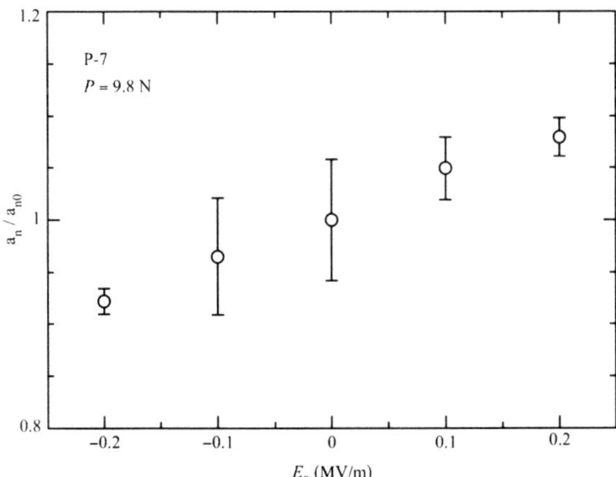

Fig. 4.20 Crack length versus electric field for IF specimen.

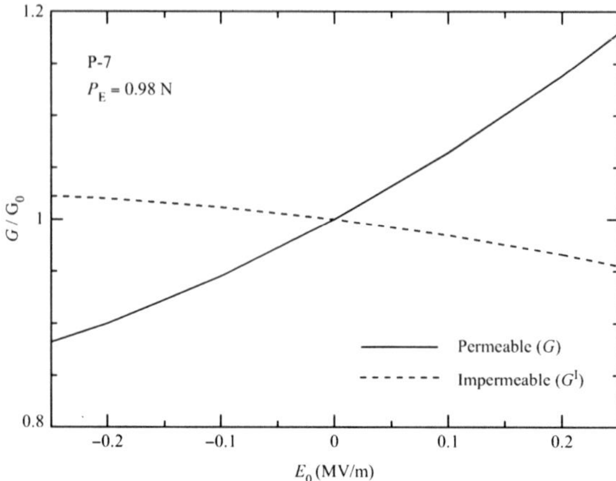

Fig. 4.21 Energy release rate versus electric field for IF specimen.

electric field. The trend for the permeable crack model is consistent with the experimental results.

4.3.2 Bending Fracture

(a) Modified small punch test

One recent experimental and analytical effort in bending fracture of piezoelectric ceramics is described in [16] and [50], using the modified small punch (MSP) test technique. The piezoelectric ceramic P-7 is selected for the experiment. The coercive electric field E_c is 0.8 MV/m. Small thin piezoceramic plate specimens of 10 by 10 by 0.5 mm used for MSP tests are sliced. Poling is done along the axis of the 0.5 mm dimension. Using a 10 kN screw-driven test machine, all MSP tests are conducted. To generate electric fields E_0, a power supply for voltages up to 1.25 kV/dc is used to apply positive and negative electric fields of 0.2, 0.4, 0.8, and 1.0 MV/m. The punch and the specimen holder, designed for MSP tests, are shown in Figure 4.22a. The MSP specimen holder consists of an upper and a lower die. The punch deformation is performed perpendicularly to the plate. Machine crosshead speed is 0.2 mm/min. Load P is recorded as a function of machine crosshead displacement δ, and fracture loads are measured for each set of specimens for various electric fields. For $E_0 = 0, \pm 0.4, \pm 0.8$, and ± 1.0 MV/m, four or five tests are performed. After testing, the microstructures of the fracture surfaces are examined using a confocal scanning laser microscope (CSLM). Electrodes are removed from the fractured specimens, and the surfaces of the piezoceramic plates are also observed by optical microscope.

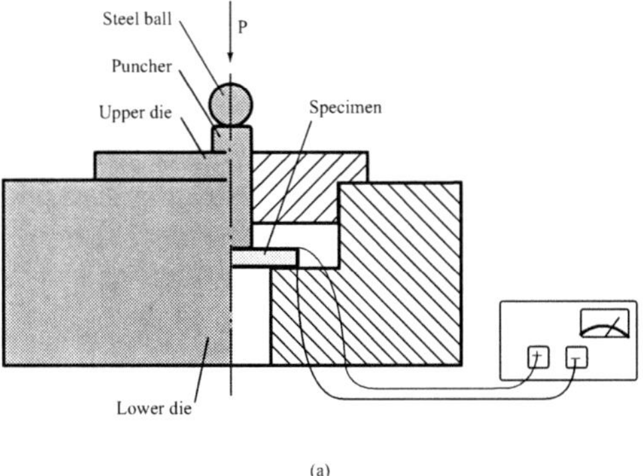

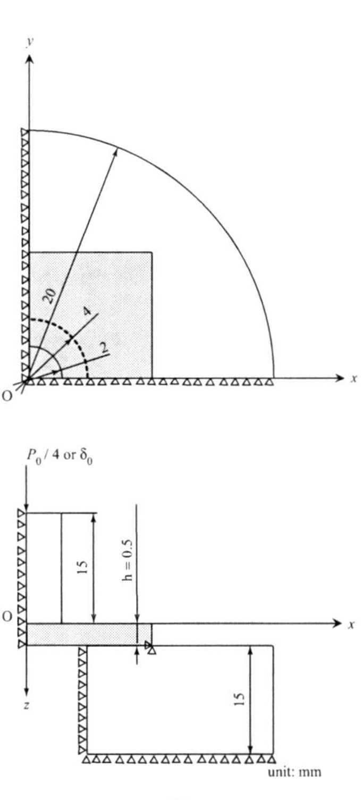

Fig. 4.22 Schematic representation of the MSP tests: (a) testing setup; (b) finite element model.

The three-dimensional model of the MSP specimen with load and boundary conditions is shown in Figure 4.22b. A rectangular Cartesian coordinate system (x, y, z) is used with the z-axis coinciding with the poling direction. The specimen is modeled with ANSYS solid eight-node hexahedral elements. The contact between the specimen and the lower die is modeled using contact elements. A mechanical load is produced by the application of either a prescribed force P_0 or a prescribed displacement δ_0 along the z-direction. For electrical loads, a positive or negative electric potential ϕ_0 is added on the surface, $z = h = 0.5$ mm; h is the specimen thickness. The surface $z = 0$ is grounded. Because of the double symmetry of the body and loading only one quarter of the body is modeled. The switching criterion of Equation (4.9) is checked for every element and for every possible polarization direction to see if switching will occur. The spontaneous polarization P^{s} and strain γ^{s} are assigned representative values of 0.3 C/m^2 and 0.004, respectively.

Figure 4.23 shows the fracture initiation loads P_{c} under different electric fields $E_0 = -\phi_0/h$ obtained from the experiment. Negative electric fields reduce the fracture initiation load, whereas positive electric fields increase it. When the electric field level in the negative direction reaches the coercive field $E_0 = -0.8$ MV/m, the direction of polarization of the stress-free sample switches. At $E_0 = -1.0$ MV/m, the polarization has reversed and is now aligned with the negative electric field. Also, the effective coercive field under applied stress changes as a function of mechanical load. The behavior of the fracture initiation load in the range $E_0 = -0.4$ MV/m to $E_0 = -1.0$ MV/m is very complicated because of all these switching phenomena. Decrease in the fracture initiation load at $E_0 = 1.0$ MV/m is attributed to dielectric breakdown and irreversible damage of the piezoelectric ceramics.

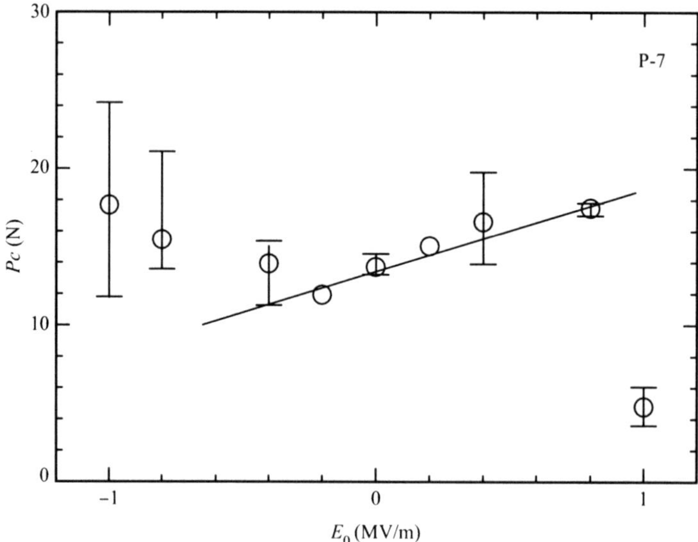

Fig. 4.23 Fracture initiation load versus electric field for MSP specimen.

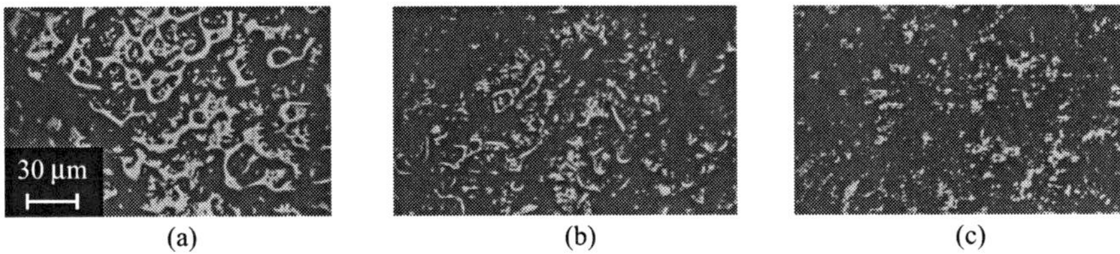

(a)   (b)   (c)

Fig. 4.24 Fracture features of MSP specimens at (a) $E_0 = +1.0$ MV/m, (b) 0.0 V/m, and (c) -1.0 MV/m.

CSLM micrographs of fracture surfaces are shown in Figure 4.24 for the MSP specimens tested at (a) $E_0 = +1.0$ MV/m, (b) $E_0 = 0.0$ MV/m, and (c) $E_0 = -1.0$ MV/m. The MSP specimen at $E_0 = -1.0$ MV/m had a relatively flat fracture surface. Under $E_0 = +1.0$ MV/m, the fracture surface appeared rougher with more intergranular fracture. It has been reported that the intergranular fracture is produced by purely electrical loading [51, 52] or dielectric breakdown [53]. The samples that experience breakdown are found to possess conduction channels terminated with craters at both sides of the specimen surfaces. The microstructure of the fractured specimen for $E_0 = +1.0$ MV/m, observed at the rim of the crater, is shown in Figure 4.25.

Figure 4.26 presents the critical MSP energy $E_{\mathrm{MSP}}^{\mathrm{c}}$ including 180° and 90° switching effects (open triangle) for various electric fields E_0. The load displacement curves were drawn up to the average fracture initiation load P_{c} using the finite element method, and the $E_{\mathrm{MSP}}^{\mathrm{c}}$ was calculated from the area under the curve (energy to failure). It is interesting to note that the MSP energies for $E_0 = +0.8$ MV/m and $E_0 = -0.8$ MV/m have very nearly the same values. A similar phenomenon is observed for $E_{\mathrm{MSP}}^{\mathrm{c}}$ without the 90° switching effect (solid circle), and the 90° switching has no effect on the critical MSP energy. Figure 4.27 shows the maximum strain energy density $W_{\mathrm{MSP}}^{\mathrm{m}}$ versus E_0. The $W_{\mathrm{MSP}}^{\mathrm{m}}$ as a function of applied load is computed using the average fracture initiation load P_c via finite element analysis without the 90° switching effect. The W_{MSP}^{m} occurs at the observed crack initiation location.

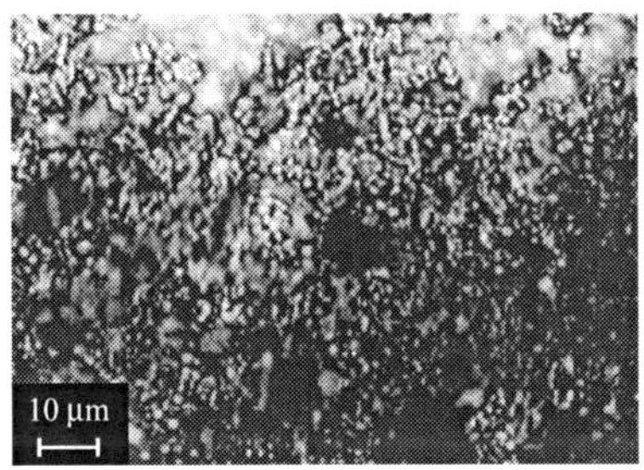

Fig. 4.25 Optical micrograph of the surface of MSP specimen tested at $E_0 = +1.0$ MV/m.

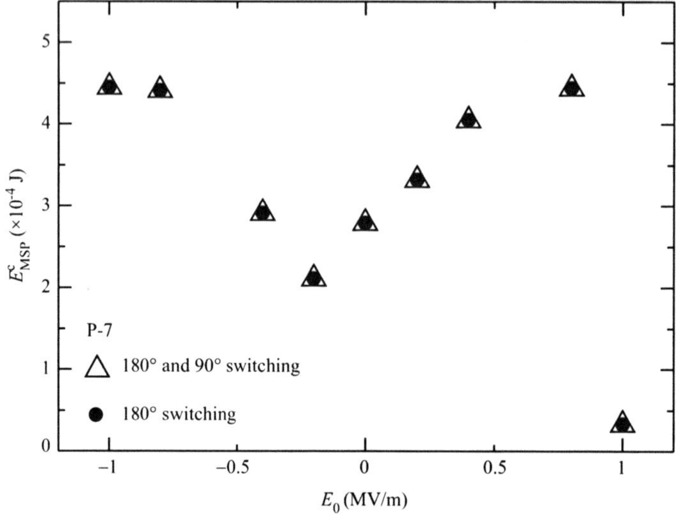

Fig. 4.26 Critical MSP energy versus electric field for MSP specimen.

(b) Other related tests

The three-point bending tests were conducted on the piezoelectric ceramics, where the poling is the width direction, under electric field, for example [54].

The specimen and jigs were immersed in a silicone oil bath, and electric fields were applied parallel and antiparallel to the poling. The three-point

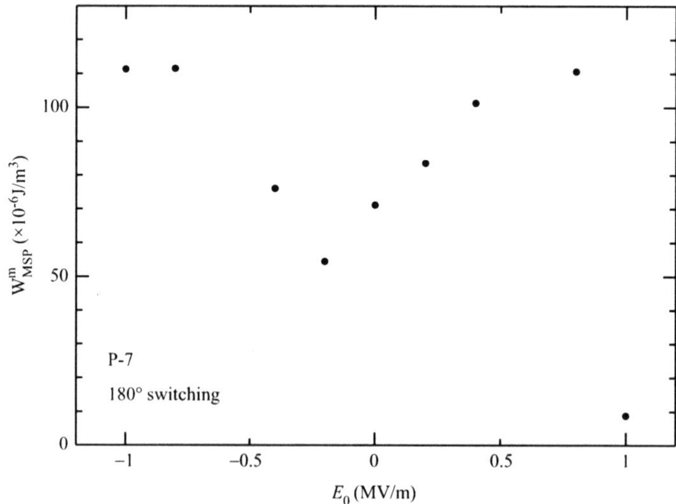

Fig. 4.27 Maximum strain energy density versus electric field for MSP specimen.

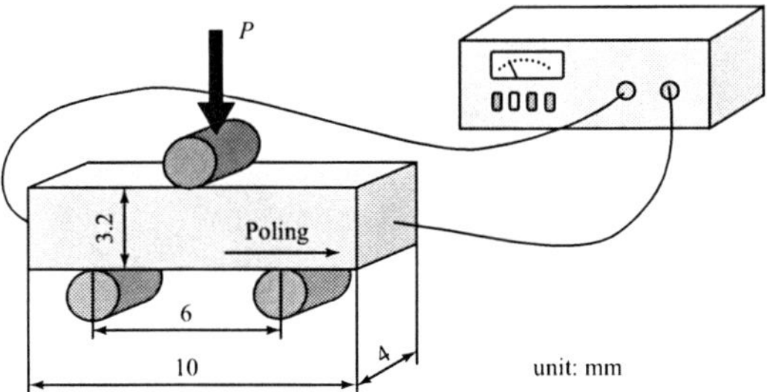

Fig. 4.28 Schematic representation of the three-point bending tests.

bending tests were also carried out in [55] using the piezoelectric samples poled in the length direction. The three-point bending specimen is depicted in Figure 4.28. The specimen of width 3.2 mm, thickness 4 mm, and length 10 mm is commercially supplied PZT-841 (American Piezo Ceramics), and has a span of 6 mm. The samples were placed in a silicone oil, and the load P was applied until specimen failure. The bending strength was then obtained from beam theory. Using the same specimens, the three-point bending tests were further performed in an air environment [56]. The bending strength was evaluated based on finite element analysis.

4.3.3 Fracture of Cracked Specimens

(a) **Single-edge precracked-beam test**

There has been significant effort aimed at the understanding of piezoelectric fracture in cracked specimens. Here, we consider single-edge precracked-beam (SEPB) methods, for example. The testing setup is illustrated in Figure 4.29a. The size of the specimens used in [57] is 5 mm thick, 5 mm wide, and 15 mm long. Poling is done along the axis of the 15mm dimension. Vickers indents are introduced using a commercial microhardness testing machine. At least 11 indents are placed at the midspan on the polished surface of the specimen along a line, with indent diagonals normal to the edges of the specimen bar. Indented specimens are carefully aligned and centered on a steel bridge-anvil. The specimens are compressed until a precrack is formed, and unloaded immediately after "pop-in" to avoid additional slow crack extension. The crack thus produced has initial length of a. The precracked specimens are loaded (load P) to failure in a three-point flexure apparatus with a support span of 13 mm, and the fracture loads P_c are measured for each set of specimens for various electric fields E_0. To generate electric fields, a power supply for voltages up to 125 kV/dc is used.

Plane strain finite element calculations are made to determine the energy release rate for the cracked piezoelectric specimens [57, 58]. The specimen

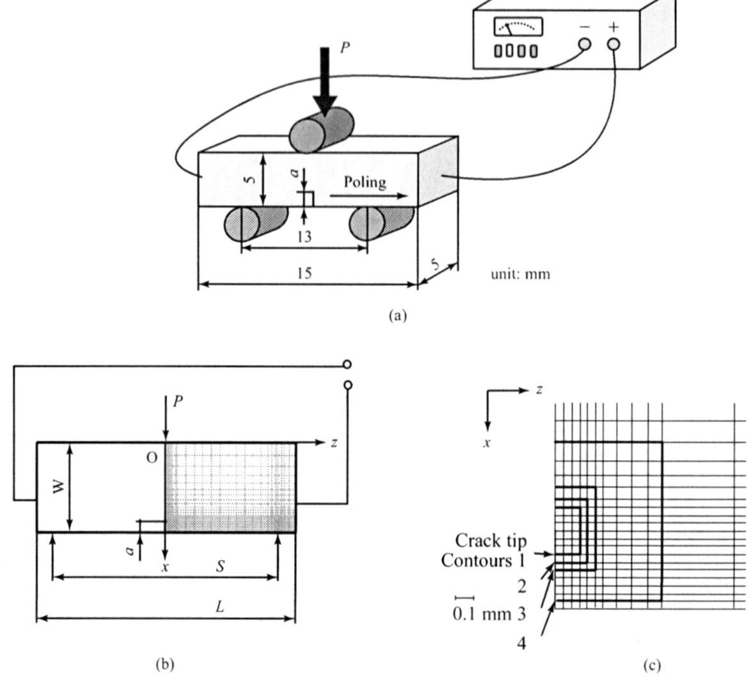

Fig. 4.29 Schematic representation of the SEPB tests: (a) testing set-up; (b) finite element model; (c) detail of crack-tip mesh.

and loading geometries are shown in Figure 4.29b. Let the coordinate axes x and z be chosen such that the y-axis coincides with the thickness direction. The z-axis is oriented parallel to the poling direction. The three-point flexure specimen with a span S is a beam of width W and length L containing a crack of length a. A mechanical load is produced by the application of a prescribed force P at $x = 0, z = 0$ along the x-direction. For electrical loads, an electric potential $\phi_0/2$ is applied at the edge $0 \leq x \leq W, z = L/2$. The electric potential at the edge $0 \leq x \leq W, z = -L/2$ is $-\phi_0/2$. Because of symmetry, only the right half of the model is used in the finite element analysis. In the analysis, the energy release rate is computed using the J-integral approach. For the calculation of G, four contours are defined in the finite element mesh (see Figure 4.29c).

Table 4.4 shows the fracture loads P_c for hard piezoelectric ceramics PCM-80 (Panasonic Electronic Devices, Japan) with a precrack of length

Table 4.4 SEPB test results

	E_0 (MV/m)	−0.2	−0.1	0	+0.1	+0.2
PCM-80	P_c (N)	150	150	150	146	147
	G_c (J/m^2)	7.67	7.67	7.67	7.26	7.36

Table 4.5 Material properties of PCM-80

	Elastic Stiffnesses $(\times 10^{10}$ N/m$^2)$					Piezoelectric Coefficients (C/m$^2)$			Dielectric Constants $(\times 10^{-10}$ C/Vm)	
	c_{11}^E	c_{12}^E	c_{13}^E	c_{33}^E	c_{44}^E	e_{31}	e_{33}	e_{15}	ε_{11}^S	ε_{33}^S
PCM-80	17.0	10.6	11.5	16.5	3.05	-5.99	15.6	13.7	95.2	68.4

$a = 0.5$ mm under $E_0 = -\phi_0/L = 0$, ± 0.1, ± 0.2 MV/m obtained from the experiment; material properties are listed in Table 4.5. Averaged values of three or four data are presented. Also listed are the corresponding critical energy release rates G_c. The critical energy release rate is calculated from the finite element analysis using the permeable crack model ($a = 0.5$ mm). Average values of four contours are listed. The fracture load and critical energy release rate for the hard piezoelectric ceramics are little dependent on E_0.

Table 4.6 lists the energy release rates for the permeable, impermeable, open, and nonlinear electrically discharging cracks of length $a = 0.5$ mm under $P = 100$ N and $E_0 = 0.5$ MV/m. The results obtained by the four contours and the average values (Avg.) are listed, and the values in parentheses are the contribution from the crack interior. The values of G for each of these contours are practically identical. Also, the contribution to G from the open and discharging crack interior, $-2H^c(x^\Gamma)u_z^+(x^\Gamma)$ in Equation (4.110), is very small. In the energy release rates for the piezoelectric specimens, negligible differences are noted between the permeable and discharging crack models [59]. Figure 4.30 shows the dependence of the energy release rate G for the permeable, open, impermeable, and discharging crack models on E_0 under $P = 100$ N for $a = 0.5$ mm. Average values of four contours are presented. The energy release rates for the permeable and discharging crack models are independent of E_0. In the impermeable and open crack models, applying the electric field in either direction decreases the energy release rate. A negative energy release rate is also produced under large electric fields.

Table 4.6 Energy release rate for SEPB specimen under $P = 100$ N and $E_0 = 0.5$ MV/m[a]

	G (N/m^2)			
	Permeable	Impermeable	Open	Discharging
Contour 1	3.28	1.58	2.84 (-1.12×10^{-2})	3.28 (-8.49×10^{-6})
Contour 2	3.27	1.64	2.96 (-1.29×10^{-2})	3.27 (-9.81×10^{-6})
Contour 3	3.26	1.70	3.07 (-1.49×10^{-2})	3.26 (-1.01×10^{-5})
Contour 4	3.26	1.70	3.09 (-1.96×10^{-2})	3.26 (-1.59×10^{-5})
Avg.	3.27	1.66	2.99 (-1.47×10^{-2})	3.27 (-1.11×10^{-5})

[a]Values in parentheses are the results of $-2H^c(x^\Gamma)u_z^+(x^\Gamma)$ in Equation (4.110).

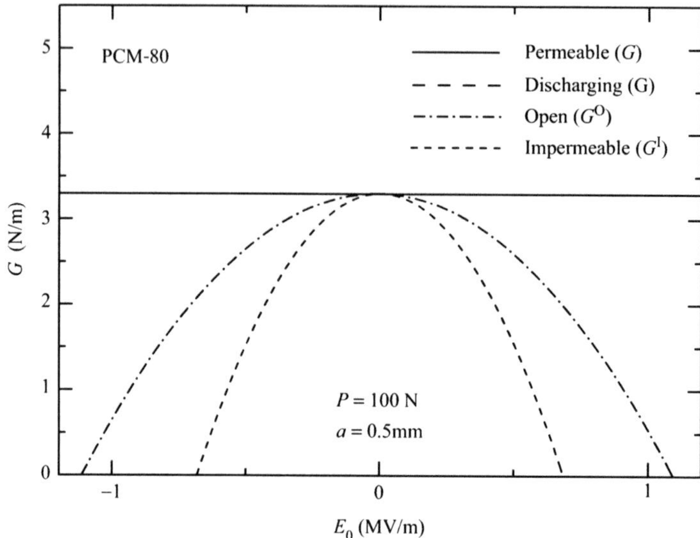

Fig. 4.30 Energy release rate versus electric field for SEPB specimen.

According to the fracture mechanics interpretation, a negative energy release rate would correspond to a crack that could absorb energy due to crack extension. Because this would exclude the fracture in piezoelectric ceramics under electric fields, in contradiction with the experimental observations, the parameters for the impermeable and open crack models have questionable physical significance. Therefore, the electrical boundary conditions (4.79) and (4.95) are not appropriate for a slit crack in piezoelectric ceramics.

(b) DT (double torsion) test

Let us now consider a composite piezoelectric double torsion (DT) specimen described by the Cartesian coordinate system (x, y, z), as shown in Figure 4.31 [60]. The piezoelectric samples of width $W_p = 5$ mm, thickness 5 mm, and length 30 mm are cut. The specimen is produced by first poling a 5 mm wide piezoelectric beam and then bonding it between two wider brass beams of width 7.5 mm, thickness 5 mm, and length 30 mm with high-strength epoxy. The Young's modulus and Poisson's ratio of brass are 100.6 GPa and 0.35, respectively. A side groove of depth 2.5 mm and width 1 mm is machined in the piezoelectric ceramics. Before testing, a thin notch is cut in the end of the piezoelectric ceramics to a depth of 2.5 mm, and a length of $a = 5$ mm. The specimens are loaded by concentrated loads $P/2$ at $x = 0$ mm, $y = 2.5$ mm, and $z = \pm 2$ mm. The moment arm and distance for the loading machine are fixed at 5.5 mm and 2 mm, respectively. To generate electric fields E_0, a power supplier that can produce up to 1.25 kV in dc is used.

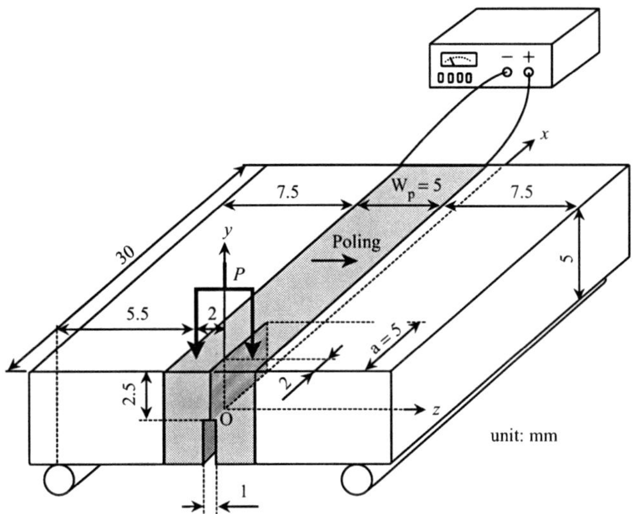

Fig. 4.31 Schematic representation of the DT tests.

Loads that caused fracture are measured for each set of specimens for various electric fields.

Three-dimensional finite element calculations are made to determine the fracture mechanics parameters for the composite DT specimens. A mechanical load is produced by the application of prescribed forces $P/2$ at $x = 0$ mm, $y = 2.5$ mm, and $z = \pm 2$ mm. For electrical loads, an electric potential $\phi_0/2$ is applied at the interface $z = W_p/2$. The electric potential at the interface $z = -W_p/2$ is $-\phi_0/2$. Because of symmetry, only the right half of the model was used in the finite element analysis.

Figure 4.32 shows the measured fracture loads P_c of P-7 and C-91 under various values of electric field $E_0 = -\phi_0/W_p$. A positive electric field increases the fracture load, and a negative one decreases it. Hence, the crack opens less under a positive electric field than under a negative electric field. These experimentally observed phenomena contradict the results of three-point bending and IF tests. It is suggested that the negative electric field puts the DT specimen under tension (near the crack-tip) and the positive electric field exerts a compressive stress (near the crack-tip). The growth mechanism for a crack in the DT specimen is quite different from that in the conventional fracture tests with SEPB and Vickers' indentation specimens. The fracture loads depend on the material properties.

Figure 4.33 shows the dependence of G for the permeable crack model without and with switching effect under $P = 205$ N on E_0 for brass/soft P-7/brass ($a = 5$ mm) at $y = 0$ mm, normalized by values for $E_0 = 0$ V/m. Also shown is the result for the impermeable crack model without switching effect. For a given load, positive E_0 decreases the energy release rate for the permeable crack model, and negative E_0 has an opposite effect.

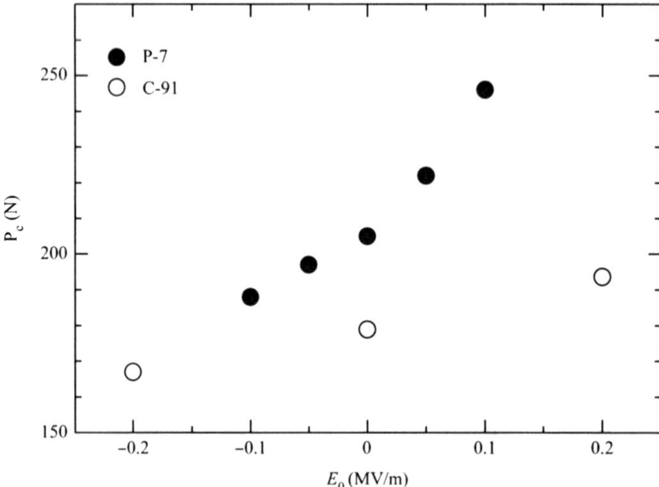

Fig. 4.32 Fracture load versus electric field for DT specimen.

The increase in P_c with increasing positive E_0 is attributed to the decrease of G with increasing positive E_0 under a constant load. A monotonically increasing negative E_0 causes polarization switching. After E_0 reaches about -0.4 MV/m, polarization switching in a local region leads to an unexpected decrease in G for the permeable crack model.

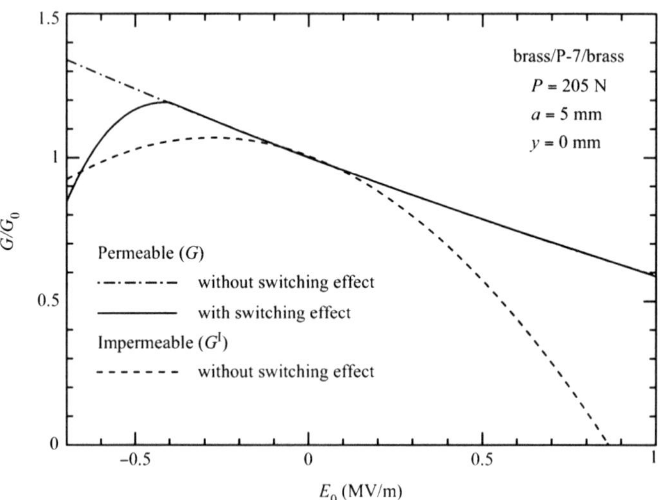

Fig. 4.33 Energy release rate versus electric field for DT specimen.

(c) Other related tests

Compact tension (CT) tests were conducted in [15] to study the effect of electric field on the fracture toughness of piezoelectric ceramics PZT-841. A notch was created normal to the poling and electric field directions.

4.4 Fatigue Test and Analysis

In this section, we review the results on the fatigue behavior of piezoelectric ceramics. Electric-field-induced fatigue crack growth in piezoelectric ceramics due to indentation was investigated in [51]. For piezoelectric ceramics (PZT-5) with a short notch, cyclic electric-field-induced fatigue cracking was studied in [61]. A notch was created normal to the poling and electric field directions. The results showed that under low electric fields, the emergence and growth of microcracks is the major fatigue mechanism that impedes the growth of the main crack, whereas the main crack is the only mode of fatigue cracking under high electric fields. The fatigue crack growth data were also presented in [62] for ferroelectric materials (PZT-5H and PLZT 8/65/35) with a V-shaped notch subjected to purely cyclic electrical loading. A V-shaped notch was normal to the poling and electric field directions. It was found that the rate of advance of cracks decreases with increasing cycle number, finally resulting in arrest. On the other hand, the crack growth under cyclic mechanical load in piezoelectric ceramics (Soft PZT151; PI Ceramics, Germany) was investigated in [63]. A V-shaped notch was oriented parallel to the poling direction, and the specimens were loaded in four-point bending without an electric field.

For the piezoelectric ceramics under both mechanical and electrical loads, static fatigue tests were carried out in [64] using the SEPB specimens (see Figure 4.29a). The material used was hard PCM-80, and the crack was created perpendicular to the poling direction. Time-to-failure under different mechanical loads and electric fields (parallel to the poling) were obtained from the experiment. A finite element analysis was also made, and the applied energy release rate for the permeable crack model was calculated. The effect of electric fields on the applied energy release rate versus lifetime curve was then examined.

The experimental data for the average time-to-failure t_f of two to four measurements in PZT ceramics under static mechanical load P and applied electric field $E_0 = +0.1, 0$ and -0.1 MV/m are listed in Table 4.7. The most important conclusion is that the lifetimes for the piezoelectric specimens under positive electric field are much shorter than the failure times of specimens under negative electric field for the same mechanical load level.

Cyclic crack growth tests were conducted in [58] on the piezoelectric ceramics under constant electric fields using the SEPB method. Piezoelectric

Table 4.7 Static fatigue test results

	E_0 (MV/m)		-0.1	0	$+0.1$
PCM-80	t_f (sec)	$P = 140$ (N)	203	108	43
		135	1573	776	361

samples PCM-80 were tested, subject to constant amplitude sinusoidal loads at a frequency of 1 or 10 Hz, using three-point bending apparatus (see Figure 4.29a). The load ratio R, defined as the ratio of minimum load $P_{\min}$ to maximum load $P_{\max}$ of the fatigue cycle N, was 0.2 or 0.5. The maximum load $P_{\max}$ was 140 N in all of the experiments. To determine crack growth rate da/dN, fatigue crack length a was monitored during the test using the digital microscope video camera. A finite element analysis was also used to calculate the maximum energy release rate $G_{\max}$.

Cyclic fatigue crack propagation (da/dN) data under $E_0 = +0.1, 0$, and -0.1 MV/m are plotted in Figure 4.34 as a function of the maximum energy release rate $G_{\max}$, at 1 Hz with $R = 0.5$. Also shown are the results at 1 Hz with $R = 0.2$ for $E_0 = +0.1$ and 0 MV/m. At $R = 0.5$, crack growth rates exhibited a local minimum resulting in V-shaped growth rate behavior. Fatigue crack propagation results after the minimum da/dN show that a positive (negative) electric field increases (decreases) the crack growth rate. At $R = 0.2$, the crack growth rate decreased strongly with increasing $G_{\max}$. The cracks were retarded during cyclic loading at $R = 0.2$ in the piezoelectric

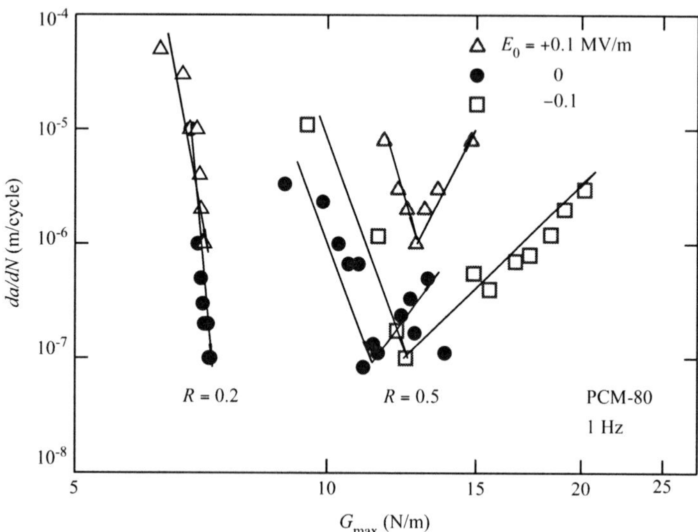

Fig. 4.34 Crack growth rate versus maximum energy release rate for SEPB specimen.

ceramics under $E_0 = +0.1$ and 0 MV/m, and no additional crack propagation was detected. The piezoelectric fatigue cracks exhibit features such as wedging [51], grain bridging [65], strain hysteresis [66], and microcrack nucleation [61], followed by crack arrest.

Dynamic fatigue or slow crack growth in piezoelectric ceramics was studied under electromechanical loading by a combined numerical-experimental approach [67]. Constant load-rate testing was conducted in three-point flexure using the PCM-80 SEPB specimens (see Figure 4.29a) under zero and positive electric fields, and the effects of electric field and loading rate (from 0.05 to 1 Ns^{-1}) on the fracture load and crack propagation were examined. A finite element analysis was also employed to calculate the energy release rate. Crack propagation velocity versus energy release rate curves at various loading rates were then estimated based on the finite element analysis using measured data. It was found that the piezoelectric ceramics under positive electric field have low dynamic fatigue or slow crack growth resistance, compared to those under no electric field. Also, the crack propagation velocity increased at first with increasing the energy release rate, reaching a peak, and then tended to decrease at a higher energy release rate before final failure.

4.5 Summary and Future Research Directions

Fracture and localized polarization switching are among the properties that limit the use of piezoelectric materials as sensors and actuators in smart material systems and structures technology. This chapter shows that electric field and mechanical loading influence the crack behavior in piezoelectric materials. We also show that, in modeling cracks in piezoelectric materials, the impermeable and open crack face assumptions can lead to significant errors and the permeable crack face is a better assumption. Piezoelectric materials could have a massive impact on various industries by reducing maintenance requirements, and increasing safety and product lifetime. At present, there remains a need for efficient numerical methods and models for predicting basic macroscopic material response while simultaneously accounting for microscale phenomena, such as domain switching [22, 35] and domain wall motion [68, 69]. Also, one needs a good understanding of nonlinear piezoelectric fracture behavior due to the microscale nature of the material. In addition, functional grading of materials could effectively reduce the magnitude of internal electromechanical fields or obtain an optimal pattern of fields in devices for a given design application [70–72]. Therefore, research on crack mechanics of functional graded piezoelectric materials is required. Although some works focus on functional graded piezoelectric materials with cracks, very little work is available for practical use.

The combined influence of electromechanical loading requires more research to provide deeper understanding of piezoelectric fracture for future demanding micro/nano device applications. There is a great interest in accurate multiscale computational methods and piezoelectric material models for linking atomistic, domain, and macroscale behavior. We believe that future research will concentrate on this area.

References

[1] Parton, V. Z.: Fracture mechanics of piezoelectric materials. *Acta Astro.* **3**, 671–683 (1976)

[2] Shindo, Y., Ozawa, E., Nowacki, J. P.: Singular stress and electric fields of a cracked piezoelectric strip. *Int. J. Appl. Electromagn. Mater.* **1**, 77–87 (1990)

[3] Deeg, W. F. J.: The analysis of dislocation, crack, and inclusion problems in piezoelectric solids. Ph.D. thesis, Stanford University (1980)

[4] Pak, Y. E.: Crack extension force in a piezoelectric material. *ASME J. Appl. Mech.* **57**, 647–653 (1990)

[5] Sosa, H. A.: On the fracture mechanics of piezoelectric solids. *Int. J. Solids Struct.* **29**, 2613–2622 (1992)

[6] Hao, T.-H., Shen, Z.-Y.: A new electric boundary condition of electric fracture mechanics and its applications. *Eng. Fract. Mech.* **47**, 793–802 (1994)

[7] Xu, X.-L., Rajapakse, R. K. N. D.: On a plane crack in piezoelectric solids. *Int. J. Solids Struct.* **38**, 7643–7658 (2001)

[8] Landis, C. M.: Energetically consistent boundary conditions for electromechanical fracture. *Int. J. Solids Struct.* **41**, 6291–6315 (2004)

[9] Ou, Z.-C., Chen, Y.-H.: Re-examination of the PKHS crack model in piezoelectric materials. *Euro. J. Mech.* A **26**, 659–675 (2007)

[10] Qin, Q. H.: *Fracture Mechanics of Piezoelectric Materials.* WIT Press, Southampton, Boston (2001)

[11] Zhang, T.-Y., Zhao, M. H., Tong, P.: Fracture of piezoelectric ceramics. *Adv. Appl. Mech.* **38**, 147–289 (2002)

[12] Chen, Y.-H., Hasebe, N.: Current understanding on fracture behaviors of ferroelectric/piezoelectric. *J. Intell. Mater. Syst. Struct.* **16**, 673–687 (2005)

[13] Park, S. B., Sun, C. T.: Fracture criteria for piezoelectric ceramics. *J. Am. Ceram. Soc.* **78**, 1475–1480 (1995)

[14] Narita, F., Shindo, Y., Horiguchi, K. Electroelastic fracture mechanics of piezoelectric ceramics. In: Shindo, Y. (ed.) *Mechanics of Electromagnetic Material Systems and Structures*, pp. 89–101, WIT Press, Southampton, Boston (2003)

[15] Fu, R., Zhang, T.Y.: Effect of an applied electric field on the fracture toughness of lead zirconate titanate ceramics. *J. Am. Ceram. Soc.* **83**, 1215–1218 (2000)

[16] Shindo, Y., Narita, F., Horiguchi, K., Magara, Y., Yoshida, M.: Electric fracture and polarization switching properties of piezoelectric ceramic PZT studied by the modified small punch test. *Acta Mater.* **51**, 4773–4782 (2003)

[17] Tiersten, H. F.: *Linear Piezoelectric Plate Vibrations.* Plenum, New York (1969)

[18] Maugin, G. A.: *Continuum Mechanics of Electromagnetic Solids.* Elsevier Science, Amsterdam (1988)

[19] Saito, Y., Takao, H., Tani, T., Nonoyama, T., Takatori, K., Homma, T., Nagaya, T, Nakamura, M.: Lead-free piezoceramics. *Nature* **432**, 84–87 (2004)

[20] Guo, Y., Kakimoto, K., Ohsato, H.: Phase transitional behavior and piezoelectric properties of $(Na_{0.5}K_{0.5})NbO_3$-$LiNbO_3$ ceramics. *Appl. Phys. Lett.* **85**, 4121–4123 (2004)

[21] Chen, P. J., Tucker, T. J.: One dimensional polar mechanical and dielectric responses of the ferroelectric ceramic PZT 65/35 due to domain switching. *Int. J. Eng. Sci.* **19**, 147–158 (1981)

[22] Hwang, S. C., Lynch, C. S., McMeeking, R. M.: Ferroelectric/ferroelastic interactions and a polarization switching model. *Acta Metall.* Mater. **43**, 2073–2084 (1995)

[23] Kamlah, M., Tsakmakis, C: Phenomenological modeling of the non-linear electromechanical coupling in ferroelectrics. *Int. J. Solids Struct.* **36**, 669–695 (1999)

[24] Huber, J. E., Fleck, N. A.: Multi-axial electrical switching of a ferroelectric: Theory versus experiment. *J. Mech. Phys. Solids* **49**, 785–811 (2001)

[25] Sun, C. T., Achuthan, A., Domain-switching criteria for ferroelectric materials subjected to electrical and mechanical loads. *J. Am. Ceram. Soc.* **87**, 395–400 (2004)

[26] Loge, R. E., Suo, Z.: Nonequilibrium thermodynamics of ferroelectric domain evolution. *Acta Mater.* **44**, 3429–3438 (1996)

[27] Sneddon, I. N.: *Fourier Transforms.* McGraw-Hill, New York (1951)

[28] Sosa, H. A., Khutoryansky, N.: New development concerning piezoelectric materials with defects. *Int. J. Solids Struct.* **33**, 3399–3414 (1996)

[29] Shindo, Y., Watanabe, K., Narita, F.: Electroelastic analysis of a piezoelectric ceramic strip with a central crack. *Int. J. Eng. Sci.* **38**, 1–19 (2000)

[30] Sih, G. C.: *Mechanics of Fracture Initiation and Propagation.* Kluwer Academic, The Netherlands (1991)

[31] Zuo, J. Z., Sih, G. C.: Energy density theory formulation and interpretation of cracking behavior for piezoelectric ceramics. *Theor. Appl. Fract. Mech.* **34**, 17–33 (2000)

[32] Lin, S., Narita, F., Shindo, Y.: Comparison of energy release rate and energy density criteria for a piezoelectric layered composite with a crack normal to interface. *Theor. Appl. Fract. Mech.* **39**, 229–243 (2003)

[33] Pak, Y. E.: Linear electro-elastic fracture mechanics of piezoelectric materials. *Int. J. Fract.* **54**, 79–100 (1992)

[34] Wang, B. L., Mai, Y.-W.: On the electrical boundary conditions on the crack surfaces in piezoelectric ceramics. *Int. J. Eng. Sci.* **41**, 633–652 (2003)

[35] Shindo, Y., Narita, F., Saito, F.: Electroelastic intensification and domain switching near plane strain crack in rectangular piezoelectric material. *J. Mech. Mater. Struct.* **2**, 1525–1540 (2007)

[36] Pak, Y. E., Tobin, A.: On electric field effects in fracture of piezoelectric materials. In: Lee, J. S., Maugin, G. A., Shindo, Y. (eds.) *Mechanics of Electromagnetic Materials and Structures*, AMD-Vol.161, MD-Vol.42, pp. 51–62. ASME (1993)

[37] Shindo, Y., Oka, M., Horiguchi, K.: Analysis and testing of indentation fracture behavior of piezoelectric ceramics under an electric field. *ASME J. Eng. Mater. Tech.* **123**, 293–300 (2001)

[38] Shindo, Y., Ozawa, E.: Dynamic analysis of a cracked piezoelectric material. In: *Proc. IUTAM Symposium on Mechanical Modellings of New Electromagnetic Materials*, pp. 297–304 (1990)

[39] Shindo, Y., Narita, F., Ozawa, E.: Impact response of a finite crack in an orthotropic piezoelectric ceramic. *Acta Mech.* **137**, 99–107 (1999)

[40] Weichert, D., Schulz, M.: *J*-integral concept for multiphase materials. *Comput. Mater. Sci.* **1**, 241–248 (1993)

[41] Shindo, Y., Narita, F., Mikami, M.: Electroelastic fracture mechanics of piezoelectric layered composites. *J. Intell. Mater. Syst. Struct.* **16**, 573–582 (2005)

[42] Yoshida, M., Narita, F., Shindo, Y., Karaiwa, M., Horiguchi, K.: Electroelastic field concentration by circular electrodes in piezoelectric ceramics. *Smart Mater. Struct.* **12**, 972–978 (2003)

[43] Hayashi, K., Shindo, Y., Narita, F.: Displacement and polarization switching properties of piezoelectric laminated actuators under bending. *J. Appl. Phys.* **94**, 4603–4607 (2003)

[44] Narita, F., Shindo, Y., Hayashi, K.: Bending and polarization switching of piezoelectric laminated actuators under electromechanical loading. *Comput. Struct.* **83**, 1164–1170 (2005)

[45] McMeeking, R. M.: Crack tip energy release rate for a piezoelectric compact tension specimen. *Eng. Fract. Mech.* **64**, 217–244 (1999)

[46] McMeeking, R. M.: The energy release rate for a Griffith crack in a piezoelectric material. *Eng. Fract. Mech.* **71**, 1149–1163 (2004)

[47] Kalyanam, S., Sun, C. T.: Modeling of electrical boundary condition and domain switching in piezoelectric materials. *Mech. Mater.* **37**, 769–784 (2005)

[48] Sun, C. T., Park, S. B.: Measuring fracture toughness of piezoceramics by Vickers indentation under the influence of electric fields. *Ferroelectrics* **248**, 79–95 (2000)

[49] Lin, S., Narita, F., Shindo, Y.: Electroelastic analysis of a penny-shaped crack in a piezoelectric ceramic under mode I loading. *Mech. Res. Comm.* **30**, 371–386 (2003)

[50] Magara, Y., Narita, F., Shindo, Y., Karaiwa, M.: Finite element analysis of electric fracture properties in modified small punch testing of piezoceramic plates. *J. Appl. Phys.* **95**, 4303–4309 (2004)

[51] Cao, H., Evans, A. G.: Electric-field-induced fatigue crack growth in piezoelectrics. *J. Am. Ceram. Soc.* **77**, 1783–1786 (1994)

[52] Shang, J. K., Tan, X.: A maximum strain criterion for electric-field-induced fatigue crack propagation in ferroelectric ceramics. *Mater. Sci. Eng. A.* **301**, 131–139 (2001)

[53] Liu, B., Fang, D., Hwang, K. C.: Electric-field-induced fatigue crack growth in ferroelectric ceramics. *Mater. Lett.* **54**, 442–446 (2002)

[54] Makino, H., Kamiya, N.: Effect of DC electric field on mechanical properties of piezoelectric ceramics. *Jpn. J. Appl. Phys.* **33**, 5323–5327 (1994)

[55] Fu, R., Zhang, T. Y.: Effects of an applied electric field on the modulus of rupture of poled lead zirconate titanate ceramics. *J. Am. Ceram. Soc.* **81**, 1058–1060 (1998)

[56] Fu, R., Zhang, T. Y.: Influence of temperature and electric field on the bending strength of lead zirconate titanate ceramics. *Acta Mater.* **48**, 1729–1740 (2000)

[57] Shindo, Y., Murakami, H., Horiguchi, K., Narita, F.: Evaluation of electric fracture properties of piezoelectric ceramics using the finite element and single-edge precracked-beam methods. *J. Am. Ceram. Soc.* **85**, 1243–1248 (2002)

[58] Narita, F., Shindo, Y., Saito, F.: Cyclic fatigue crack growth in three-point bending PZT ceramics under electromechanical loading. *J. Am. Ceram. Soc.* **90**, 2517–2524 (2007)

[59] Narita, F., Shindo, Y., Hirama, M.: Electric delayed fracture and localized polarization switching of cracked piezoelectric ceramics in three-point bending. *Int. J. Damage Mech.*, in press

[60] Shindo, Y., Narita, F., Mikami, M.: Double torsion testing and finite element analysis for determining the electric fracture properties of piezoelectric ceramics. *J. Appl. Phys.* **97**, 114109 (2005)

[61] Fang, D., Liu, B., Sun, C. T.: Fatigue crack growth in ferroelectric ceramics driven by alternating electric fields. *J. Am. Ceram. Soc.* **87**, 840–846 (2004)

[62] Shieh, J., Huber, J. E., Fleck, N. A.: Fatigue crack growth in ferroelectrics under electrical loading. *J. Euro. Ceram. Soc.* **26**, 95–109 (2006)

[63] Salz, C. R. J., Hoffman, M., Westram, I., Rodel, J.: Cyclic fatigue crack growth in PZT under mechanical loading. *J. Am. Ceram. Soc.* **88**, 1331–1333 (2005)

[64] Shindo, Y., Narita, F., Saito, F.: Static fatigue behavior of cracked piezoelectric ceramics in three-point bending under electric fields. *J. Euro. Ceram. Soc.* **27**, 3135–3140 (2007)

[65] Lynch, C. S., Yang, W., Collier, L., Suo, Z., McMeeking, R. M.: Electric field induced cracking in ferroelectric ceramics. *Ferroelectrics* **166**, 11–30 (1995)

[66] Weitzing, H., Schneider, G. A., Steffens, J., Hammer, M., Hoffmann, M. J.: Cyclic fatigue due to electric loading in ferroelectric ceramics. *J. Euro. Ceram. Soc.* **19**, 1333–1337 (1999)

[67] Shindo, Y., Narita, F., Hirama, M.: Dynamic fatigue of cracked piezoelectric ceramics under electromechanical loading: Three-point bending test and finite element analysis. *J. Mech. Mater. Struct.*, in press

[68] Li, S., Bhalla, A. S., Newnham,R. E., Cross, L. E.: Quantitative evaluation of extrinsic contribution to piezoelectric coefficient d_{33} in ferroelectric PZT ceramics. *Mater. Lett.* **17**, 21–26 (1993)

[69] Narita, F., Shindo, Y., Mikami, M.: Analytical and experimental study of nonlinear bending response and domain wall motion in piezoelectric laminated actuators under AC electric fields. *Acta Mater.* **53**, 4523–4529 (2005)

[70] Taya, M., Almajid, A. A., Dunn, M., Takahashi, H.: Design of bimorph piezo-composite actuators with functionally graded microstructure. *Sens. Actuators A* **107**, 248–260 (2003)

[71] Qiu, J., Tani, J., Ueno, T., Morita, T., Takahashi, H., Du, H.: Fabrication and high durability of functionally graded piezoelectric bending actuators. *Smart Mater. Struct.* **12**, 115–121 (2003)

[72] Shindo, Y., Narita, F., Nakagawa, J.: Nonlinear bending characteristics and sound pressure level of functionally graded piezoelectric actuators by AC electric fields: simulation and experiment. *Smart Mater. Struct.* **16**, 2296–2301 (2007)

Chapter 5
Boundary Element Method

Qing-Hua Qin

5.1 Introduction

In Chapter 2 Green's functions in piezoelectric materials were described. Applications of these Green's functions to the boundary element method (BEM) are discussed in this chapter. In contrast to the finite element method (FEM), BEM involves only discretization of the boundary of the structure due to the governing differential equation being satisfied exactly inside the domain leading to a relatively smaller system size with sufficient accuracy. This is an important advantage over domain-type solutions such as FEM or the finite difference method. During the past two decades several BEM techniques have been successfully developed for analyzing structure performance with piezoelectric materials [1–4]. Lee and Jiang [1] derived the boundary integral equation of piezoelectric media by the method of weighted residuals for plane piezoelectricity. Lu and Mahrenholtz [5] presented a variational boundary integral equation for the same problem. Ding, Wang, and Chen [6] developed a boundary integral formulation that is efficient for analyzing crack problems in piezoelectric material. Rajapakse [7] discussed three boundary element methods (direct boundary method, indirect boundary element method, and fictitious stress-electric charge method) in coupled electroelastic problems. Xu and Rajapakse [8] and Rajapakse and Xu [9] extended the formulations in [7, 8] to the case of piezoelectric solids with various defects (cavities, inclusions, cracks, etc.). Liu and Fan [10] established a boundary integral equation in a rigorous way and addressed the question of degeneration for problems of cracks and thin shelllike structures. Pan [11] derived a single-domain BE formulation for 2D static crack problems. Denda and Lua [12] developed a BEM formulation using Stroh's formalism to derive

Qing-Hua Qin
Department of Engineering Australian National University, Canberra, ACT 0200, Australia, e-mail: Qinghua.qin@anu.edu.au

J. Yang (ed.), *Special Topics in the Theory of Piezoelectricity*, DOI: 10.1007/978-0-387-89498-0_5, 137
© Springer Science + Business Media, LLC 2009

the fundamental solution but did not show any numerical results. Davi and Milazzo [13] used the known subdomain method to formulate a multidomain BEM, well suited for crack problems, by modeling crack faces as boundaries of the different subdomains. Groh and Kuna [14] developed a direct collocation boundary element code with a subdomain technique for analyzing crack problems and calculating stress intensity factors. Khutoryaansky, Sosa, and Zu [15] introduced a BE formulation for time-dependent problems of linear piezoelectricity. A brief review of this field can be found in [4].

5.2 Boundary Integral Formulations

5.2.1 Governing Equations

In this section, the theory of piezoelectricity presented in Chapter 1 is briefly summarized for deriving the corresponding boundary integral equation. Under the condition of a static deformation, the governing equations for a linear and generally anisotropic piezoelectric solid consist of

(i) Equilibrium equations

$$T_{ij,j} + \tilde{f}_i = 0, \quad D_{i,i} = \rho_e, \tag{5.1}$$

where $\tilde{f}_i = \rho_0 f_i$;
(ii) Constitutive relations

$$T_{ij} = c_{ijkl}S_{kl} - e_{mij}E_m, \quad D_k = e_{kij}S_{ij} + \varepsilon_{mk}E_m; \tag{5.2}$$

(iii) Elastic strain-displacement and electric field-potential relations

$$S_{ij} = \frac{(u_{i,j} + u_{j,i})}{2}, \quad E_i = -\phi_{,i}; \tag{5.3}$$

(iv) Boundary conditions

$$\left.\begin{array}{ll} u_i = \bar{u}_i & \text{on} \quad S_u \\ t_i = T_{ij}n_j = \bar{t}_i & \text{on} \quad S_T \end{array}\right\}, \qquad \left.\begin{array}{ll} \phi = \bar{\phi} & \text{on} \quad S_\phi \\ D_n = D_i n_i = -\bar{\sigma} & \text{on} \quad S_D \end{array}\right\}. \tag{5.4}$$

5.2.2 Boundary Integral Equation

Several approaches have been used in the literature to establish boundary integral equations of piezoelectric materials, such as the weighted residual approach [1], the variational approach [5], and Betti's reciprocity theorem [11, 12]. A brief discussion of Betti's reciprocity theorem which is used in later sections of this chapter is given here.

With the reciprocity theorem, we consider two electroelastic states, namely [12, 15]

$$\text{State 1:} \quad \{U_I^{(1)}\}^T = \{u_1^{(1)}, u_2^{(1)}, u_3^{(1)}, \phi^{(1)}\}^T;$$

$$\{\tilde{f}_I^{(1)}\}^T = \{\tilde{f}_1^{(1)}, \tilde{f}_2^{(1)}, \tilde{f}_3^{(1)}, \rho_e^{(1)}\}^T;$$

$$\{t_I^{(1)}\}^T = \{t_1^{(1)}, t_2^{(1)}, t_3^{(1)}, D_n^{(1)}\}^T \tag{5.5}$$

$$\text{State 2:} \quad \{U_I^{(2)}\}^T = \{u_1^{(2)}, u_2^{(2)}, u_3^{(2)}, \phi^{(2)}\}^T;$$

$$\{\tilde{f}_I^{(2)}\}^T = \{\tilde{f}_1^{(2)}, \tilde{f}_2^{(2)}, \tilde{f}_3^{(2)}, -\rho_e^{(2)}\}^T;$$

$$\{t_I^{(2)}\}^T = \{t_1^{(2)}, t_2^{(2)}, t_3^{(2)}, D_n^{(2)}\}^T. \tag{5.6}$$

The first state represents the solution to piezoelectric problems with finite domains and general loading conditions; the second state is of an artificial nature and represents the fundamental solution to the case of a fictitious infinite body subjected to a point force or a point charge. Furthermore, we introduce the compatible field $\{\widehat{S}_{ij}, \widehat{u}_i, \widehat{E}_i, \widehat{\phi}\}$ that satisfies Equation (5.3). The principle of virtual work is given by

$$\int_V \tilde{f}_I^{(1)} \widehat{U}_I dV + \int_S t_I^{(1)} \widehat{U}_I dS = \int_V (T_{ij}^{(1)} \widehat{S}_{ij} - D_i^{(1)} \widehat{E}_i) dV. \tag{5.7}$$

On the other hand, for a linear piezoelectric solid, we can show that the following reciprocal property of Betti type holds:

$$T_{ij}^{(1)} S_{ij}^{(2)} - D_i^{(1)} E_i^{(2)} = T_{ij}^{(2)} S_{ij}^{(1)} - D_i^{(2)} E_i^{(1)}. \tag{5.8}$$

By substituting Equation (5.8) into (5.7), the following reciprocal relation can be obtained,

$$\int_V \tilde{f}_I^{(1)} U_I^{(2)} dV + \int_S t_I^{(1)} U_I^{(2)} dS = \int_V \tilde{f}_I^{(2)} U_I^{(1)} dV + \int_S t_I^{(2)} U_I^{(1)} dS. \tag{5.9}$$

To convert Equation (5.9) into a boundary integral equation, assume that State 1 is the actual solution for a body V with the boundary S and State 2 is the solution for a fictitious infinite body subjected to a point force at $\widehat{\mathbf{X}}$ in the x_m direction with no bulk charge distribution (i.e., $\rho_e^{(2)} = 0$), namely

$$\tilde{f}_i^{(2)}(\mathbf{x}) = \delta(\mathbf{x} - \widehat{\mathbf{x}})\delta_{im}, \qquad (m = 1, 2, 3). \tag{5.10}$$

The displacement, electric potential, stress, and electric displacement induced by the above-mentioned point force were discussed in Chapter 2. Using the solution given in Chapter 2, the variables $u^{(2)}, \phi^{(2)}, T_{ij}^{(2)}$, and $D_i^{(2)}$

can be written in the form

$$u_i^{(2)} = u_{ij}^*(\mathbf{x}, \widehat{\mathbf{x}})e_j, \qquad \phi^{(2)} = \phi_j^*(\mathbf{x}, \widehat{\mathbf{x}})e_j = u_{4j}^*(\mathbf{x}, \widehat{\mathbf{x}})e_j,$$

$$T_{ij}^{(2)} = \Sigma_{ijm}^*(\mathbf{x}, \widehat{\mathbf{x}})e_m = [c_{ijkl}u_{km,l}^*(\mathbf{x}, \widehat{\mathbf{x}}) - e_{lij}u_{4m,l}^*(\mathbf{x}, \widehat{\mathbf{x}})]e_m,$$

$$D_i^{(2)} = D_{im}^*(\mathbf{x}, \widehat{\mathbf{x}})e_m = [e_{ikl}u_{km,l}^*(\mathbf{x}, \widehat{\mathbf{x}}) + \kappa_{il}u_{4m,l}^*(\mathbf{x}, \widehat{\mathbf{x}})]e_m, \qquad (5.11)$$

where u_{ij}^* represents the displacement in the jth direction at field point $\mathbf{x}$ due to a point force in the ith direction applied at the source point $\widehat{\mathbf{x}}$ interior to S, and u_{4i}^* denotes the ith displacement at $\mathbf{x}$ due to a point electric charge at $\widehat{\mathbf{x}}$. Making use of Equation (5.4), the traction and surface charge can be obtained as

$$t_i^{(2)} = t_{im}^*(\mathbf{x}, \widehat{\mathbf{x}})e_m = \Sigma_{ijm}^*(\mathbf{x}, \widehat{\mathbf{x}})n_j e_m,$$

$$D_n^{(2)} = t_{4m}^*(\mathbf{x}, \widehat{\mathbf{x}})e_m = D_{im}^*(\mathbf{x}, \widehat{\mathbf{x}})n_i e_m. \qquad (5.12)$$

Substituting all the above quantities associated with State 2 into Equation (5.9) yields

$$u_i(\widehat{\mathbf{x}}) = \int_V u_{Ji}^*(\mathbf{x}, \widehat{\mathbf{x}})\tilde{f}_J(\mathbf{x})dV(\mathbf{x}) - \int_S t_{Ji}^*(\mathbf{x}, \widehat{\mathbf{x}})u_J(\mathbf{x})dS(\mathbf{x})$$

$$+ \int_S u_{Ji}^*(\mathbf{x}, \widehat{\mathbf{x}})t_J(\mathbf{x})dS(\mathbf{x}), \qquad (5.13)$$

where $i = 1, 2, 3$, $J = 1 - 4$, $u_4 = -\phi$, $\tilde{f}_4 = \rho_e$, and $t_4 = D_n$.

Next, assume that State 2 of the fictitious infinite body is subjected to a point charge at $\widehat{\mathbf{x}}$ with no body force distribution; that is,

$$-\rho_e^{(2)}(\mathbf{x}) = \delta(\mathbf{x} - \widehat{\mathbf{x}}), \qquad \tilde{f}_i^{(2)}(\mathbf{x}) = 0 \quad (i = 1, 2, 3) \qquad (5.14)$$

The resulting displacement and electric potential induced by this point charge are given by

$$u_i^{(2)} = u_{i4}^*(\mathbf{x}, \widehat{\mathbf{x}}),$$

$$\phi^{(2)} = \phi_4^*(\mathbf{x}, \widehat{\mathbf{x}}) = u_{44}^*(\mathbf{x}, \widehat{\mathbf{x}}), \qquad (5.15)$$

where u_{i4}^* and u_{44}^* denote, respectively, the ith displacement and electric potential at $\mathbf{x}$ due to a point electric charge at $\widehat{\mathbf{x}}$.

Substituting the solution in Equation (5.15) into (5.2) and (5.4), we have

$$T_{ij}^{(2)} = \Sigma_{ij4}^*(\mathbf{x}, \widehat{\mathbf{x}}) = c_{ijkl}u_{k4,l}^*(\mathbf{x}, \widehat{\mathbf{x}}) - e_{lij}u_{44,l}^*(\mathbf{x}, \widehat{\mathbf{x}}),$$

$$D_i^{(2)} = D_{i4}^*(\mathbf{x}, \widehat{\mathbf{x}}) = e_{ikl}u_{k4,l}^*(\mathbf{x}, \widehat{\mathbf{x}}) + \kappa_{il}u_{44,l}^*(\mathbf{x}, \widehat{\mathbf{x}}),$$

$$t_i^{(2)} = t_{i4}^*(\mathbf{x}, \widehat{\mathbf{x}}) = \Sigma_{ij4}^*(\mathbf{x}, \widehat{\mathbf{x}})n_j, \qquad D_n^{(2)} = t_{44}^*(\mathbf{x}, \widehat{\mathbf{x}}) = D_{i4}^*(\mathbf{x}, \widehat{\mathbf{x}})n_i, \qquad (5.16)$$

where t_{ij}^* are related to u_{ij}^* by Equations (5.2) and (5.4).

Substituting Equations $(5.14)-(5.16)$ into (5.9), we obtain

$$\phi(\widehat{\mathbf{x}}) = \int_V u^*_{J4}(\mathbf{x}, \widehat{\mathbf{x}})\tilde{f}_J(\mathbf{x})dV(\mathbf{x}) - \int_S t^*_{J4}(\mathbf{x}, \widehat{\mathbf{x}})u_J(\mathbf{x})dS(\mathbf{x})$$
$$+ \int_S u^*_{J4}(\mathbf{x}, \widehat{\mathbf{x}})t_J(\mathbf{x})dS(\mathbf{x}) . \tag{5.17}$$

Combining Equation (5.13) with (5.17), we have

$$u_I(\widehat{\mathbf{x}}) = \int_V u^*_{JI}(\mathbf{x}, \widehat{\mathbf{x}})\tilde{f}_J(\mathbf{x})dV(\mathbf{x}) - \int_S t^*_{JI}(\mathbf{x}, \widehat{\mathbf{x}})u_J(\mathbf{x})dS(\mathbf{x})$$
$$+ \int_S u^*_{JI}(\mathbf{x}, \widehat{\mathbf{x}})t_J(\mathbf{x})dS(\mathbf{x}) , \tag{5.18}$$

where $I, J = 1\text{--}4$.

Making use of Equations (5.2) and (5.18), the corresponding stresses and electric displacements are expressed as

$$\Pi_{iJ}(\widehat{\mathbf{x}}) = \int_V D^*_{KiJ}(\mathbf{x}, \widehat{\mathbf{x}})\tilde{f}_K(\mathbf{x})dV(\mathbf{x}) - \int_S S^*_{KiJ}(\mathbf{x}, \widehat{\mathbf{x}})u_K(\mathbf{x})dS(\mathbf{x})$$
$$+ \int_S D^*_{KiJ}(\mathbf{x}, \widehat{\mathbf{x}})t_K(\mathbf{x})dS(\mathbf{x}) , \tag{5.19}$$

where

$$\Pi_{iJ} = \begin{cases} \sigma_{ij}, & J \leq 3, \\ D_i & J = 4 \end{cases} \tag{5.20}$$

$$S^*_{KiJ}(\mathbf{x}, \widehat{\mathbf{x}}) = E_{iJMn}\frac{\partial t^*_{MK}(\mathbf{x}, \widehat{\mathbf{x}})}{\partial x_n} \tag{5.21}$$

$$D^*_{KiJ}(\mathbf{x}, \widehat{\mathbf{x}}) = E_{iJMn}\frac{\partial u^*_{MK}(\mathbf{x}, \widehat{\mathbf{x}})}{\partial x_n} \tag{5.22}$$

with

$$E_{iJMn} = \begin{cases} c_{ijmn}, & J, \ M \leq 3, \\ e_{nij}, & J \leq 3, \ M = 4, \\ e_{imn}, & J = 4, \ M \leq 3, \\ -\varepsilon_{in}, & J = 4, \ M = 4. \end{cases} \tag{5.23}$$

The integral representation formula for the generalized traction components can be obtained from Equations (5.4) and (5.19) as

$$t_J(\widehat{\mathbf{x}}) = \int_V V^*_{IJ}(\mathbf{x}, \widehat{\mathbf{x}})\tilde{f}_I(\mathbf{x})dV(\mathbf{x}) - \int_S W^*_{IJ}(\mathbf{x}, \widehat{\mathbf{x}})u_I(\mathbf{x})dS(\mathbf{x})$$
$$+ \int_S V^*_{IJ}(\mathbf{x}, \widehat{\mathbf{x}})t_I(\mathbf{x})dS(\mathbf{x}) , \tag{5.24}$$

where

$$V_{IJ}^*(\mathbf{x}, \widehat{\mathbf{x}}) = D_{IkJ}^*(\mathbf{x}, \widehat{\mathbf{x}})n_k(\widehat{\mathbf{x}}), \qquad W_{IJ}^*(\mathbf{x}, \widehat{\mathbf{x}}) = S_{IkJ}^*(\mathbf{x}, \widehat{\mathbf{x}})n_k(\widehat{\mathbf{x}}). \tag{5.25}$$

It can be seen from Equations (5.13), (5.17), and (5.24) that to obtain the fields at internal points, the boundary data of traction, displacement, electric potential, and the normal component of electric displacement need to be known throughout the boundary S. For this purpose, we can examine the limiting forms of Equations (5.13), (5.17), and (5.24) as $\widehat{\mathbf{x}}$ approaches the boundary. To properly circumvent the singular behavior when $\mathbf{x}$ approaches $\widehat{\mathbf{x}}$, Chen and Lin [16] assumed a singular point $\widehat{\mathbf{x}}$ on the boundary surrounded by a small hemispherical surface of radius ε, say S_ε, centered at the point $\widehat{\mathbf{x}}$ with $\varepsilon \to 0$. Because the asymptotic behavior of Green's function in piezoelectric solids at $r = |\mathbf{x} - \widehat{\mathbf{x}}| \to 0$ is mathematically similar to that of uncoupled elasticity, Equations (5.13), (5.17), and (5.24) can be rewritten as [16]

$$c_{ki}u_k(\widehat{\mathbf{x}}) = \int_V u_{Ji}^*(\mathbf{x}, \widehat{\mathbf{x}})\tilde{f}_J(\mathbf{x})dV(\mathbf{x}) - \int_S t_{Ji}^*(\mathbf{x}, \widehat{\mathbf{x}})u_J(\mathbf{x})dS(\mathbf{x})$$
$$+ \int_S u_{Ji}^*(\mathbf{x}, \widehat{\mathbf{x}})t_J(\mathbf{x})dS(\mathbf{x}) \tag{5.26}$$

$$b\phi(\widehat{\mathbf{x}}) = \int_V u_{J4}^*(\mathbf{x}, \widehat{\mathbf{x}})\tilde{f}_J(\mathbf{x})dV(\mathbf{x}) - \int_S t_{J4}^*(\mathbf{x}, \widehat{\mathbf{x}})u_J(\mathbf{x})dS(\mathbf{x})$$
$$+ \int_S u_{J4}^*(\mathbf{x}, \widehat{\mathbf{x}})t_J(\mathbf{x})dS(\mathbf{x}) \tag{5.27}$$

$$c_{ki}t_k(\widehat{\mathbf{x}}) = \int_V V_{Ji}^*(\mathbf{x}, \widehat{\mathbf{x}})\tilde{f}_J(\mathbf{x})dV(\mathbf{x}) - \int_S W_{Ji}^*(\mathbf{x}, \widehat{\mathbf{x}})u_J(\mathbf{x})dS(\mathbf{x})$$
$$+ \int_S V_{Ji}^*(\mathbf{x}, \widehat{\mathbf{x}})t_J(\mathbf{x})dS(\mathbf{x}) \tag{5.28}$$

$$bD_n(\widehat{\mathbf{x}}) = \int_V V_{J4}^*(\mathbf{x}, \widehat{\mathbf{x}})\tilde{f}_J(\mathbf{x})dV(\mathbf{x}) - \int_S W_{J4}^*(\mathbf{x}, \widehat{\mathbf{x}})u_J(\mathbf{x})dS(\mathbf{x})$$
$$+ \int_S V_{J4}^*(\mathbf{x}, \widehat{\mathbf{x}})t_J(\mathbf{x})dS(\mathbf{x}), \tag{5.29}$$

where $\widehat{\mathbf{x}} \in S$, and the coefficient c_{ki} and b are defined as

$$c_{ki}(\widehat{\mathbf{x}}) = \delta_{ki} + \lim_{\varepsilon \to 0} \int_{S_\varepsilon} t_{ki}^*(\mathbf{x}, \widehat{\mathbf{x}})dS(\mathbf{x}) \tag{5.30}$$

$$b(\widehat{\mathbf{x}}) = 1 + \lim_{\varepsilon \to 0} \int_{S_\varepsilon} t_{44}^*(\mathbf{x}, \widehat{\mathbf{x}})dS(\mathbf{x}). \tag{5.31}$$

In the field of BEM, the coefficients c_{ki} and b are usually known as boundary shape coefficients: $c_{ii}(\widehat{\mathbf{x}}) = b(\widehat{\mathbf{x}}) = 1$ if $\widehat{\mathbf{x}} \in \Omega$, $c_{ii}(\widehat{\mathbf{x}}) = b(\widehat{\mathbf{x}}) = 1/2$ if $\widehat{\mathbf{x}}$ is on the smooth boundary [17]. Using the concept of boundary shape coefficients, Equation (5.18) can be rewritten as

$$c_{KI}u_K(\widehat{\mathbf{x}}) = \int_V u^*_{JI}(\mathbf{x},\widehat{\mathbf{x}})\tilde{f}_J(\mathbf{x})dV(\mathbf{x}) - \int_S t^*_{JI}(\mathbf{x},\widehat{\mathbf{x}})u_J(\mathbf{x})dS(\mathbf{x})$$

$$+ \int_S u^*_{JI}(\mathbf{x},\widehat{\mathbf{x}})t_J(\mathbf{x})dS(\mathbf{x})\,, \tag{5.32}$$

where $\widehat{\mathbf{x}} \in \Gamma$, and $c_{K4} = c_{4K} = b\delta_{K4}$.

It is more convenient to work with matrices rather than continue with indicial notation. To this effect the generalized displacement $\mathbf{U}$, traction $\mathbf{T}$, body force $\mathbf{b}$, and boundary shape coefficients $\mathbf{C}$ are defined as [6]

$$\mathbf{U} = \begin{Bmatrix} u_1 \\ u_2 \\ u_3 \\ u_4 \end{Bmatrix}, \qquad \mathbf{T} = \begin{Bmatrix} t_1 \\ t_2 \\ t_3 \\ t_4 \end{Bmatrix}, \qquad \tilde{\mathbf{f}} = \begin{Bmatrix} \tilde{f}_1 \\ \tilde{f}_2 \\ \tilde{f}_3 \\ \tilde{f}_4 \end{Bmatrix}, \qquad \mathbf{C} = \begin{bmatrix} c_{11} & c_{12} & c_{13} & 0 \\ c_{21} & c_{22} & c_{23} & 0 \\ c_{31} & c_{32} & c_{33} & 0 \\ 0 & 0 & 0 & c_{44} \end{bmatrix}^T, \tag{5.33}$$

where $c_{44} = b$.

Similarly, the fundamental solution coefficients can be defined in matrix form as

$$\mathbf{U}^* = \begin{bmatrix} u^*_{11} & u^*_{12} & u^*_{13} & u^*_{14} \\ u^*_{21} & u^*_{22} & u^*_{23} & u^*_{24} \\ u^*_{31} & u^*_{32} & u^*_{33} & u^*_{34} \\ u^*_{41} & u^*_{42} & u^*_{43} & u^*_{44} \end{bmatrix}^T, \qquad \mathbf{T}^* = \begin{bmatrix} t^*_{11} & t^*_{12} & t^*_{13} & t^*_{14} \\ t^*_{21} & t^*_{22} & t^*_{23} & t^*_{24} \\ t^*_{31} & t^*_{32} & t^*_{33} & t^*_{34} \\ t^*_{41} & t^*_{42} & t^*_{43} & t^*_{44} \end{bmatrix}^T, \tag{5.34}$$

where $t^*_{ij}(i = 1,2,3)$ represents the traction in the jth direction at a field point $\mathbf{x}$ due to a unit point load acting at the source point $\widehat{\mathbf{x}}$, and t^*_{4j} denotes the jth traction at $\mathbf{x}$ due to a unit electric charge at $\widehat{\mathbf{x}}$. Using the matrix notation defined in Equations (5.33) and (5.34), the integral equation (5.32) can be written in matrix form as

$$\mathbf{C}\mathbf{U} = \int_V \mathbf{U}^*\tilde{\mathbf{f}}dV - \int_S \mathbf{T}^*\mathbf{U}dS + \int_S \mathbf{U}^*\mathbf{T}dS\,. \tag{5.35}$$

Using Equation (5.35), the generalized displacement at a point, say point $\widehat{\mathbf{x}}_i$, can be obtained by enforcing a point load at the same point. In this case Equation (5.35) becomes

$$\mathbf{C}(\widehat{\mathbf{x}}_i)\mathbf{U}(\widehat{\mathbf{x}}_i) = \int_V \mathbf{U}^*(\widehat{\mathbf{x}}_i,\mathbf{x})\tilde{\mathbf{f}}(\mathbf{x})dV(\mathbf{x}) - \int_S \mathbf{T}^*(\widehat{\mathbf{x}}_i,\mathbf{x})\mathbf{U}(\mathbf{x})dS(\mathbf{x})$$

$$+ \int_S \mathbf{U}^*(\widehat{\mathbf{x}}_i,\mathbf{x})\mathbf{T}(\mathbf{x})dS(\mathbf{x})\,. \tag{5.36}$$

5.2.3 Boundary Element Equation

To obtain a weak solution of Equation (5.36), as in the usual BEM, the boundary S and the domain V in Equation (5.36) are divided into K boundary elements and M internal cells, respectively. Boundary displacements, tractions, electric potential, and electric displacement are written in terms of their values at a series of nodal points. After discretization is performed by use of various kinds of BE (e.g., constant elements, linear elements, or higher-order elements), the boundary integral equation (5.36) becomes a set of linear algebraic equations including boundary variables $\mathbf{U}$ and $\mathbf{T}$. Once the boundary conditions are applied the linear algebraic equations can be solved to obtain all the unknown values. For a particular boundary element e, the variables $\mathbf{U}$ and $\mathbf{T}$ can be approximated in terms of the shape function $\mathbf{N}$ in the form [17]

$$\mathbf{U} = \sum_{i=1}^{Q} N_i \mathbf{U}^{ei} = \mathbf{N}\mathbf{U}^e, \quad \mathbf{T} = \sum_{i=1}^{Q} N_i \mathbf{T}^{ei} = \mathbf{N}\mathbf{T}^e, \tag{5.37}$$

where N_i is the shape function associated with node i and is discussed in the next two sections, $\mathbf{U}^{ei}$ and $\mathbf{T}^{ei}$ are, respectively, the values of $\mathbf{U}$ and $\mathbf{T}$ at node i of the element, and Q is the number of nodes of the element. The interpolation function $\mathbf{N}$ is a $4 \times Q$ array of shape functions; that is,

$$\mathbf{N} = \begin{bmatrix} N_1 & 0 & 0 & 0 & N_2 & \cdots & N_Q & 0 & 0 & 0 \\ 0 & N_1 & 0 & 0 & 0 & \cdots & 0 & N_Q & 0 & 0 \\ 0 & 0 & N_1 & 0 & 0 & \cdots & 0 & 0 & N_Q & 0 \\ 0 & 0 & 0 & N_1 & 0 & \cdots & 0 & 0 & 0 & N_Q \end{bmatrix} \tag{5.38}$$

for a three-dimensional electroelastic problem. Substituting Equation (5.37) into (5.36) yields

$$\mathbf{C}^i \mathbf{U}^i + \sum_{j=1}^{K} \left\{ \int_{S_j} \mathbf{T}^{*(i)} \mathbf{N} dS \right\} \mathbf{U}^j = \sum_{j=1}^{K} \left\{ \int_{S_j} \mathbf{U}^{*(i)} \mathbf{N} dS \right\} \mathbf{T}^j$$

$$+ \sum_{s=1}^{M} \left\{ \int_{V_s} \mathbf{U}^{*(i)} \tilde{\mathbf{f}} dV \right\}, \tag{5.39}$$

where K is the number of boundary elements, $\mathbf{C}^i = \mathbf{C}(\widehat{\mathbf{x}}_i)$, $\mathbf{U}^i = \mathbf{U}(\widehat{\mathbf{x}}_i)$, $\mathbf{U}^{*(i)} = \mathbf{U}^*(\widehat{\mathbf{x}}_i, \mathbf{x})$, $\mathbf{T}^{*(i)} = \mathbf{T}^*(\widehat{\mathbf{x}}_i, \mathbf{x})$, and S_j is the surface of a "j" element.

Once the integrals in Equation (5.39) have been carried out, they can be further written as

$$\mathbf{C}^i \mathbf{U}^i + \sum_{j=1}^{K} \overline{\mathbf{H}}_{ij} \mathbf{U}^j = \sum_{j=1}^{K} \mathbf{G}_{ij} \mathbf{T}^j + \sum_{s=1}^{M} B_{is}, \tag{5.40}$$

where the inference matrices $\overline{\mathbf{H}}_{ij}$ and $\mathbf{G}_{ij}$ are evaluated by

$$\overline{\mathbf{H}}_{ij} = \sum_t \int_{S_t} \mathbf{T}^*(\widehat{\mathbf{x}}_i, \mathbf{x})\mathbf{N}_q(\mathbf{x})dS(\mathbf{x}),$$

$$\mathbf{G}_{ij} = \sum_t \int_{S_t} \mathbf{U}^*(\widehat{\mathbf{x}}_i, \mathbf{x})\mathbf{N}_q(\mathbf{x})dS(\mathbf{x}), \tag{5.41}$$

where the summation extends to all the elements to which node j belongs and q is the number of the order of node j within element t. The pseudo-loading component $\mathbf{B}_{is}$ is defined as

$$\mathbf{B}_{is} = \int_{V_s} \mathbf{U}^*(\widehat{\mathbf{x}}_i, \mathbf{x})\tilde{\mathbf{f}}(\mathbf{x})dV(\mathbf{x}). \tag{5.42}$$

If we define $\mathbf{H}_{ij} = \overline{\mathbf{H}}_{ij} + \delta_{ij}\mathbf{C}(\widehat{\mathbf{x}}_i)$ and apply Equation (5.40) to the K nodes "j" the following system of equations is obtained,

$$\mathbf{HU} = \mathbf{GT} + \mathbf{B}. \tag{5.43}$$

The vectors $\mathbf{U}$ and $\mathbf{T}$ represent all the values of generalized displacements and tractions before applying boundary conditions. These boundary conditions can be introduced by collecting the unknown terms to the left-hand side and the known terms to the right-hand side of Equation (5.43). This gives the final system of equations; that is,

$$\mathbf{EX} = \mathbf{R}. \tag{5.44}$$

By solving the above system the vector $\mathbf{X}$ of boundary variables is fully determined.

5.3 One-Dimensional Elements

5.3.1 Shape Functions

One-dimensional elements are used to model the boundary of the plane domain of a problem. Within a particular element e (see Figure 5.1), field variables can be interpolated using constant, linear, and higher-order polynomials. In this section, however, we focus on BEM formulations with linear elements.

Consider a linear element as shown in Figure 5.2, in which ξ is a dimensionless coordinate whose value equals zero at the center and ± 1 at the ends.

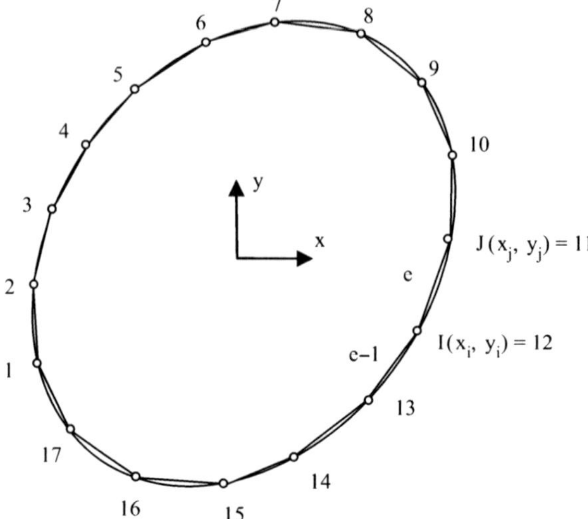

Fig. 5.1 One-dimensional element.

It is easy to verify that the Cartesian coordinates of a point on the element shown in Figure 5.2 are related to the dimensional coordinate ξ by the relation

$$x(\xi) = \frac{1-\xi}{2}x_i + \frac{1+\xi}{2}x_j = N_1(\xi)x_i + N_2(\xi)x_j$$

$$y(\xi) = \frac{1-\xi}{2}y_i + \frac{1+\xi}{2}y_j = N_1(\xi)y_i + N_2(\xi)y_j \,, \tag{5.45}$$

where N_1 and N_2 are the two shape functions of the element. Similarly, the field variables $\mathbf{U}$ and $\mathbf{T}$ of this element can be approximated in terms of the same shape functions as

$$\mathbf{U}(\xi) = \frac{1-\xi}{2}\mathbf{U}^i + \frac{1+\xi}{2}\mathbf{U}^j = N_1(\xi)\mathbf{U}^i + N_2(\xi)\mathbf{U}^j$$

$$\mathbf{T}(\xi) = \frac{1-\xi}{2}\mathbf{T}^i + \frac{1+\xi}{2}\mathbf{T}^j = N_1(\xi)\mathbf{T}^i + N_2(\xi)\mathbf{T}^j \,. \tag{5.46}$$

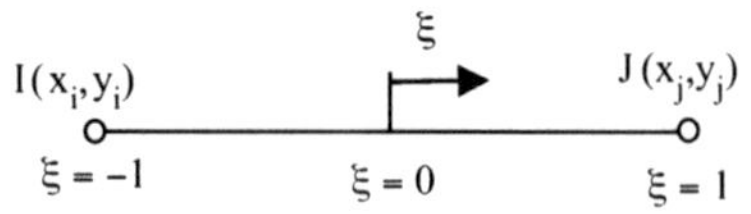

Fig. 5.2 Linear element with Cartesian and dimensional coordinates.

Substituting Equation (5.46) into (5.41) and noting the node i shown in Figure 5.1, we have

$$\overline{\mathbf{H}}_{ki} = \int_{S_{e-1}} \mathbf{T}^*(\widehat{\mathbf{x}}_k, \mathbf{x})\mathbf{N}_2(\mathbf{x})dS(\mathbf{x}) + \int_{S_c} \mathbf{T}^*(\widehat{\mathbf{x}}_k, \mathbf{x})\mathbf{N}_1(\mathbf{x})dS(\mathbf{x})$$

$$\mathbf{G}_{ki} = \int_{S_{e-1}} \mathbf{U}^*(\widehat{\mathbf{x}}_k, \mathbf{x})\mathbf{N}_2(\mathbf{x})dS(\mathbf{x}) + \int_{S_c} \mathbf{U}^*(\widehat{\mathbf{x}}_k, \mathbf{x})\mathbf{N}_1(\mathbf{x})dS(\mathbf{x}) \,. \quad (5.47)$$

5.3.2 Differential Geometry

In evaluating Equation (5.47), the information on the unit vector normal to a line element is required. The best way to determine the unit vector is by using vector algebra. To this end, consider a one-dimensional element as shown in Figure 5.3 [18]. The tangential vector in the direction of ξ can be obtained by the differentiation of Equation (5.45) as

$$\mathbf{v}_\xi = \nu_x\mathbf{i} + \nu_y\mathbf{j} = \frac{d}{d\xi}x(\xi)\mathbf{i} + \frac{d}{d\xi}y(\xi)\mathbf{j}\,, \quad (5.48)$$

where $\mathbf{i}$ and $\mathbf{j}$ are, respectively, the unit vectors in the x- and y-directions.

A vector normal to the line element e in Figure 5.3, $\mathbf{v}_n$, may then be obtained by taking the cross-product of $\mathbf{v}_\xi$ with a unit vector in the z-direction $(\mathbf{v}_\xi = \{0, 0, 1\}^T)$:

$$\mathbf{v}_n = \mathbf{v}_\xi \times \mathbf{v}_z = \left\{\begin{matrix} \frac{dx}{d\xi} \\ \frac{dy}{d\xi} \\ 0 \end{matrix}\right\} \times \left\{\begin{matrix} 0 \\ 0 \\ 1 \end{matrix}\right\} = \left\{\begin{matrix} \frac{dy}{d\xi} \\ -\frac{dx}{d\xi} \\ 0 \end{matrix}\right\}\,. \quad (5.49)$$

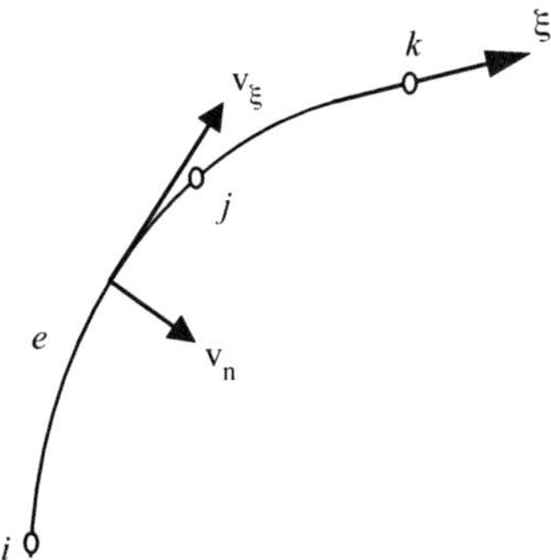

Fig. 5.3 Normal and tangential vectors for one-dimensional element [18].

The length of the vector $\mathbf{v}_n$ is equal to

$$v_n = |\mathbf{v}_n| = \sqrt{\left(\frac{dy}{d\xi}\right)^2 + \left(-\frac{dx}{d\xi}\right)^2} \tag{5.50}$$

which is also the length of $d\xi$ and hence the Jacobian of the coordinate transformation from the Cartesian coordinate system to the intrinsic coordinate ξ. Therefore the unit vector normal to the line element is given by

$$\mathbf{n} = \frac{\mathbf{v}_n}{|\mathbf{v}_n|} = \frac{1}{\sqrt{\left(\frac{dy}{d\xi}\right)^2 + \left(-\frac{dx}{d\xi}\right)^2}} \left\{ \begin{array}{c} \frac{dy}{d\xi} \\ -\frac{dx}{d\xi} \\ 0 \end{array} \right\}. \tag{5.51}$$

5.4 Two-Dimensional Elements

5.4.1 Shape Functions

For modeling the boundary of three-dimensional problems, two-dimensional elements are used. For illustration we take an 8-node isoparametric element shown in Figure 5.4 as an example. The shape functions $Ni(i=1\text{–}8)$ for the 8-node isoparametric element take the form [18]

$$N_1 = \frac{1}{4}(1-\xi)(1-\eta)(-\xi-\eta-1),$$

$$N_2 = \frac{1}{4}(1+\xi)(1-\eta)(\xi-\eta-1), \tag{5.52}$$

$$N_3 = \frac{1}{4}(1+\xi)(1+\eta)(\xi+\eta-1),$$

$$N_4 = \frac{1}{4}(1-\xi)(1+\eta)(-\xi+\eta-1), \tag{5.53}$$

$$N_5 = \frac{1}{2}(1-\xi^2)(1-\eta), \qquad N_6 = \frac{1}{2}(1+\xi)(1-\eta^2), \tag{5.54}$$

$$N_7 = \frac{1}{2}(1-\xi^2)(1+\eta), \qquad N_8 = \frac{1}{2}(1-\xi)(1-\eta^2). \tag{5.55}$$

These shape functions have the property that N_i is equal to unity at node i and zero at all other nodes. In the isoparametric formulation, both the geometry and the variable fields are approximated with the same shape

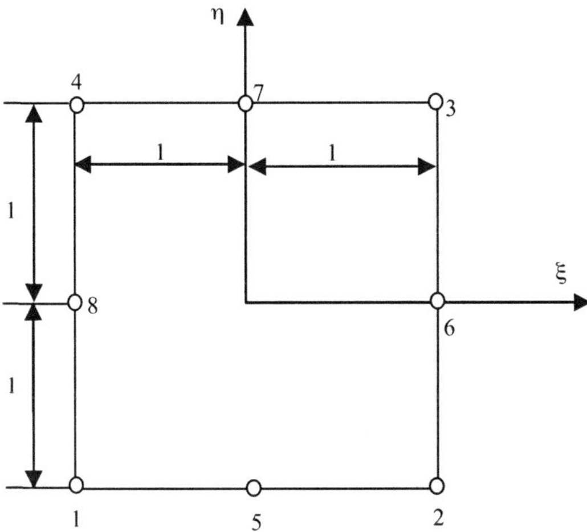

Fig. 5.4 Eight-node isoparametric element.

functions as

$$\mathbf{x}(\xi, \eta) = \sum_{i=1}^{8} N_i(\xi, \eta)\mathbf{x}^i \tag{5.56}$$

$$\mathbf{U}(\xi, \eta) = \sum_{i=1}^{8} N_i(\xi, \eta)\mathbf{U}^i \tag{5.57}$$

$$\mathbf{T}(\xi, \eta) = \sum_{i=1}^{8} N_i(\xi, \eta)\mathbf{T}^i \, . \tag{5.58}$$

To illustrate how to evaluate Equation (5.41) using Equations (5.57) and (5.58), consider node 11 in Figure 5.5. Substituting Equations (5.57) and (5.58) into (5.41), we obtain

$$\overline{\mathbf{H}}_{ki} = \int_{S_{e1}} \mathbf{T}^*(\widehat{\mathbf{x}}_k, \mathbf{x})\mathbf{N}_3(\mathbf{x})dS(\mathbf{x}) + \int_{S_{e2}} \mathbf{T}^*(\widehat{\mathbf{x}}_k, \mathbf{x})\mathbf{N}_4(\mathbf{x})dS(\mathbf{x})$$
$$+ \int_{S_{e3}} \mathbf{T}^*(\widehat{\mathbf{x}}_k, \mathbf{x})\mathbf{N}_1(\mathbf{x})dS(\mathbf{x}) + \int_{S_{e4}} \mathbf{T}^*(\widehat{\mathbf{x}}_k, \mathbf{x})\mathbf{N}_2(\mathbf{x})dS(\mathbf{x})$$
$$\tag{5.59}$$

$$\mathbf{G}_{ki} = \int_{S_{e1}} \mathbf{U}^*(\widehat{\mathbf{x}}_k, \mathbf{x})\mathbf{N}_3(\mathbf{x})dS(\mathbf{x}) + \int_{S_{e2}} \mathbf{U}^*(\widehat{\mathbf{x}}_k, \mathbf{x})\mathbf{N}_4(\mathbf{x})dS(\mathbf{x})$$
$$+ \int_{S_{e3}} \mathbf{U}^*(\widehat{\mathbf{x}}_k, \mathbf{x})\mathbf{N}_1(\mathbf{x})dS(\mathbf{x}) + \int_{S_{e4}} \mathbf{U}^*(\widehat{\mathbf{x}}_k, \mathbf{x})\mathbf{N}_2(\mathbf{x})dS(\mathbf{x}) \, . $$
$$\tag{5.60}$$

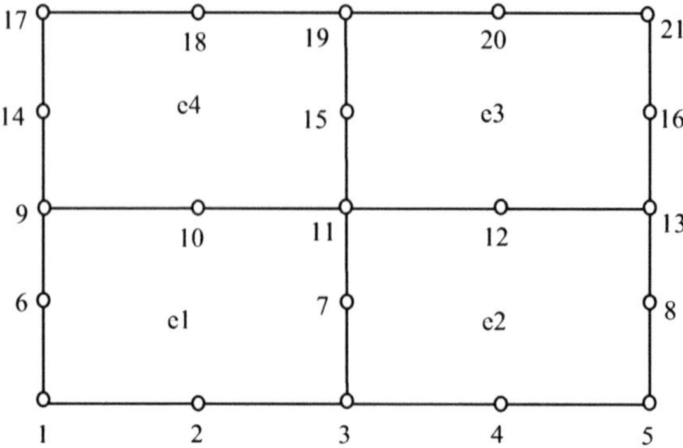

Fig. 5.5 Node 11 belongs to four elements.

5.4.2 Differential Geometry

The unit vector **n** normal to a surface can be obtained in a way similar to that in the one-dimensional element. Consider a two-dimensional element as shown in Figure 5.6. The tangential vectors $\mathbf{v}_\xi$ in the ξ-direction and $\mathbf{v}_\eta$ in the η-direction are obtained by differentiating Equation (5.56) [18]:

$$\mathbf{v}_\xi = \frac{\partial \mathbf{x}}{\partial \xi} = \sum_{i=1}^{8} \frac{\partial N_i}{\partial \xi} \mathbf{x}^i,$$

$$\mathbf{v}_\eta = \frac{\partial \mathbf{x}}{\partial \eta} = \sum_{i=1}^{8} \frac{\partial N_i}{\partial \eta} \mathbf{x}^i. \tag{5.61}$$

The vector normal to the surface in Figure 5.6, $\mathbf{v}_n$ can then be determined by taking the cross-product of $\mathbf{v}_\xi$ and $\mathbf{v}_\eta$:

$$\mathbf{v}_n = \mathbf{v}_\xi \times \mathbf{v}_\eta = \left\{ \begin{array}{c} v_{nx} \\ v_{my} \\ v_{nz} \end{array} \right\} = \left\{ \begin{array}{c} \frac{dx}{d\xi} \\ \frac{dy}{d\xi} \\ \frac{dz}{d\xi} \end{array} \right\} \times \left\{ \begin{array}{c} \frac{dx}{d\eta} \\ \frac{dy}{d\eta} \\ \frac{dz}{d\eta} \end{array} \right\} \left\{ \begin{array}{c} \frac{dy}{d\xi}\frac{dz}{d\eta} - \frac{dy}{d\eta}\frac{dz}{d\xi} \\ \frac{dz}{d\xi}\frac{dx}{d\eta} - \frac{dz}{d\eta}\frac{dx}{d\xi} \\ \frac{dx}{d\xi}\frac{dy}{d\eta} - \frac{dx}{d\eta}\frac{dy}{d\xi} \end{array} \right\}. \tag{5.62}$$

The unit normal vector n is then given by

$$\mathbf{n} = \frac{\mathbf{v}_n}{|\mathbf{v}_n|}, \tag{5.63}$$

where

$$v_n = |\mathbf{v}_n| = \sqrt{v_{nx}^2 + v_{ny}^2 + v_{nz}^2} \tag{5.64}$$

is the Jacobian of the transformation from the Cartesian coordinate system to the intrinsic coordinate system (ξ, η).

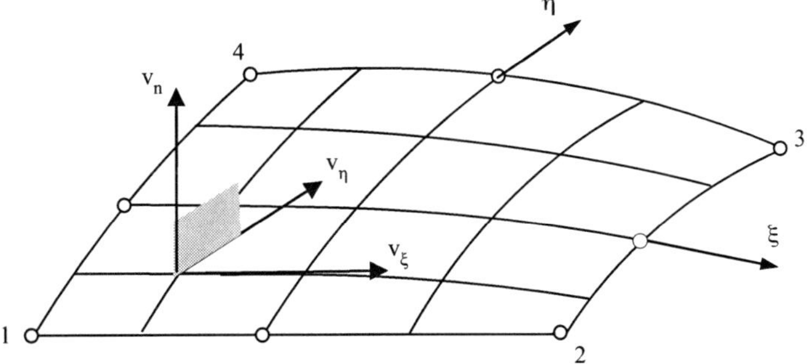

Fig. 5.6 Vectors $\mathbf{v}_n, \mathbf{v}_\xi$, and $\mathbf{v}_\eta$ in two-dimensional elements [18].

5.5 Numerical Integration over Elements

Generally, the analytical solution to the integrals in Equation (5.41) is very difficult, and numerical integration over the element is thus required. Because Gaussian integration formulae are popular, simple, and very accurate for a given number of points, they are adopted in this chapter.

5.5.1 One-Dimensional Elements

In the numerical treatment of one-dimensional elements the integral to be evaluated is written as

$$I_1 = \int_{-1}^{1} f(\xi)d\xi \approx \sum_{i-1}^{n} f(\xi_i)w_i , \tag{5.65}$$

where ξ_i is the coordinate of the ith integration point, w_i is the associated weighting factor, and n is the total number of integration points, which can be found from the BEM textbooks [19]. Notice that ξ_i values are symmetric with respect to $\xi = 0, w_i$ being the same for the two symmetric values.

When we apply the Gaussian integration formulae (5.65) to the expression (5.41), the limitations of the integrals need to be converted from S_t to $[-1,1]$ and the relationship between dS and the increment of intrinsic coordinate $d\xi$. Noting Equation (5.50), this relationship is given by

$$dS = v_n d\xi = J d\xi = \sqrt{\left(\frac{dx}{d\xi}\right)^2 + \left(\frac{dy}{d\xi}\right)^2}\, d\xi, \tag{5.66}$$

where J is the Jacobian of the intrinsic transformation. Substituting Equations (5.65) and (5.66) into (5.41) and using the relation (5.45) yields

$$\overline{\mathbf{H}}_{ij} = \sum_t \left[\int_{-1}^{+1} \mathbf{T}^*(\widehat{\mathbf{x}}_i, \mathbf{x}(\xi)) \mathbf{N}_q(\mathbf{x}(\xi)) J_t(\xi)\, d\xi \right]$$

$$\approx \sum_t \left[\sum_{k=1}^{n} w_k \mathbf{T}^*(\widehat{\mathbf{x}}_i, \mathbf{x}(\xi_k)) \mathbf{N}_q(\mathbf{x}(\xi_k)) J_t(\xi_k) \right] \tag{5.67}$$

$$\mathbf{G}_{ij} = \sum_t \left[\int_{-1}^{+1} \mathbf{U}^*(\widehat{\mathbf{x}}_i, \mathbf{x}(\xi)) \mathbf{N}_q(\mathbf{x}(\xi)) J_t(\xi)\, d\xi \right]$$

$$\approx \sum_t \left[\sum_{k=1}^{n} w_k \mathbf{U}^*(\widehat{\mathbf{x}}_i, \mathbf{x}(\xi_k)) \mathbf{N}_q(\mathbf{x}(\xi_k)) J_t(\xi_k) \right], \tag{5.68}$$

where n is the number of Gaussian sampling points employed in the Gaussian numerical integration, and $\mathbf{U}^*(\widehat{\mathbf{x}}_i, \mathbf{x}(\xi))$ and $\mathbf{T}^*(\widehat{\mathbf{x}}_i, \mathbf{x}(\xi))$ are the fundamental solutions at $\mathbf{x}(\xi)$ for a source at point $\widehat{\mathbf{x}}_i$.

5.5.2 Two-Dimensional Elements

Two-dimensional integration formulation is obtained by combining expression (5.65) in the form

$$I_2 = \int_{-1}^{1} \int_{-1}^{1} f(\xi, \eta) d\xi d\eta \cong \sum_{j=1}^{n} \sum_{i=1}^{n} f(\xi_i, \eta_i) w_i w_j. \tag{5.69}$$

The relationship between dS and the increment of intrinsic coordinates $d_\xi d_\eta$ is given by

$$dS = v_n d\xi d\eta = J(\xi, \eta) d\xi d\eta = \sqrt{v_{nx}^2 + v_{ny}^2 + v_{nz}^2}\, d\xi d\eta. \tag{5.70}$$

Substituting Equations (5.69) and (5.70) into (5.41) and using the relation (5.56), we have

$$\overline{\mathbf{H}}_{ij} = \sum_t \left[\int_{-1}^{+1} \mathbf{T}^*(\widehat{\mathbf{x}}_i, \mathbf{x}(\xi, \eta)) \mathbf{N}_q(\mathbf{x}(\xi, \eta)) J_t(\xi, \eta)\, d\xi d\eta \right]$$

$$\approx \sum_t \left[\sum_{l=1}^{n} \sum_{k=1}^{n} w_k w_l \mathbf{T}^*(\widehat{\mathbf{x}}_i, \mathbf{x}(\xi_k, \eta_l)) \mathbf{N}_q(\mathbf{x}(\xi_k, \eta_l)) J_t(\xi_k, \eta_l) \right] \tag{5.71}$$

$$\overline{\mathbf{G}}_{ij} = \sum_t \left[\int_{-1}^{+1} \mathbf{U}^*(\widehat{\mathbf{x}}_i, \mathbf{x}(\xi, \eta)) \mathbf{N}_q(\mathbf{x}(\xi, \eta)) J_t(\xi, \eta) \, d\xi d\eta \right]$$

$$\approx \sum_t \left[\sum_{l=1}^{n} \sum_{k=1}^{n} w_k w_l \mathbf{U}^*(\widehat{\mathbf{x}}_i, \mathbf{x}(\xi_k, \eta_l)) \mathbf{N}_q(\mathbf{x}(\xi_k, \eta_l)) J_t(\xi_k, \eta_l) \right] \quad (5.72)$$

5.6 Treatment of Singular Integrals

The accuracy of BEM for piezoelectric problems is critically dependent upon proper evaluation of the boundary integrals. The integrals (5.41) and (5.42) present a singular behavior of the order $O(1/r)$ and $O(1/r^2)$ for the generalized displacement and traction fundamental solutions, where r is the distance from a source point to the element under evaluation. The discussion which follows illustrates the basic procedure in treating singular integrals, taken from the results in [20, 21].

5.6.1 Nearly Singular Integrals [20]

Nearly singular integrals are evaluated using the standard Gauss quadrature formulae (5.69) in a special element subdivision scheme [22]. In this scheme, the element is subdivided into M and N subelements in the directions ξ and η, respectively. The Gauss quadrature formulae are applied in each element, obtaining the value of the integral as the sum of all individual contributions from the subdivisions. The criteria for choosing the number of subdivisions M and N are based on the size of the element of integration and its relative distance to the collocation point. These criteria are formulated in such a way that, for a given element, the closer the source point is to the element, the greater is the number of subdivisions.

5.6.2 Weakly Singular Integrals

Integrals of the kernels $\mathbf{U}^*$ in Equation (5.41) show a weak singularity of the type $O[\ln(z_K - z_{K0})]$ when the source point and the field point are either coincident or a short distance apart in comparison with the size of the element, which can be dealt with by a method of transformation in which the Jacobian of the transformation cancels out the singularity or weakens its effect, allowing a regular quadrature formula to give accurate results. For example, Rizzo and Shippy [23] used a Cartesian to polar transformation to integrate weak singular kernels. A polar coordinate system (r, θ) is introduced, originating at the singular node, such that the element of area $d\xi d\eta$ in the Cartesian system (ξ, η) becomes $rdrd\theta$ in the polar coordinate system. The additional r in the integrand cancels the $O(1/r)$ singularity.

5.6.3 Nonhypersingular Integrals [21]

The kernels T* appearing in Equation (5.41) show a strong singularity of $O[1/(z_K - \widehat{z}_K)]$ as $\widehat{\mathbf{x}} \to \mathbf{x}$, where $(z_K - \widehat{z}_K)$ is defined in Equation (5.75) below. Integration of such kernels over the element S_j that contains the source point $\widehat{\mathbf{x}}$ can be achieved as follows.

It is obvious that integrals of the type

$$\int_{S_j} \mathbf{T}^*(\widehat{\mathbf{x}}_j, \mathbf{x})\mathbf{N}_q(\mathbf{x})dS(\mathbf{x}) \tag{5.73}$$

contain the basic integral as

$$I_K = \int_{\Gamma_j} [p_K n_1(\mathbf{x}) - n_2(\mathbf{x})]\frac{1}{z_K - \widehat{z}_K}\mathbf{N}_q(\mathbf{x})dS(\mathbf{x}) \qquad (K = 1-4) \tag{5.74}$$

where n_1, n_2 are the components of the external unit normal to the boundary at the observation point $\mathbf{x}$ (see Figure 5.7) and p_K is the material's eigenvalues [2]. Define

$$r_K = z_K - \widehat{z}_K = (x_1 - \widehat{x}_1) + p_K(x_2 - \widehat{x}_2). \tag{5.75}$$

It follows that

$$\frac{dr_K}{dS} = \frac{dr_K}{dx_1}\frac{dx_1}{dS} + \frac{dr_K}{dx_2}\frac{dx_2}{dS} = -n_2 + p_K n_1. \tag{5.76}$$

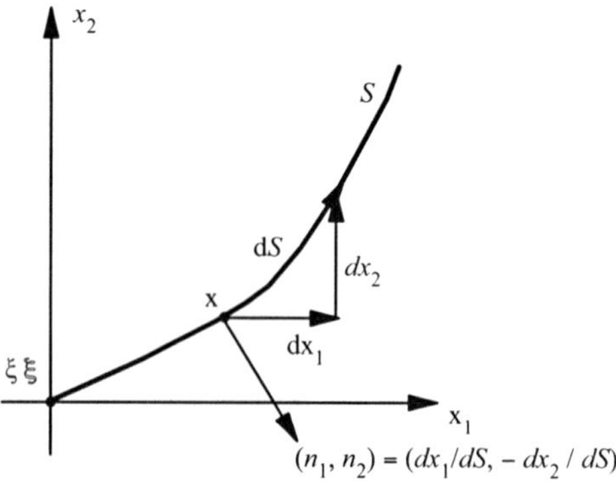

Fig. 5.7 Outward unit normal at boundary point $\mathbf{x}$.

Equation (5.76) is the key for all the transformations proposed below, and this illustrates that the Jacobian dr_K/dS of the coordinate transformation that maps the geometry of the boundary element S_j onto the complex plane r_K is included in the fundamental solution itself for the piezoelectric case.

Making use of Equations (5.76) and (5.74) can be rewritten as

$$I_K = \int_{S_j} \frac{1}{r_K} \mathbf{N}_q(\mathbf{x}) dr_K \tag{5.77}$$

which can be decomposed into the sum of a regular integral plus a singular integral with a known analytical solution

$$I_K = \int_{\Gamma_j} \frac{1}{r_K}(\mathbf{N}_q(\mathbf{x}) - 1) dr_K + \int_{\Gamma_j} \frac{1}{r_K} dr_K . \tag{5.78}$$

Integration of the kernels D^*_{KiJ} and V^*_{IJ} in Equations (5.19) and (5.24) can be achieved in a similar way as for $\mathbf{T}^*$ kernels because they contain singularities of the same type when $\widehat{\mathbf{x}} \to \mathbf{x}$. From Equation (5.22), we have the singular integral of the type [21]

$$I'_K = \int_{\Gamma_j} [p_K n_1(\widehat{\mathbf{x}}) - n_2(\widehat{\mathbf{x}})] \frac{1}{r_K} \mathbf{N}_q(\mathbf{x}) d\Gamma(\mathbf{x}) \quad (K = 1\text{--}4) \tag{5.79}$$

which can be regularized as follows,

$$I'_K = \int_{\Gamma_j} [p_K n_1(\widehat{\mathbf{x}}) - n_2(\widehat{\mathbf{x}}) - \frac{dr_K}{d\Gamma}] \frac{1}{r_K} \mathbf{N}_q(\mathbf{x}) d\Gamma(\mathbf{x})$$
$$+ \int_{\Gamma_j} \frac{1}{r_K} \mathbf{N}_q(\mathbf{x}) dr_K \tag{5.80}$$

The first integral in Equation (5.80) is regular and the second integral can be easily evaluated.

5.6.4 Hypersingular Integrals [21]

Note that the integration of W^*_{IJ} in Equation (5.24) has a hypersingularity of the order $O(1/r^2)$ as $\mathbf{x} \to \widehat{\mathbf{x}}$. From Equation (5.21) it follows that the hypersingular integral in Equation (5.24) is of the form

$$I''_K = \int_{\Gamma_j} [p_K n_1(\mathbf{x}) - n_2(\mathbf{x})] \frac{1}{r_K^2} \mathbf{N}_q(\mathbf{x}) d\Gamma(\mathbf{x}) = \int_{\Gamma_j} \frac{1}{r_K^2} \mathbf{N}_q(\mathbf{x}) dr_K \quad (K = 1\text{--}4) . \tag{5.81}$$

As indicated in [21], the integral (5.81) can be again decomposed into the sum of a regular integral plus singular integrals with known analytical solutions by using Equation (5.76) and the first two terms of the series expansion of the shape function N_q at $\widehat{\mathbf{x}}$, considered as a function of the complex space variable r_K

$$N_q(r_K) = N_q|_{r_K=0} + \left.\frac{dN_q}{dr_K}\right|_{r_K=0} r_K + 0(r_K^2) \approx N_{q0} + N'_{q0}r_K . \qquad (5.82)$$

Thus, I''_K can be written as

$$I''_K = \int_{\Gamma_j} \frac{1}{r_K^2}\mathbf{N}_q(\mathbf{x})dr_K = \int_{\Gamma_j} \frac{1}{r_K^2}[\mathbf{N}_q(\mathbf{x}) - N_{q0} - N'_{q0}r_K)dr_K$$

$$+ N_{q0}\int_{\Gamma_j} \frac{1}{r_K^2}\mathbf{N}_q(\mathbf{x})dr_K + N'_{q0}\int_{\Gamma_j} \frac{1}{r_K}\mathbf{N}_q(\mathbf{x})dr_K . \qquad (5.83)$$

The first integral in (5.83) is regular and the other two can be easily evaluated analytically.

5.7 Evaluation of Domain Integrals

It is noted that the boundary integral equation (5.36) contains a domain integral due to the existence of body forces and body charges. The simplest way of computing the domain integral in Equation (5.36) is by subdividing the region into a series of internal cells, on each of which a numerical integration scheme such as Gauss quadrature can be applied. The use of cells to evaluate domain integrals implies an internal discretization, which considerably increases the amount of data needed to run the corresponding program. Hence the method is an obvious disadvantage and diminishes the elegance and computational efficiency of BEM which relies on the transformation of domain integrals into boundary ones. This can be overcome using the dual reciprocity method [24, 25]. The method may, theoretically, be used with any type of fundamental solution and does not need internal cells. The discussion in this section follows the developments in [17, 26].

5.7.1 Dual Reciprocity Formulation

The dual reciprocity formulation is derived by weighting the inhomogeneous differential equation [26],

$$\Xi_{JK}\hat{U}^l_{KN} + b^l_{JN} = 0 , \qquad (5.84)$$

with the fundamental solution u^*_{MJ} leading to

$$\int_V u^*_{MJ} b^l_{JN} dV = C_{MJ} \hat{U}^l_{JN} + \int_S (t^*_{MJ} \hat{U}^l_{JN} - u^*_{MJ} \hat{T}^l_{JN}) dS , \qquad (5.85)$$

where $\Xi_{JK} = E_{iJKm} \partial_i \partial m$ is the elliptic operator of piezoelectricity, C_{MJ} is defined in Equation (5.33), u^*_{MJ} and t^*_{MJ} are given in Equation (5.34), and $\hat{T}^l_{JN} = E_{iJKq} \hat{U}^l_{KN} n_i$ is the corresponding traction field. Consequently, the dual reciprocity method requires the use of a series of particular solutions $\hat{U}^l_{KM}$, where the number of $\hat{U}^l_{KM}$ is equal to the total number of nodes in the problem with collocation points. If there are W boundary nodes and L internal nodes, there will be $W + L$ values of $\hat{U}^l_{KM}$. Now the source term $\tilde{f}_J$ is approximated by a series of tensor functions b^l_{JN} and unknown coefficients α^l_N,

$$\tilde{f}_J \approx \sum_{l=1}^{W+L} b^l_{JN} \alpha^l_N , \qquad (5.86)$$

where the approximation functions b^l_{JN} are linked to the particular $\hat{U}^l_{KN}$ through the relationship (5.84). Substituting the approximation (5.86) into Equation (5.36) and making use of Equation (5.85), the following equation is obtained

$$\begin{aligned}
\mathbf{C}(\hat{\mathbf{x}}_i)\mathbf{U}(\hat{\mathbf{x}}_i) = &\int_S \mathbf{U}^*(\hat{\mathbf{x}}_i, \mathbf{x})\mathbf{T}(\mathbf{x}) dS(\mathbf{x}) - \int_S \mathbf{T}^*(\hat{\mathbf{x}}_i, \mathbf{x})\mathbf{U}(\mathbf{x}) dS(\mathbf{x}) \\
&+ \sum_{j=1}^{W+L} \left(\mathbf{C}(\hat{\mathbf{x}}_i)\hat{\mathbf{U}}^j - \int_S \mathbf{U}^*(\hat{\mathbf{x}}_i, \mathbf{x})\hat{\mathbf{T}}^j dS(\mathbf{x}) \right. \\
&\left. + \int_S \mathbf{T}^*(\hat{\mathbf{x}}_i, \mathbf{x})\hat{\mathbf{U}}^j dS(\mathbf{x}) \right) \alpha^j .
\end{aligned} \qquad (5.87)$$

After discretization and integrating over each element using interpolation equation (5.37), Equation (5.87) becomes

$$\mathbf{C}^i \mathbf{U}^i + \sum_{j=1}^{K} \overline{\mathbf{H}}_{ij} \mathbf{U}^j = \sum_{j=1}^{K} \mathbf{G}_{ij} \mathbf{T}^j + \sum_{k=1}^{W+L} \left(\mathbf{C}^i \hat{\mathbf{U}}^{ik} + \sum_{j=1}^{K} \overline{\mathbf{H}}_{ij} \hat{\mathbf{U}}^{jk} - \sum_{j=1}^{K} \mathbf{G}_{ij} \hat{\mathbf{T}}^{jk} \right) \alpha^k , \qquad (5.88)$$

where $\overline{\mathbf{H}}_{ij}$ are $\mathbf{G}_{ij}$ defined in Equation (5.41), i is a source node, j a boundary element, and k the collocation points of the dual reciprocity scheme.

The contribution for all "i" nodes can be written together in matrix form as

$$\mathbf{HU} = \mathbf{GT} + (\mathbf{H}\hat{\mathbf{U}} - \mathbf{G}\hat{\mathbf{T}})\alpha . \qquad (5.89)$$

5.7.2 Coefficients α

The coefficients α can be calculated by point allocation, that is, by forcing approximation (5.86) to be exact at $W + L$ collocation points. This leads to the following system equations:

$$
\begin{bmatrix}
\mathbf{b}_1(\mathbf{x}_1) & \mathbf{b}_2(\mathbf{x}_1) & \cdots & \mathbf{b}_P(\mathbf{x}_1) \\
\mathbf{b}_1(\mathbf{x}_2) & \mathbf{b}_2(\mathbf{x}_2) & \cdots & \mathbf{b}_P(\mathbf{x}_2) \\
\vdots & \vdots & \ddots & \vdots \\
\mathbf{b}_1(\mathbf{x}_P) & \mathbf{b}_2(\mathbf{x}_P) & \cdots & \mathbf{b}_P(\mathbf{x}_P)
\end{bmatrix}
\begin{bmatrix}
\alpha_1 \\ \alpha_2 \\ \vdots \\ \alpha_P
\end{bmatrix}
=
\begin{bmatrix}
\tilde{\mathbf{f}}_1 \\ \tilde{\mathbf{f}}_2 \\ \vdots \\ \tilde{\mathbf{f}}_P
\end{bmatrix},
\qquad (5.90)
$$

where $P = W + L, \mathbf{b}_i(\mathbf{x}_j) = \mathbf{b}_j(\mathbf{x}_i) = \mathbf{b}(\left|\mathbf{x}_i - \mathbf{x}_j\right|)$. Then, the coefficient vector α can be expressed in terms of nodal values of the generalized body force vector $\tilde{\mathbf{f}}$ as

$$
\begin{bmatrix}
\alpha_1 \\ \alpha_2 \\ \vdots \\ \alpha_P
\end{bmatrix}
=
\begin{bmatrix}
\mathbf{b}_1(\mathbf{x}_1) & \mathbf{b}_2(\mathbf{x}_1) & \cdots & \mathbf{b}_P(\mathbf{x}_1) \\
\mathbf{b}_1(\mathbf{x}_2) & \mathbf{b}_2(\mathbf{x}_2) & \cdots & \mathbf{b}_P(\mathbf{x}_2) \\
\vdots & \vdots & \ddots & \vdots \\
\mathbf{b}_1(\mathbf{x}_P) & \mathbf{b}_2(\mathbf{x}_P) & \cdots & \mathbf{b}_P(\mathbf{x}_P)
\end{bmatrix}^{-1}
\begin{bmatrix}
\tilde{\mathbf{f}}_1 \\ \tilde{\mathbf{f}}_2 \\ \vdots \\ \tilde{\mathbf{f}}_P
\end{bmatrix}.
\qquad (5.91)
$$

5.7.3 Particular Solutions $\hat{U}$ and Approximation Functions b

Due to the complexity of the governing differential equation it is very difficult to derive particular solutions in closed form. Several methods for approximate calculation of particular solutions have been presented in the literature [27–29]. Using a simple approach as presented in [17], instead of choosing radial basis functions for the approximation functions $\mathbf{b}$ and solving the differential equation (5.84) to obtain the particular solutions, we can simply choose $\hat{U}^l_{KN} = \hat{U}^l_{KN}(r_l)$ and calculate the corresponding body force term b^l_{JN} using Equation (5.84). Using this approach, Kogl and Gaul [26] selected the following particular solution,

$$
\hat{U}^l_{KN} = \delta_{KN}(r^2 + r^3),
\qquad (5.92)
$$

which yields the derivatives

$$
\hat{U}^l_{KN,q} = \delta_{KN}(2r + 3r^2)r_{,q}
\qquad (5.93)
$$

$$
\hat{U}^l_{KN,qi} = \delta_{KN}[(2 + 3r)\delta_{qi} + 3rr_{,q}r_{,i}].
\qquad (5.94)
$$

Then b^l_{JN} can be obtained through use of Equations (5.84) and (5.94).

5.8 Multidomain Problems

The discussion in the previous sections of this chapter is suitable for problems with a single solution domain only, as the fundamental solution used assumes that material properties do not change inside the domain being analyzed. If the solution domain is made up piecewise of different materials the problem can be solved by multidomain BEM [13, 18]. The basic idea is to consider a number of regions that are connected to each other, much like pieces of a puzzle. Each region is treated in the same way as discussed previously, but can now be assigned different material properties. Because at the interfaces between the regions both $\mathbf{U}$ and $\mathbf{T}$ are not known, the number of unknowns is increased and additional equations are required to solve the problem. These equations can be obtained from the conditions of equilibrium and compatibility at the region interfaces. The multidomain approach presented in [13] can be adopted in implementation of the method. It is based on the division of the origin domain into homogeneous subregions (see Figure 5.8) so that Equation (5.43) still holds for each single subdomain, and we can write

$$\mathbf{H}^{(i)}\mathbf{U}^{(i)} - \mathbf{G}^{(i)}\mathbf{T}^{(i)} = \mathbf{B}^{(i)}, \qquad (i = 1, 2, \ldots, J), \qquad (5.95)$$

where J is the number of subregions and the superscript (i) indicates quantities associated with the ith subregion. To obtain the solution it is necessary to restore domain unity by enforcing generalized displacement and traction continuity conditions along the interfaces between contiguous subdomains. We now introduce a partition of the linear algebraic system

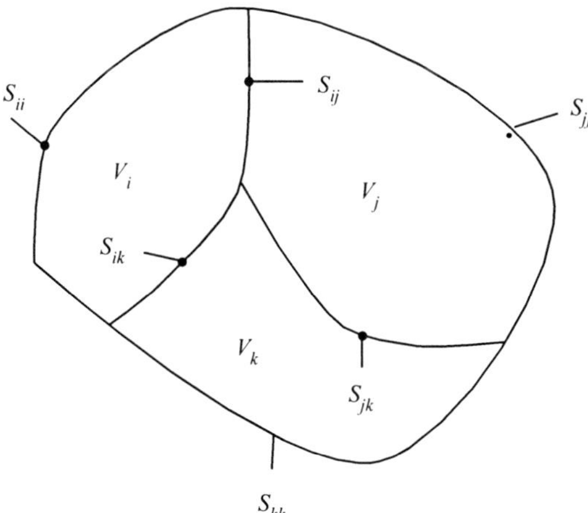

Fig. 5.8 Multidomain configuration.

given by Equation (5.95) in such a way that the generic vector can be written as [13]

$$y^{(i)} = \left\{ y_{S_{il}}^{(i)} \quad \cdots \quad y_{S_{iJ}}^{(i)} \right\}^{T}, \tag{5.96}$$

where the vector $y_{S_{iJ}}^{(i)}$ collects the components of $y^{(i)}$ associated with the nodes belonging to the interface S_{ij} between the ith and jth subdomain, with the convention that S_{ii} stands for the external boundary of the ith subdomain (see Figure 5.8). Based on this arrangement, the interface compatibility and equilibrium conditions are given by [13]

$$\mathbf{U}_{S_{ij}}^{(i)} = \mathbf{U}_{S_{ij}}^{(j)}, \qquad \mathbf{T}_{S_{ij}}^{(i)} = -\mathbf{T}_{S_{ij}}^{(j)}, \qquad (i = 1, \ldots, J-1; j = i+1, \ldots, J). \tag{5.97}$$

It should be noted that if the ith and jth subdomain have no common boundary, $y_{S_{ij}}^{(i)}$ is a zero-order vector, and Equation (5.97) is no longer valid. The system of Equation(5.95) and the interface continuity conditions (5.97) provide a set of relationships that, together with the external boundary conditions, allows derivation of the electroelastic solution in terms of generalized displacement and traction on the boundary of each subdomain. It should be mentioned that the multidomain approach described here is suitable for modeling general fracture problems in piezoelectric media [13].

5.9 Numerical Examples

To illustrate the application of the element model described above, three examples are presented. The first example deals with a piezoelectric column subjected to tension at its two ends; the second treats an infinite piezoelectric solid with a horizontal finite crack; and the third illustrates the behavior of crack-tip fields in a skew-cracked rectangular panel.

5.9.1 A Piezoelectric Column Under Uniaxial Tension [6]

In this example, a piezoelectric prism under simple extension is considered (see Figure 5.9). The size of the prism is $2a \times 2a \times 2b$. The corresponding boundary conditions are given by

$$T_{zz} = p, \qquad T_{xz} = T_{yz} = D_z = 0, \quad \text{for} \quad z = \pm b,$$
$$\phi = 0, \quad \text{for} \quad z = 0$$
$$T_{xx} = T_{xz} = D_x = 0, \quad \text{for} \quad x = \pm a$$
$$T_{yy} = T_{yz} = D_y = 0, \quad \text{for} \quad y = \pm a.$$

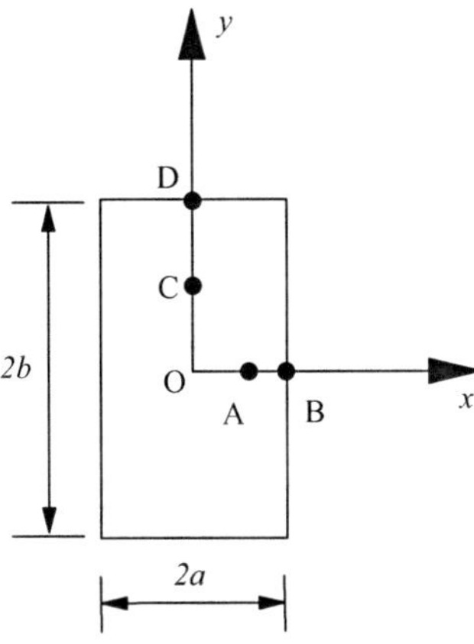

Fig. 5.9 Geometry of the piezoelectric prism.

In the calculation, $2a = 3$ m, $2b = 10$ m, and $p = 100$ Nm^2 are assumed and 32 elements are used [6]. The material considered is PZT$-$4 whose material parameters are

$$c_{11} = 13.9 \times 10^{10} \text{ NM}^{-2},$$

$$c_{12} = 7.78 \times 10^{10} \text{ NM}^{-2},$$

$$c_{13} = 7.43 \times 10^{10} \text{ NM}^{-2}$$

$$c_{33} = 11.5 \times 10^{10} \text{ NM}^{-2},$$

$$c_{44} = 2.56 \times 10^{10} \text{ NM}^{-2},$$

$$e_{15} = 12.7 \text{ Cm}^{-2} \qquad e_{31} = -5.2 \text{ Cm}^{-2},$$

$$e_{33} = 15.1 \text{ Cm}^{-2}, \qquad \varepsilon_{11} = 730\varepsilon_0, \qquad \varepsilon_{33} = 635\varepsilon_0,$$

where $\varepsilon_0 = 8.854 \times 10^{-12}$ C^2/Nm^2. Table 5.1 lists the displacements and electric potential at points A(2, 0), B(3, 0), C(0, 5), and D(0, 10) using BEM, and comparison is made with analytical results. It is evident that the BEM results are in good agreement with the analytical results even when only six boundary elements are used in the calculation [6].

Table 5.1 $u_1, u_2,$ and ϕ of BEM results and comparison with exact solution [6]

Point		A(2, 0)	B(3, 0)	C(0, 5)	D(0, 10)
BEM	$u_1 \times 10^{10}$(m)	-0.9665	-1.4502	0	0
	$u_2 \times 10^{9}$(m)	0	0	0.4997	1.0004
	ϕ(V)	0	0	0.6879	1.3762
Exact [6]	$u_1 \times 10^{10}$(m)	-0.9672	-1.4508	0	0
	$u_2 \times 10^{9}$(m)	0	0	0.5006	1.0011
	ϕ(V)	0	0	0.6888	1.3775

5.9.2 A Horizontal Finite Crack in an Infinite Piezoelectric Medium

The second example is a finite horizontal crack along the x-direction in an infinite PZT-4 medium under a uniform far-field stress or electric displacement. To effectively evaluate crack-tip fields, Pan [11] constructed the following crack-tip element with its tip at $\zeta = -1$, where ζ is the intrinsic coordinate in a quarter-point element defined in Figure 5.10,

$$\triangle \mathbf{U} = \sum_{k=1}^{3} \Phi_k \triangle \mathbf{U}^k \,, \tag{5.98}$$

where the superscript $k(k = 1, 2, 3)$ denotes the the generalized relative crack displacement (GRCD) at nodes $\zeta = -2/3, 0, 2/3$, respectively. ζ is the boundary natural coordinate in a quarter-point element defined in Figure 5.10. The quarter-point quadratic crack-tip element is obtained by setting [21]

$$\zeta = 2\sqrt{\frac{r}{L}} - 1 \,, \tag{5.99}$$

where r is the distance from a field point to the crack-tip, and L the element length (see Figure 5.10). In the element, the collocation points NC1, NC2, and NC3 for the quarter-point element are located at $\zeta_1 = -3/4$, $\zeta_2 = 0$, and $\zeta_3 = 3/4$, respectively (see Figure 5.10). In such a case, the distance r from the collocation nodes of the quarter-point element to the crack-tip follows from Equation (5.99)

$$r_1 = \frac{L}{64} \quad \text{at NC1,}$$

$$r_2 = \frac{L}{4} \quad \text{at NC2,}$$

$$r_3 = \frac{49L}{64} \quad \text{at NC3.} \tag{5.100}$$

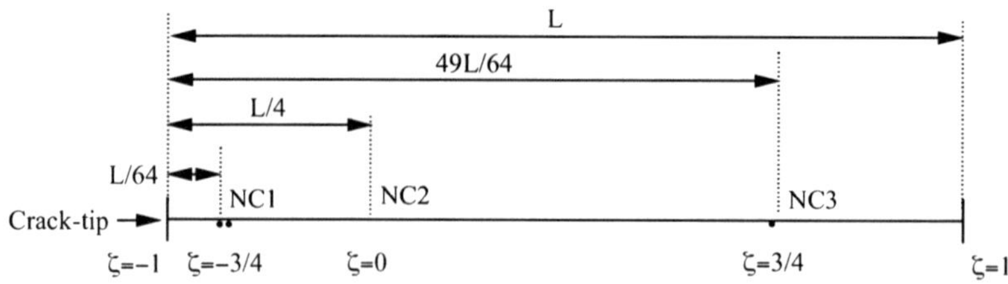

Fig. 5.10 Configuration of a quarter-point element.

Making use of the boundary natural coordinate ζ, the shape functions Φ_k in Equation (5.98) can be defined by [11]

$$\Phi_1 = \frac{3\sqrt{3}}{8}\sqrt{\zeta+1}[-5+18(\zeta+1)-9(\zeta+1)^2],$$

$$\Phi_2 = \frac{1}{4}\sqrt{\zeta+1}[5-8(\zeta+1)+3(\zeta+1)^2],$$

$$\Phi_3 = \frac{3\sqrt{3}}{8\sqrt{5}}\sqrt{\zeta+1}[1-4(\zeta+1)+3(\zeta+1)^2].\tag{5.101}$$

For calculation of the generalized stress intensity factors (GSIF), Pan [11] employed the extrapolation method of the GRCD, which requires an analytical relation between the generalized displacement and the GSIF. This relation can be expressed as

$$\triangle\mathbf{U}(r) = 2\sqrt{\frac{2r}{\pi}}(i\mathbf{AB}^{-1})\mathbf{K},\tag{5.102}$$

where the matrices $\mathbf{A}$ and $\mathbf{B}$ are matrices associated with material proper ties [2], and $\mathbf{K}$ is defined by [21],

$$\mathbf{K} = \left\{\begin{matrix} K_{II} \\ K_I \\ K_{IV} \end{matrix}\right\} = 2\sqrt{\frac{2\pi}{L}}(\mathrm{Re}[\mathbf{B}])^{-1}\left\{\begin{matrix} \triangle u_1^{NC1} \\ \triangle u_2^{NC1} \\ \triangle\phi^{NC1} \end{matrix}\right\}.\tag{5.103}$$

In this example, Pan [11] used 20 discontinuous quadratic elements described above to discretize the crack surface which has a length of $2a$ (=1m). Tables 5.2 and 5.3 list the GRCD caused by a far-field stress $T_{yy}(=1 \text{ N/m}^2)$ and a far-field electric displacement $D_y(=1 \text{ C/m}^2)$, and comparison is made with analytical results. It is obvious that a far-field stress induces a nonzero

Table 5.2 GRCD caused by a far-field $T_{yy}(= 1 \text{ N/m}^2)$[11]

	$\triangle u_y(10^{-12}\text{ m})$		$\triangle\phi(10^{-1}\text{ V})$	
x(m)	BEM	Analytical	BEM	Analytical
0.492	0.032	0.032	0.040	0.040
0.425	0.094	0.093	0.116	0.116
0.358	0.124	0.124	0.154	0.154
0.292	0.144	0.144	0.179	0.179
0.225	0.158	0.158	0.197	0.197
0.158	0.168	0.168	0.210	0.210
0.092	0.174	0.174	0.217	0.217
0.025	0.177	0.177	0.221	0.221

Table 5.3 GRCD caused by a far-field $D_y (= 1 \ \mathrm{C/m^2})$[11]

x(m)	$\triangle\phi(10^8 \ \mathrm{V})$		$\triangle u_y(10^{-1} \ \mathrm{m})$	
	BEM	Analytical	BEM	Analytical
0.492	0.161	0.160	0.040	0.040
0.425	0.466	0.465	0.116	0.116
0.358	0.616	0.616	0.154	0.154
0.292	0.717	0.717	0.179	0.179
0.225	0.789	0.789	0.197	0.197
0.158	0.838	0.838	0.210	0.210
0.092	0.868	0.868	0.217	0.217
0.025	0.882	0.882	0.221	0.221

$\triangle\phi$ even though the corresponding K_{IV} is zero. Similarly, a far-field electric displacement can induce a nonzero $\triangle u_y$.

5.9.3 A Rectangular Piezoelectric Solid with a Central Inclined Crack [13]

The third example is a rectangular piezoelectric solid with a central crack ($a = 0.1$ m) inclined $\theta = 45°$ with respect to the positive x-direction. The ratios of crack length to width and of height to width are $a/w = 0.2$ and $h/w = 2$, respectively (see Figure 5.11). The analysis is carried out for the

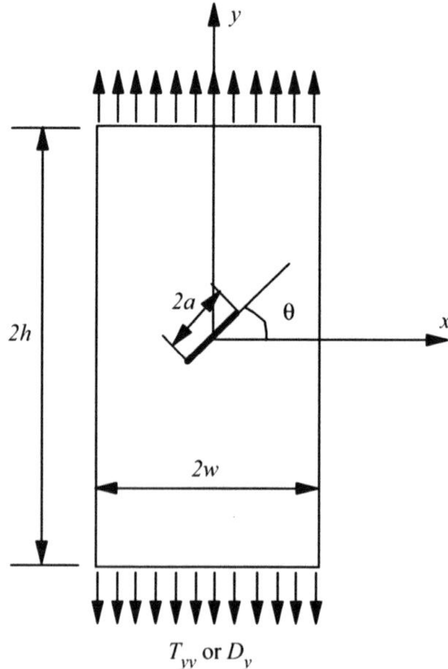

Fig. 5.11 Finite rectangular solid with an inclined crack under uniform tension or electric displacement in the y-direction.

Table 5.4 GSIF for cracked rectangle loaded by σ_y or D_y

	Ref	$K_I/T_{yy}\sqrt{\pi a}$	$K_{II}/T_{yy}\sqrt{\pi a}$	$K_{IV}/D^*\sqrt{\pi a}$
Loaded by σ_y	[11]	0.5303	0.5151	-2.97×10^{-12}
	[11]	0.5292	0.5163	-2.79×10^{-12}
	Ref	$K_I/T^*\sqrt{\pi a}$	$K_{II}/T^*\sqrt{\pi a}$	$K_{IV}/D_y\sqrt{\pi a}$
Loaded by D_y	[11]	-1.42×10^6	1.69×10^5	-0.7278
	[13]	-1.44×10^6	1.64×10^5	-0.7283

rectangle loaded by a uniform tension and electric displacement applied in the y-direction. Table 5.4 lists the normalized GSIF for the two loading conditions considered and the results obtained are compared with those given by Pan [11], who used 10 discontinuous quadratic elements on the crack surfaces and 32 quadratic elements on the outside boundaries in his analysis. Tables 5.5 and 5.6 give the corresponding GRCD. In Table 5.4, D^* is a nominal electric displacement expressed in C/m^2 with its amplitude equal to that of T_{yy} expressed in N/m^2, and T^* is a nominal stress expressed in N/m^2 with its amplitude equal to that of Dy expressed in C/m^2.

It can be seen from Table 5.4 that an electric load of $D_y(= 1\ C/m^2)$ can produce very large mechanical stress intensity factors. Conversely, the electric displacement intensity factor due to mechanical loads is usually negligible. This phenomenon indicates clearly that crack initiation criteria based on a single stress intensity factor (SIF) cannot be simply extended to the piezoelectric case. It is also found from Tables 5.5 and 5.6 that a mechanical load can induce relative crack electric potential and, conversely, an electric load can give rise to relative crack displacement.

Table 5.5 GRCD for cracked rectangle caused by a far-field $T_{yy}(= 1\ N/m^2)$ [13]

$x = y\ (10^{-1}\ m)$	$\triangle u_x(10^{-13}\ m)$	$\triangle u_y(10^{-11}\ m)$	$\triangle\phi(10^{-2}\ V)$
0.477	0.256	-0.190	0.232
0.424	0.279	-0.206	0.252
0.371	0.299	-0.219	0.268
0.318	0.315	-0.230	0.281
0.265	0.328	-0.239	0.292
0.212	0.339	-0.246	0.300
0.159	0.347	-0.251	0.307
0.106	0.352	-0.255	0.312
0.053	0.356	-0.257	0.314

Table 5.6 GRCD for cracked rectangle caused by a far-field $D_y(= 1 \text{ C/m}^2)$ [13]

$x = y\,(10^{-1}\text{ m})$	$\triangle u_x(10^{-5}\text{ m})$	$\triangle u_y(10^{-2}\text{ m})$	$\triangle\phi(10^8\text{ V})$
0.477	0.353	0.232	0.093
0.424	0.371	0.252	0.100
0.371	0.385	0.268	0.107
0.318	0.391	0.281	0.112
0.265	0.406	0.292	0.116
0.212	0.410	0.300	0.120
0.159	0.416	0.307	0.123
0.106	0.420	0.312	0.125
0.053	0.431	0.314	0.126

References

[1] Lee JS, Jiang LZ (1994) A boundary integral formulation and 2D fundamental solution for piezoelectric media. *Mech Res Comm* 21: 47–54

[2] Qin QH (2001) *Fracture Mechanics of Piezoelectric Materials.* WIT Press, Southampton

[3] Qin QH (2004) Material Properties of Piezoelectric Composites by BEM and Homogenization Method. *Composite Struc* 66: 295–299

[4] Qin QH (2007) *Green's Function and Boundary Elements in Multifield Materials.* Elsevier, Oxford

[5] Lu P, Mahrenholtz O (1994) A variational boundary element formulation for piezoelectricity. *Mech Res Comm* 21: 605–611

[6] Ding HJ, Wang GP, Chen WQ (1998) Boundary integral formulation and 2D fundamental solutions for piezoelectric media. *Comp Meth Appl Mech Eng* 158: 65–80

[7] Rajapakse RK (1997) Boundary element methods for piezoelectric solids. *Procs of SPIE, Mathematics and Control in Smart Structures* 3039: 418–428

[8] Xu XL, Rajapakse RKND (1998) Boundary element analysis of piezoelectric solids with defects. *Composites Part B: Eng* 29: 655–669

[9] Rajapakse RKND, Xu XL (2001) Boundary element modeling of cracks in piezoelectric solids. *Eng Anal Boun Elem* 25: 771–781

[10] Liu YJ, Fan H (2001) On the conventional boundary integral formulation for piezoelectric solids with defects or of thin shapes. *Eng Anal Boun Elem* 25: 77–91

[11] Pan E (1999) A BEM analysis of fracture mechanics in 2D anisotropic piezoelectric solids. *Eng Anal Boun Elem* 23: 67–76

[12] Denda M, Lua J (1999) Development of the boundary element method for 2D piezoelectricity. *Composites: Part B* 30: 699–707

[13] Davi G, Milazzo A (2001) Multidomain boundary integral formulation for piezoelectric materials fracture mechanics. *Int J Solids Struct* 38: 7065–7078

[14] Groh U, Kuna M (2005) Efficient boundary element analysis of cracks in 2D piezoelectric structures. *Int J Solids Struct* 42: 2399–2416

[15] Khutoryaansky N, Sosa H, Zu WH (1995) Approximate Green's functions and a boundary element method for electroelastic analysis of active materials. *Comput Struct* 66: 289–299

[16] Chen T, Lin FZ (1995) Boundary integral formulations for three–dimensional anisotropic piezoelectric solids. *Comput Mech* 15: 485–496

[17] Schclar NA (1994) *Anisotropic Analysis Using Boundary Elements.* Computational Mechanics, Southampton

[18] Beer G (2001) *Programming the Boundary Element Method.* John Wiley & Sons, Chichester

[19] Stroud AH, Secrest D (1966) *Gaussian Quadrature Formulas.* Prentice-Hall, New York

[20] Diego N (2002) *Thermoelastic Fracture Mechanics Using Boundary Elements.* WIT Press, Southampton

[21] Garcia-Sanchez F, Saez A, Dominguez J (2005) Anisotropic and piezoelectric materials fracture analysis by BEM. *Comput Struct* 83: 804–820

[22] Lachat JC, Watson JO (1976) Effective numerical treatment of boundary integral equations: A formulation for three-dimensional elastostatics. *Int J Num Meth Eng* 10: 991–1005

[23] Rizzo FJ, Shippy DJ (1977) An advanced boundary integral equation method for three-dimensional thermoelasticity. *Int J Num Meth Eng* 11: 1753–1768

[24] Nardini D, Brebbia CA (1982) A new approach to free vibration analysis using boundary elements. In: *Boundary Element Methods in Engineering.* Springer, Berlin

[25] Partridge PW, Brebbia CA, Wrobel LC (1992) *The Dual Reciprocity Boundary Element Method.* Computational Mechanics, Southampton

[26] Kogl M, Gaul L (2000) A boundary element method for transient piezoelectric analysis. *Eng Anal Boun Elements* 24: 591–598

[27] Atkinson KE (1985) The numerical evaluation of particular solutions for Poisson's equation. *IMAJ Numer Anal* 5: 319–338

[28] Golberg MA (1995) The numerical evaluation of particular solutions in the BEM-a review. *Bound Elem Commun* 6: 99–106

[29] Grundemann H (1989) A general procedure transferring domain integrals onto boundary integrals in BEM. *Eng Anal Boun Elem* 6: 214–222

Chapter 6
Waves in Strained/Polarized Media

Olivian Simionescu-Panait

Dedicated to the memory of my Professor Eugen Soós

6.1 Introduction

The problems related to electroelastic materials subject to incremental fields superposed on initial mechanical and electric fields have attracted considerable attention recently, due their complexity and to multiple applications (see [3, 5, 10, 31–33]). The basic equations of the theory of piezoelectric bodies subject to infinitesimal deformations and fields superposed on large initial mechanical and electric fields were described by Eringen and Maugin in their well-known monograph [6]. An useful development of the equations of electromagnetism in material continua may be found in [30]. As regards the description of mechanics of a continuum medium we refer to the classical textbook of Malvern [12].

The chapter is divided into four parts. The first one presents the fundamental equations of incremental fields superposed on large static deformation and electric fields. Following the paper [2], we derive the balance equations, constitutive equations, and boundary conditions for this problem, using the updated Lagrangean description. We analyse the important special case of homogeneous initial state and nonpolarisable environment. In this framework we obtain the dynamic and static energy balance, we present the static and dynamic local stability criteria, we derive the conditions of plane harmonic wave propagation, and we define the characteristic surfaces.

In the second part we analyse the propagation conditions of plane harmonic waves in various types of crystals subject to initial electromechanical fields. Following the papers [15–18, 27] we derive the propagation conditions for isotropic solids, cubic crystals, and 6 mm-type crystals, we show the electrostrictive effect, we define and analyse the generalised anisotropy

Olivian Simionescu-Panait
Department of Geometry, Bucharest University, Str. Academiei 14, Bucharest 010014, Romania, e-mail: o_simionescu@yahoo.com

J. Yang (ed.), *Special Topics in the Theory of Piezoelectricity*, DOI: 10.1007/978-0-387-89498-0_6, 169
© Springer Science + Business Media, LLC 2009

factor and the coupling coefficients, and we demonstrate the influence of initial fields on the shape of slowness surfaces. From a mathematical point of view, the problem reduces to the spectral properties of the acoustic tensor, supposed symmetric and positive definite.

The third part presents the problem of attenuated wave propagation in isotropic solids and in cubic crystals subject to initial electromechanical initial fields (see papers [19–24]). Here the acoustic tensor is no longer symmetric and positive definite in the general case, but we find particular forms of the initial electric field in order to solve the spectral problem.

Finally, following the papers [25, 26], we study the coupling conditions for propagation of planar guided waves in a piezoelectric semi-infinite plane subject to initial fields and we obtain the decomposition of the corresponding mechanical and electrical boundary conditions. The mathematical problem is linked to the spectral properties of the acoustic tensor, having differential operators of second-order with complex coefficients, as components.

6.2 Basic Equations

In this section we present the fundamental equations describing the behaviour of a nonmagnetisable elastic dielectric material, which conducts neither heat nor electricity. Thus, we assume the quasi-electrostatic approximation of field equations and of boundary conditions, as described in monograph [6]. We follow here the paper [2], with some variations. A detailed analysis of the problem may be found there.

6.2.1 The Quasi-Electrostatic Approximation of Balance Equations

We assume the material to be an hyperelastic dielectric, which is nonmagnetisable and conducts neither heat nor electricity. We use the quasi-electrostatic approximation of the equations of balance. Let B_R be the reference configuration, in which at time $t = 0$ the body is undeformed and free of all fields and B_t the present (current) configuration. Let V_R and V_t be the geometric domains associated with B_R and B_t. The material position (reported to B_R) and the spatial position (reported to B_t) of an arbitrary particle X of the body are denoted by $\mathbf{X}$ (resp. $\mathbf{x}$). The first case is known as *Lagrangean description*, and the second as *Eulerian description* of the deformation.

Let ρ_R and ρ be the mass densities of the body, referred to the configurations B_R and B_t. If J is the determinant of the deformation gradient $\mathbf{F}$ from B_R to B_t, then we obtain the *mass conservation law* (in Lagrangean description) in the form:

$$\rho J = \rho_R. \qquad (6.1)$$

Similarly, let q_R and q be the volumetric electric charge densities, reported to B_R and B_t. Then, we determine the *charge conservation law* (in Lagrangean description) as

$$q J = q_R. \qquad (6.2)$$

Let $\mathbf{T}$ be the electromechanical stress tensor of Cauchy type and $\mathbf{S}$ its symmetric part. We suppose that the latter tensor is derived from Helmholtz free energy as in nonlinear elasticity, but now depending on electromechanical variables, as follows. Let $\mathbf{E}$, $\mathbf{P}$, and $\mathbf{D}$ be the electric field, the electric polarization and the electric displacement vectors, respectively. Then, we suppose the following relations (see [6]),

$$\mathbf{T} = \mathbf{S} - \mathbf{P} \otimes \mathbf{E}, \qquad \mathbf{D} = \mathbf{E} + \mathbf{P}. \qquad (6.3)$$

The balance equations in the quasi-electrostatic approximation have the form (see [6]):

$$\rho \dot{\mathbf{v}} = \mathrm{div}_{\mathbf{x}} \mathbf{T} + \rho \mathbf{f} + q \mathbf{E} + (\mathbf{P} \cdot \nabla_{\mathbf{x}}) \mathbf{E},$$
$$\mathrm{div}_{\mathbf{x}} \mathbf{D} = q, \qquad \mathrm{rot}_{\mathbf{x}} \mathbf{E} = 0. \qquad (6.4)$$

Here $\mathbf{v}$ is the velocity field vector, $\mathbf{f}$ represents the mechanical body force, and q is the volumetric charge density. A superposed dot is used to denote the material time derivative. In Equations (6.4) we used the Eulerian description, the fields involved depending on spatial coordinate $\mathbf{x}$ and on time t. Here one finds the electrostatic form of Coulomb and Faraday laws. The first differential relation is derived from the momentum balance.

Furthermore, the jump conditions and the electromechanical surface stress vector $\mathbf{t_n}$, defined on the boundary ∂V_t, are given by:

$$\mathbf{n} \cdot [\mathbf{D}] = w, \qquad \mathbf{n} \times [\mathbf{E}] = \mathbf{0},$$
$$\mathbf{t_n} = \mathbf{T} \mathbf{n} = (\mathbf{S} - \mathbf{P} \otimes \mathbf{E}) \mathbf{n} \quad \text{on} \quad \partial V_t. \qquad (6.5)$$

Here w represents the surface charge density, $\mathbf{n}$ the exterior normal unit vector to the boundary, and $[\phi] = \phi^+ - \phi^-$ is the jump of the field ϕ across the boundary. From now on, we denote simply by ϕ the inside limit value ϕ^-.

The previous field equations and boundary conditions may be expressed in Lagrangean description using electromechanical stress tensors of Piola–Kirchhoff type. They are related to the reference configuration B_R and are defined by the relations:

$$\boldsymbol{\Theta} = J \mathbf{F}^{-1} \mathbf{S}, \qquad \boldsymbol{\Pi} = J \mathbf{F}^{-1} \mathbf{S} \mathbf{F}^{-T}, \qquad \boldsymbol{\Theta} = \boldsymbol{\Pi} \mathbf{F}^{T}. \qquad (6.6)$$

Here $\boldsymbol{\Theta}$ and $\boldsymbol{\Pi}$ are the nominal electromechanical stress tensors of Piola–Kirchhoff type related to the symmetric part $\mathbf{S}$ of the electromechanical stress tensor $\mathbf{T}$ of Cauchy type. It is obvious that $\boldsymbol{\Pi}$ is a symmetric tensor.

Similarly, we introduce the Lagrangean version of vectors $\mathbf{P}$, $\mathbf{E}$, and $\mathbf{D}$, as follows:

$$\boldsymbol{\mathcal{P}} = J\mathbf{F}^{-1}\mathbf{P}, \qquad \boldsymbol{\mathcal{E}} = \mathbf{F}^{T}\mathbf{E},$$
$$\boldsymbol{\mathcal{D}} = J\mathbf{F}^{-1}\mathbf{D} = J\mathbf{F}^{-1}\mathbf{E} + \boldsymbol{\mathcal{P}}. \tag{6.7}$$

Finally, using the first relation (6.3) we obtain the associated material version of tensor $\mathbf{T}$:

$$\boldsymbol{\mathcal{T}} = J\mathbf{F}^{-1}\mathbf{T} = \boldsymbol{\Theta} - \boldsymbol{\mathcal{P}} \otimes \mathbf{E}. \tag{6.8}$$

Consequently, we derive from the momentum balance the Lagrangean form of the balance equations:

$$\rho_R\ddot{\mathbf{u}} = \mathrm{div}_{\mathbf{X}}\boldsymbol{\mathcal{T}} + \rho_R\mathbf{f} + q_R\mathbf{E} + (\boldsymbol{\mathcal{P}} \cdot \nabla_{\mathbf{X}})\mathbf{E}, \qquad \mathrm{div}_{\mathbf{X}}\boldsymbol{\mathcal{D}} = q_R, \qquad \mathrm{rot}_{\mathbf{X}}\boldsymbol{\mathcal{E}} = \mathbf{0}. \tag{6.9}$$

Here $\mathbf{u}$ is the displacement vector from B_R to B_t. The differential operators are associated with the reference configuration, and the various fields involved depend on the material coordinate $\mathbf{X}$ and on time t.

Furthermore, the jump conditions and the electromechanical surface stress vector $\mathbf{t_N}$, reported to the material configuration, are given by

$$\mathbf{N} \cdot [\boldsymbol{\mathcal{D}}] = w_R, \qquad \mathbf{N} \times [\boldsymbol{\mathcal{E}}] = \mathbf{0}, \qquad \mathbf{t_N} = \boldsymbol{\mathcal{T}}\mathbf{N} = (\boldsymbol{\Theta} - \boldsymbol{\mathcal{P}} \otimes \mathbf{E})\mathbf{N} \quad \text{on}\, \partial V_R. \tag{6.10}$$

Here $\mathbf{N}$ is the exterior normal unit vector to the boundary ∂V_R and w_R is surface charge density per material surface area.

The previous balance equations are supplemented by the constitutive equations (see [6]):

$$\boldsymbol{\Pi} = \frac{\partial \mathcal{H}}{\partial \mathbf{G}}, \qquad \boldsymbol{\mathcal{P}} = -\frac{\partial \mathcal{H}}{\partial \boldsymbol{\mathcal{E}}}, \tag{6.11}$$

where

$$\rho_R\psi = \mathcal{H}(\mathbf{G}, \boldsymbol{\mathcal{E}}) \tag{6.12}$$

is the electromechanical Helmholtz free energy, and

$$\mathbf{G} = \frac{1}{2}(\mathbf{C} - \mathbf{1}) = \frac{1}{2}(\mathbf{F}^{T}\mathbf{F} - \mathbf{1}) \tag{6.13}$$

is the Green strain tensor.

6.2.2 Small Deformation and Electric Fields Superposed on Large Static Deformation and Electric Fields

In this part we describe the behaviour of incremental electromechanical fields superposed on large initial electromechanical fields. In [6] one obtains these equations using the field equations corresponding to the spatial (Eulerian) description of the deformation. The perturbed electromechanical surface forces, as well as the surface and volumetric charges, are taken in the perturbed current configuration. In [2], these quantities are referred to the initially deformed static configuration, which is supposed as being known (i.e., the updated Lagrangean description of the deformation). We follow the second approach.

To describe this situation we use three different configurations: the *reference configuration* B_R in which at time $t = 0$ the body is undeformed and free of all fields; and the *initial configuration* $\overset{\circ}{B}$ in which the body is deformed statically and carries the initial fields; and the *present (current) configuration* B_t obtained from $\overset{\circ}{B}$ by applying time-dependent incremental deformations and fields. Let V_R, $\overset{\circ}{V}$, and V_t be the geometric domains associated with B_R, $\overset{\circ}{B}$, and B_t.

In what follows, all the fields related to the initial configuration $\overset{\circ}{B}$ are denoted by a superposed "o". The static deformation from B_R to $\overset{\circ}{B}$ is described by the relation $\mathbf{x} = \chi(\mathbf{X})$; the associated deformation gradient is $\overset{\circ}{\mathbf{F}} = \overset{\circ}{\mathbf{F}}(\mathbf{X})$ and $\overset{\circ}{J} = \det \overset{\circ}{\mathbf{F}}(\mathbf{X})$. Thus, we obtain the mass conservation law, (resp., the charge conservation law) in the form:

$$\overset{\circ}{\rho}\,\overset{\circ}{J} = \rho_R, \qquad \overset{\circ}{q}\,\overset{\circ}{J} = q_R. \tag{6.14}$$

Furthermore, due to the relations (6.3) we derive that

$$\overset{\circ}{\mathbf{T}} = \overset{\circ}{\mathbf{S}} - \overset{\circ}{\mathbf{P}} \otimes \overset{\circ}{\mathbf{E}},$$

$$\overset{\circ}{\mathbf{D}} = \overset{\circ}{\mathbf{E}} + \overset{\circ}{\mathbf{P}}. \tag{6.15}$$

According to (6.4), the field equations for these static fields are:

$$\mathrm{div}_{\mathbf{x}}\,\overset{\circ}{\mathbf{T}} + \overset{\circ}{\rho}\overset{\circ}{\mathbf{f}} + \overset{\circ}{q}\overset{\circ}{\mathbf{E}} + (\overset{\circ}{\mathbf{P}} \cdot \nabla_{\mathbf{x}})\,\overset{\circ}{\mathbf{E}} = \mathbf{0},$$

$$\mathrm{div}_{\mathbf{x}}\,\overset{\circ}{\mathbf{D}} = \overset{\circ}{q}, \qquad \mathrm{rot}_{\mathbf{x}}\,\overset{\circ}{\mathbf{E}} = 0. \tag{6.16}$$

Here all the electromechanical fields depend on the variable $\mathbf{x}$ and on time t.

Consequently, the jump conditions and the electromechanical surface stress vector $\overset{\circ}{\mathbf{t_n}}$ at the boundary $\partial \overset{\circ}{V}$ are:

$$\overset{\circ}{\mathbf{n}} \cdot [\overset{\circ}{\mathbf{D}}] = \overset{\circ}{w}, \qquad \overset{\circ}{\mathbf{n}} \times [\overset{\circ}{\mathbf{E}}] = 0,$$

$$\overset{\circ}{\mathbf{t_n}} = \overset{\circ}{\mathbf{T}}\overset{\circ}{\mathbf{n}} = (\overset{\circ}{\mathbf{S}} - \overset{\circ}{\mathbf{P}} \otimes \overset{\circ}{\mathbf{E}})\,\overset{\circ}{\mathbf{n}} \quad \text{on}\, \partial \overset{\circ}{V}. \tag{6.17}$$

Here $\overset{\circ}{\mathbf{n}}$ is the exterior normal unit vector to the boundary $\partial \overset{\circ}{V}$.

The constitutive equations (6.11), giving the electromechanical stress tensor and the electric polarization in the statical configuration $\overset{\circ}{B}$, become:

$$\overset{\circ}{\mathbf{\Pi}} = \frac{\partial \mathcal{H}}{\partial \mathbf{G}}(\overset{\circ}{\mathbf{G}}, \overset{\circ}{\mathcal{E}}), \qquad \overset{\circ}{\mathcal{P}} = -\frac{\partial \mathcal{H}}{\partial \mathcal{E}}(\overset{\circ}{\mathbf{G}}, \overset{\circ}{\mathcal{E}}), \tag{6.18}$$

where

$$\overset{\circ}{\mathbf{\Pi}} = \overset{\circ}{J}\overset{\circ}{\mathbf{F}}{}^{-1}\overset{\circ}{\mathbf{S}}\overset{\circ}{\mathbf{F}}{}^{-T}, \qquad \overset{\circ}{\mathcal{P}} = \overset{\circ}{J}\overset{\circ}{\mathbf{F}}{}^{-1}\overset{\circ}{\mathbf{P}} \tag{6.19}$$

and

$$\overset{\circ}{\mathbf{G}} = \frac{1}{2}(\overset{\circ}{\mathbf{F}}{}^{T}\overset{\circ}{\mathbf{F}} - 1), \qquad \overset{\circ}{\mathcal{E}} = \overset{\circ}{\mathbf{F}}{}^{T}\overset{\circ}{\mathbf{E}}. \tag{6.20}$$

Now, we assume that time-dependent incremental deformations and fields are applied to the body in the initial configuration $\overset{\circ}{B}$, determining their description in the current configuration B_t. Here, all the fields referring to $\overset{\circ}{B}$ as the reference configuration, are denoted by a subscript "o". Let $\mathbf{u}(\mathbf{x}, t)$ be the small displacement from $\overset{\circ}{B}$ to B_t and let $\mathbf{F}_o = \mathbf{F}_o(\mathbf{x}, t)$ be the gradient of deformation from $\overset{\circ}{B}$ to B_t, $\overset{\circ}{B}$ being taken as the reference configuration. We define the gradient of the displacement $\mathbf{u}(\mathbf{x}, t)$ by $\mathbf{H}_o(\mathbf{x}, t)$, and $J_o(\mathbf{x}, t)$ as the determinant of $\mathbf{F}_o(\mathbf{x}, t)$. All the fields involved are regarded as functions of $\mathbf{x}$ and t, when reported to $\overset{\circ}{B}$. For simplicity, we suppress the argument $\mathbf{x}$ in the following notations. Thus, we obtain:

$$\mathbf{F}_o(t) = 1 + \mathbf{H}_o(t), \qquad \mathbf{F}(t) = \mathbf{F}_o(t)\,\overset{\circ}{\mathbf{F}}, \qquad J(t) = J_o(t)\,\overset{\circ}{J}. \tag{6.21}$$

Here $\mathbf{F}(t)$ is the deformation gradient from B_R to B_t and $J(t)$ is its determinant.

Consequently, to obtain the field equations referred to the configuration $\overset{\circ}{B}$, we introduce the following Piola–Kirchhoff-type fields.

$$\mathbf{\Theta}_o(t) = J_o\mathbf{F}_o^{-1}(t)\mathbf{S}(t) = \overset{\circ}{J}{}^{-1}\overset{\circ}{\mathbf{F}}\mathbf{\Pi}(t)\overset{\circ}{\mathbf{F}}{}^{T}\mathbf{F}_o^{T}(t),$$

$$\mathbf{\Pi}_o(t) = J_o\mathbf{F}_o^{-1}\mathbf{S}\mathbf{F}_o^{-T}, \qquad \mathbf{\Theta}_o = \mathbf{\Pi}_o\mathbf{F}_o^{T}, \tag{6.22}$$

$$\boldsymbol{\mathcal{P}}_o(t) = J_o(t)\mathbf{F}_o^{-1}(t)\mathbf{P}(t) = \overset{\circ}{J}{}^{-1}\overset{\circ}{\mathbf{F}}\,\boldsymbol{\mathcal{P}}(t),$$

$$\boldsymbol{\mathcal{E}}_o(t) = \mathbf{F}_o^T(t)\mathbf{E}(t), \tag{6.23}$$

$$\boldsymbol{\mathcal{D}}_o(t) = J_o(t)\mathbf{F}_o^{-1}(t)\mathbf{D}(t) = J_o(t)\mathbf{F}_o^{-1}(t)\mathbf{E}(t) + \boldsymbol{\mathcal{P}}_o(t), \tag{6.24}$$

$$\boldsymbol{\mathcal{T}}_o(t) = J_o(t)\mathbf{F}_o^{-1}(t)\mathbf{T}(t) = \boldsymbol{\Theta}_o(t) - \boldsymbol{\mathcal{P}}_o(t) \otimes \mathbf{E}(t). \tag{6.25}$$

Furthermore,

$$\mathbf{H}_o(0) = \mathbf{0}, \qquad \mathbf{F}_o(0) = \mathbf{1}, \qquad J_o(0) = 1. \tag{6.26}$$

Therefore,

$$\boldsymbol{\Theta}_o(0) = \overset{\circ}{\mathbf{S}}, \qquad \boldsymbol{\mathcal{P}}_o(0) = \overset{\circ}{\mathbf{P}}, \qquad \boldsymbol{\mathcal{E}}_o(0) = \overset{\circ}{\mathbf{E}}, \qquad \boldsymbol{\mathcal{D}}_o(0) = \overset{\circ}{\mathbf{D}}, \qquad \boldsymbol{\mathcal{T}}_o(0) = \overset{\circ}{\mathbf{T}}. \tag{6.27}$$

In conclusion, we obtain the balance equations in updated Lagrangean description:

$$\rho(t)J_o(t) = \overset{\circ}{\rho}, \qquad q(t)J_o(t) = q_o, \tag{6.28}$$

$$\overset{\circ}{\rho}\,\ddot{\mathbf{u}}(t) = \mathrm{div}_{\mathbf{x}}\boldsymbol{\mathcal{T}}_o(t) + \overset{\circ}{\rho}\,\mathbf{f}(t) + q_o(t)\mathbf{E}(t) + (\boldsymbol{\mathcal{P}}_o(t) \cdot \nabla_{\mathbf{x}})\mathbf{E}(t),$$

$$\mathrm{div}_{\mathbf{x}}\boldsymbol{\mathcal{D}}_o(t) = q_o(t), \qquad \mathrm{rot}_{\mathbf{x}}\boldsymbol{\mathcal{E}}_o(t) = \mathbf{0}. \tag{6.29}$$

Here $q_o(t)$ is the current volumetric electric charge density per unit material volume in the configuration $\overset{\circ}{B}$.

We find that the jump conditions and the electromechanical surface stress vector of Piola–Kirchhoff-type are given by

$$\overset{\circ}{\mathbf{n}} \cdot [\boldsymbol{\mathcal{D}}_o(t)] = w_o(t), \qquad \overset{\circ}{\mathbf{n}} \times [\boldsymbol{\mathcal{E}}_o(t)] = \mathbf{0},$$

$$\mathbf{t}_{o\mathbf{n}}(t) = \boldsymbol{\mathcal{T}}_o(t)\,\overset{\circ}{\mathbf{n}} = (\boldsymbol{\Theta}_o(t) - \boldsymbol{\mathcal{P}}_o(t) \otimes \mathbf{E}(t))\,\overset{\circ}{\mathbf{n}} \quad \text{on}\,\partial\overset{\circ}{V}, \tag{6.30}$$

where $w_o(t)$ is the current surface charge density per unit material surface area in the configuration $\overset{\circ}{B}$.

Finally, we give the constitutive relations in the form:

$$\boldsymbol{\Pi}_o(t) = \frac{\partial\mathcal{H}}{\partial\mathbf{G}}(\mathbf{G}_o(t), \boldsymbol{\mathcal{E}}_o(t)), \qquad \boldsymbol{\mathcal{P}}_o(t) = -\frac{\partial\mathcal{H}}{\partial\mathbf{E}}(\mathbf{G}_o(t), \boldsymbol{\mathcal{E}}_o(t)), \tag{6.31}$$

where

$$\mathbf{G}_o(t) = \frac{1}{2}(\mathbf{F}_o(t)^T\mathbf{F}_o(t) - \mathbf{1}). \tag{6.32}$$

Now, we define by $\mathbf{e}(t) = \mathbf{e}(\mathbf{x}, t)$ the infinitesimal perturbation of the initial applied electric field $\overset{\circ}{\mathbf{E}}$:

$$\mathbf{E}(t) = \overset{\circ}{\mathbf{E}} + \mathbf{e}(t), \tag{6.33}$$

and using (6.21) we derive the useful relation

$$\mathbf{F}(t) = \overset{\circ}{\mathbf{F}} + \mathbf{H}_o(t)\, \overset{\circ}{\mathbf{F}} . \tag{6.34}$$

In what follows, we suppose that the perturbations $\mathbf{H}_o(t)$ and $\mathbf{e}(t)$ are small, such that the products of all terms containing $\mathbf{H}_o(t)$ and $\mathbf{e}(t)$ may be neglected. In particular, we obtain

$$J_o(t) = 1 + \mathrm{tr}\mathbf{H}_o(t), \qquad \mathbf{F}_o^{-1}(t) = 1 - \mathbf{H}_o(t). \tag{6.35}$$

Henceforward, we denote by a superposed bar the small perturbation of an arbitrary field. So, we have for Green tensor

$$\mathbf{G}(t) = \overset{\circ}{\mathbf{G}} + \bar{\mathbf{G}}(t), \quad \text{where} \quad \bar{\mathbf{G}}(t) = \overset{\circ}{\mathbf{F}}{}^{T} \mathbf{g}(t)\, \overset{\circ}{\mathbf{F}}. \tag{6.36}$$

Here

$$\mathbf{g}(t) = \frac{1}{2}(\mathbf{H}_o(t) + \mathbf{H}_o^{T}(t)) \tag{6.37}$$

is the associated infinitesimal strain tensor.

Similarly, we define the perturbation of the electric field $\bar{\mathcal{E}}(t)$ by

$$\mathcal{E}(t) = \overset{\circ}{\mathcal{E}} + \bar{\mathcal{E}}(t), \quad \text{where} \quad \bar{\mathcal{E}}(t) = \overset{\circ}{\mathbf{F}}{}^{T} (\mathbf{e}(t) + \mathbf{H}_o^{T}(t)\, \overset{\circ}{\mathbf{E}}). \tag{6.38}$$

We also obtain that

$$\boldsymbol{\Theta}_o(t) = \overset{\circ}{\mathbf{S}} + \bar{\boldsymbol{\Theta}}_o(t), \qquad \boldsymbol{\Pi}(t) = \overset{\circ}{\boldsymbol{\Pi}} + \bar{\boldsymbol{\Pi}}(t),$$

$$\bar{\boldsymbol{\Theta}}_o(t) = \overset{\circ}{J}{}^{-1} \overset{\circ}{\mathbf{F}}\, \bar{\boldsymbol{\Pi}}(t)\, \overset{\circ}{\mathbf{F}}{}^{T} + \overset{\circ}{\mathbf{S}}\, \mathbf{H}_o^{T}(t). \tag{6.39}$$

The previous relation shows that the stress perturbation $\bar{\boldsymbol{\Theta}}_o(t)$ is known if the stress perturbation $\bar{\boldsymbol{\Pi}}(t)$ is known, and vice versa.

Next, we define

$$\boldsymbol{\mathcal{P}}_o(t) = \overset{\circ}{\mathbf{P}} + \bar{\boldsymbol{\mathcal{P}}}_o(t), \qquad \boldsymbol{\mathcal{P}}(t) = \overset{\circ}{\boldsymbol{\mathcal{P}}} + \bar{\boldsymbol{\mathcal{P}}}(t), \quad \text{with} \quad \bar{\boldsymbol{\mathcal{P}}}_o(t) = \overset{\circ}{J}{}^{-1} \overset{\circ}{\mathbf{F}}\, \bar{\boldsymbol{\mathcal{P}}}(t). \tag{6.40}$$

Thus, the perturbation $\bar{\boldsymbol{\mathcal{P}}}_o(t)$ is known if the perturbation $\bar{\boldsymbol{\mathcal{P}}}(t)$ is known, and vice versa.

Similarly, if we take

$$\boldsymbol{\mathcal{D}}_o(t) = \overset{\circ}{\mathbf{D}} + \bar{\boldsymbol{\mathcal{D}}}_o(t), \tag{6.41}$$

we obtain that

$$\bar{\boldsymbol{\mathcal{D}}}_o(t) = \mathbf{e}(t) + \bar{\boldsymbol{\mathcal{P}}}_o(t) + \overset{\circ}{\mathbf{E}}\,\mathrm{tr}\mathbf{H}_o(t) - \mathbf{H}_o(t)\,\overset{\circ}{\mathbf{E}}; \tag{6.42}$$

that is, to know the perturbation $\bar{\boldsymbol{\mathcal{D}}}_o(t)$ we must know the perturbation $\bar{\boldsymbol{\mathcal{P}}}_o(t)$.

We also derive that

$$\boldsymbol{\mathcal{E}}_o(t) = \overset{\circ}{\mathbf{E}} + \bar{\boldsymbol{\mathcal{E}}}_o(t), \quad \text{with} \quad \bar{\boldsymbol{\mathcal{E}}}_o(t) = \mathbf{e}(t) + \mathbf{H}_o^T(t)\,\overset{\circ}{\mathbf{E}}\,. \tag{6.43}$$

Finally, we find that

$$\boldsymbol{\mathcal{T}}_o(t) = \overset{\circ}{\mathbf{T}} + \bar{\boldsymbol{\mathcal{T}}}_o(t), \quad \text{where} \quad \bar{\boldsymbol{\mathcal{T}}}_o(t) = \bar{\boldsymbol{\Theta}}_o(t) - \bar{\boldsymbol{\mathcal{P}}}_o(t) \otimes \overset{\circ}{\mathbf{E}} - \overset{\circ}{\mathbf{P}} \otimes \mathbf{e}(t). \tag{6.44}$$

At this stage it is evident that all perturbations are known if the perturbations $\bar{\boldsymbol{\Pi}}(t)$ and $\bar{\boldsymbol{\mathcal{P}}}(t)$ are known. To obtain these perturbations we use the following constitutive equations.

$$\bar{\boldsymbol{\Pi}}(t) = \frac{\partial^2 \overset{\circ}{\mathcal{H}}}{\partial \mathbf{G} \partial \mathbf{G}}[\overset{\circ}{\mathbf{F}}^T\,\mathbf{g}(t)\,\overset{\circ}{\mathbf{F}}] + \frac{\partial^2 \overset{\circ}{\mathcal{H}}}{\partial \boldsymbol{\mathcal{E}} \partial \mathbf{G}}[\overset{\circ}{\mathbf{F}}^T\,(\mathbf{e}(t) + \mathbf{H}_o^T(t)\,\overset{\circ}{\mathbf{E}})], \tag{6.45}$$

$$\bar{\boldsymbol{\mathcal{P}}}(t) = -\frac{\partial^2 \overset{\circ}{\mathcal{H}}}{\partial \mathbf{G} \partial \boldsymbol{\mathcal{E}}}[\overset{\circ}{\mathbf{F}}^T\,\mathbf{g}(t)\,\overset{\circ}{\mathbf{F}}] - \frac{\partial^2 \overset{\circ}{\mathcal{H}}}{\partial \boldsymbol{\mathcal{E}} \partial \boldsymbol{\mathcal{E}}}[\overset{\circ}{\mathbf{F}}^T\,(\mathbf{e}(t) + \mathbf{H}_o^T(t)\,\overset{\circ}{\mathbf{E}})]. \tag{6.46}$$

Here the symbol "$\circ$" superposed on $\mathcal{H}$ indicates that the corresponding second-order derivatives of the generalized Helmholtz free energy are taken at $\overset{\circ}{\mathbf{G}}$ and $\overset{\circ}{\boldsymbol{\mathcal{E}}}$.

The perturbations of the force (resp., of charge densities) are defined by

$$\mathbf{f}(t) = \overset{\circ}{\mathbf{f}} + \bar{\mathbf{f}}(t), \qquad q_o(t) = \overset{\circ}{q} + \bar{q}(t),$$
$$w_o(t) = \overset{\circ}{w} + \bar{w}(t). \tag{6.47}$$

Concluding from the relations obtained in the last section, we derive that the incremental fields satisfy the following balance equations.

$$\overset{\circ}{\rho}\,\ddot{\mathbf{u}}(t) = \mathrm{div}_{\mathbf{x}}(\bar{\boldsymbol{\Theta}}_o(t) - \bar{\boldsymbol{\mathcal{P}}}_o(t) \otimes \overset{\circ}{\mathbf{E}} - \overset{\circ}{\mathbf{P}} \otimes \mathbf{e}(t)) + \overset{\circ}{\rho}\,\bar{\mathbf{f}}(t) + \bar{q}(t)\,\overset{\circ}{\mathbf{E}}$$
$$+ \overset{\circ}{q}\,\mathbf{e}(t) + (\bar{\boldsymbol{\mathcal{P}}}_o(t) \cdot \nabla_{\mathbf{x}})\,\overset{\circ}{\mathbf{E}} + (\overset{\circ}{\mathbf{P}} \cdot \nabla_{\mathbf{x}})\mathbf{e}(t), \tag{6.48}$$

$$\mathrm{div}_{\mathbf{x}}(\mathbf{e}(t) + \bar{\boldsymbol{\mathcal{P}}}_o(t) + \overset{\circ}{\mathbf{E}}\,\mathrm{tr}\mathbf{H}_o(t) - \mathbf{H}_o(t)\,\overset{\circ}{\mathbf{E}}) = \bar{q}(t),$$

$$\mathrm{rot}_{\mathbf{x}}(\mathbf{e}(t) + \mathbf{H}_o^T(t)\,\overset{\circ}{\mathbf{E}}) = \mathbf{0}. \tag{6.49}$$

Note that without further assumptions the differential balance equations satisfied by the incremental fields cannot be simplified.

The jump conditions for the involved incremental fields are:

$$\overset{\circ}{\mathbf{n}} \cdot [\mathbf{e}(t) + \bar{\boldsymbol{\mathcal{P}}}_o(t) + \overset{\circ}{\mathbf{E}}\,\mathrm{tr}\mathbf{H}_o(t) - \mathbf{H}_o(t)\,\overset{\circ}{\mathbf{E}}] = \bar{w}(t),$$

$$\overset{\circ}{\mathbf{n}} \times [\mathbf{e}(t) + \mathbf{H}_o^T(t)\,\overset{\circ}{\mathbf{E}}] = \mathbf{0}. \tag{6.50}$$

Finally, we obtain that the incremental electromechanical surface stress vector of Piola–Kirchhoff-type $\bar{\mathbf{t}}_{\mathbf{on}}(t)$ reduces to

$$\bar{\mathbf{t}}_{\mathbf{on}}(t) = (\bar{\boldsymbol{\Theta}}_o(t) - \bar{\boldsymbol{\mathcal{P}}}_o(t) \otimes \overset{\circ}{\mathbf{E}} - \overset{\circ}{\mathbf{P}} \otimes \mathbf{e}(t))\,\overset{\circ}{\mathbf{n}} \quad \mathrm{on}\,\partial\,\overset{\circ}{V}. \tag{6.51}$$

6.2.3 Special Cases: Homogeneous Initial State and Nonpolarisable Environment

In this part we introduce two simplifying hypotheses, essential for subsequent developments.

H1: The body is homogeneous, the initial deformation gradient $\overset{\circ}{\mathbf{F}}$ is constant in the domain V_R, and the initial applied electric field $\overset{\circ}{\mathbf{E}}$ is constant in all of space.

H2: The environment (i.e., the vacuum) of the body is not polarisable.

The second assumption is justified because the dielectric constants of electroelastic materials are significantly larger than the dielectric constant of the vacuum. Then, we have that $\overset{\circ}{\mathbf{P}} = \mathbf{0}$ and $\mathbf{e}(t) = \mathbf{0}$ in the exterior of the body $\overset{\circ}{V}$. Thus, the associated limit values on $\partial\,\overset{\circ}{V}$ satisfy the relations $\mathbf{P}^{+} = \mathbf{0}$ and $\mathbf{e}^{+}(t) = \mathbf{0}$. It is evident that the second assumption leads to an important simplification of the problem, because, by neglecting the surroundings of the body, our problem is transformed into one of a hyperelastic dielectric.

Now, if we consider the first assumption, we observe that $\overset{\circ}{\mathbf{S}}, \overset{\circ}{\mathbf{P}}, \overset{\circ}{\mathbf{E}}, \overset{\circ}{\mathbf{T}}$, and $\overset{\circ}{\mathbf{D}}$ are constant fields in the domain $\overset{\circ}{V}$. Consequently, the balance equations (6.16) take place in the assumed homogeneous state, only if

$$\overset{\circ}{\mathbf{f}} = \mathbf{0}, \qquad \overset{\circ}{q} = 0 \quad \mathrm{in}\;\overset{\circ}{V}. \tag{6.52}$$

Because the initial applied electric field is constant in the entire space, and taking into account the previous remarks, we find that the second jump condition (6.17) is satisfied if $\overset{\circ}{\mathbf{P}}$ and $\overset{\circ}{w}$ are related by

$$\overset{\circ}{\mathbf{P}} \cdot \overset{\circ}{\mathbf{n}} = -\overset{\circ}{w}, \quad \text{on} \quad \partial \overset{\circ}{V}. \tag{6.53}$$

Furthermore, the electromechanical stress vector $\overset{\circ}{\mathbf{t_n}}$ is given by the third relation (6.17).

An important consequence of assumption **H1**, together with relations (6.52) and Equations (6.48) and (6.49), is that the differential balance equations take the form:

$$\overset{\circ}{\rho}\, \ddot{\mathbf{u}}(t) = \mathrm{div}_{\mathbf{x}}(\bar{\boldsymbol{\Theta}}_o(t) - \bar{\boldsymbol{\mathcal{P}}}_o(t) \otimes \overset{\circ}{\mathbf{E}}) + \overset{\circ}{\rho}\,\bar{\mathbf{f}}(t) + \bar{q}(t)\,\overset{\circ}{\mathbf{E}}, \tag{6.54}$$

$$\mathrm{div}_{\mathbf{x}}(\mathbf{e}(t) + \bar{\boldsymbol{\mathcal{P}}}_o(t)) = \bar{q}(t), \qquad \mathrm{rot}_{\mathbf{x}}\mathbf{e}(t) = \mathbf{0}. \tag{6.55}$$

Moreover, the second assumption **H2** implies that the boundary conditions (6.50) and (6.51) reduce to:

$$\overset{\circ}{\mathbf{n}} \cdot (\mathbf{e}(t) + \bar{\boldsymbol{\mathcal{P}}}_o(t)) = -\bar{w}(t), \qquad \overset{\circ}{\mathbf{n}} \times \mathbf{e}(t) = \mathbf{0},$$

$$\bar{\mathbf{t}}_{on}(t) = (\bar{\boldsymbol{\Theta}}_o(t) - \bar{\boldsymbol{\mathcal{P}}}_o(t) \otimes \overset{\circ}{\mathbf{E}})\,\overset{\circ}{\mathbf{n}} \quad \text{on}\, \partial\, \overset{\circ}{V}. \tag{6.56}$$

The system (6.54)–(6.56) takes place whenever the initial state of the body is homogeneous and the environment of the body is not polarisable.

To complete the description of the incremental behaviour of the body, we analyse the constitutive equations (6.45) and (6.46), which give the perturbations $\bar{\boldsymbol{\Pi}}(t)$ and $\bar{\boldsymbol{\mathcal{P}}}(t)$, under the present assumptions. Moreover, if we use the relations (6.39) and (6.40), we obtain the perturbations $\bar{\boldsymbol{\Theta}}_o(t)$ and $\bar{\boldsymbol{\mathcal{P}}}_o(t)$ in the form:

$$\bar{\Theta}_{okl} = (\overset{\circ}{c}_{klmn} + \overset{\circ}{S}_{kn}\,\delta_{lm} - \overset{\circ}{e}_{nkl}\overset{\circ}{E}_m)u_{m,n} - \overset{\circ}{e}_{mkl}\,e_m \tag{6.57}$$

$$\bar{\mathcal{P}}_{ok} = (\overset{\circ}{e}_{kmn} + \overset{\circ}{\eta}_{km}\overset{\circ}{E}_n)u_{n,m} + \overset{\circ}{\eta}_{kl}\,e_l. \tag{6.58}$$

Here

$$\overset{\circ}{c}_{klmn} = \overset{\circ}{J}{}^{-1}\overset{\circ}{F}_{kp}\overset{\circ}{F}_{lq}\overset{\circ}{F}_{mr}\overset{\circ}{F}_{ns}\,\frac{\partial^2\,\overset{\circ}{\mathcal{H}}}{\partial G_{rs}\partial G_{pq}},$$

$$\overset{\circ}{e}_{mkl} = -\overset{\circ}{J}{}^{-1}\overset{\circ}{F}_{mp}\overset{\circ}{F}_{kq}\overset{\circ}{F}_{lr}\,\frac{\partial^2\,\overset{\circ}{\mathcal{H}}}{\partial \mathcal{E}_p\partial G_{qr}},$$

$$\overset{\circ}{\eta}_{kl} = -\overset{\circ}{J}{}^{-1}\overset{\circ}{F}_{km}\overset{\circ}{F}_{ln}\,\frac{\partial^2\,\overset{\circ}{\mathcal{H}}}{\partial \mathcal{E}_m\partial \mathcal{E}_n} \tag{6.59}$$

are the *instantaneous material moduli* (elastic, piezoelectric, and dielectric moduli). The constitutive relations (6.59) are valid even if the simplifying assumptions **H1** and **H2** are not satisfied.

The instantaneous material moduli possess the following symmetry properties,

$$\overset{\circ}{c}_{klmn} = \overset{\circ}{c}_{lkmn} = \overset{\circ}{c}_{klnm} = \overset{\circ}{c}_{mnkl},$$

$$\overset{\circ}{e}_{mkl} = \overset{\circ}{e}_{mlk}, \qquad \overset{\circ}{\eta}_{kl} = \overset{\circ}{\eta}_{lk}. \tag{6.60}$$

It follows that, for general anisotropy, there exist 21 independent instantaneous moduli $\overset{\circ}{c}_{klmn}$, 18 independent instantaneous moduli $\overset{\circ}{e}_{mkl}$, and 6 independent instantaneous moduli $\overset{\circ}{\eta}_{kl}$.

At this point, we introduce the incremental electromechanical stress tensor $\boldsymbol{\Sigma}$, and the incremental electric displacement vector $\boldsymbol{\Delta}$ by the relations:

$$\boldsymbol{\Sigma}(t) = \bar{\boldsymbol{\Theta}}_o(t) - \bar{\boldsymbol{P}}_o(t) \otimes \overset{\circ}{\mathbf{E}}, \qquad \boldsymbol{\Delta}(t) = \mathbf{e}(t) + \bar{\boldsymbol{P}}_o(t). \tag{6.61}$$

It follows that, according to relations (6.57) and (6.58), the constitutive relations describing the behaviour of the incremental fields, under the previous hypotheses, are:

$$\Sigma_{kl} = \overset{\circ}{\Omega}_{klmn}\, u_{m,n} - \overset{\circ}{\Lambda}_{mkl}\, e_m, \qquad \Delta_k = \overset{\circ}{\Lambda}_{kmn}\, u_{n,m} + \overset{\circ}{\epsilon}_{kl}\, e_l, \tag{6.62}$$

where

$$\overset{\circ}{\Omega}_{klmn} = \overset{\circ}{c}_{klmn} + \overset{\circ}{S}_{kn}\, \delta_{lm} - \overset{\circ}{e}_{kmn}\overset{\circ}{E}_l - \overset{\circ}{e}_{nkl}\overset{\circ}{E}_m - \overset{\circ}{\eta}_{kn}\overset{\circ}{E}_l\overset{\circ}{E}_m,$$

$$\overset{\circ}{\Lambda}_{mkl} = \overset{\circ}{e}_{mkl} + \overset{\circ}{\eta}_{mk}\overset{\circ}{E}_l, \qquad \overset{\circ}{\epsilon}_{kl} = \delta_{kl} + \overset{\circ}{\eta}_{kl} \tag{6.63}$$

are the components of the instantaneous elasticity tensor $\overset{\circ}{\boldsymbol{\Omega}}$, and the instantaneous coupling tensor $\overset{\circ}{\boldsymbol{\Lambda}}$, respectively, of the instantaneous dielectric tensor $\overset{\circ}{\boldsymbol{\epsilon}}$.

From relations (6.60) we find the symmetry relations

$$\overset{\circ}{\Omega}_{klmn} = \overset{\circ}{\Omega}_{nmlk}, \qquad \overset{\circ}{\epsilon}_{kl} = \overset{\circ}{\epsilon}_{lk}. \tag{6.64}$$

Moreover, we see that $\overset{\circ}{\Omega}_{klmn}$ is not symmetric according to indices (k, l) and (m, n) and $\overset{\circ}{\Lambda}_{mkl}$ is not symmetric relative to indices (k, l). It follows that, generally, there are 45 independent instantaneous elastic moduli $\overset{\circ}{\Omega}_{klmn}$, 27 independent instantaneous coupling moduli $\overset{\circ}{\Lambda}_{mkl}$, and 6 independent instantaneous dielectric moduli $\overset{\circ}{\epsilon}_{kl}$. These moduli are constant parameters depending on the considered hyperelastic material, and on the initial electric and mechanical applied fields. We note at this stage that, even if this problem is linearized, the solution depends nonlinearly on the initial applied electric field.

In this framework, using the incremental electromechanical stress tensor $\boldsymbol{\Sigma}$ and the incremental electric displacement vector $\boldsymbol{\Delta}$, we derive from (6.54) and (6.55) the differential balance equations in the form:

$$\overset{\circ}{\rho}\,\ddot{\mathbf{u}}(t) = \mathrm{div}_{\mathbf{x}}\boldsymbol{\Sigma} + \overset{\circ}{\rho}\,\bar{\mathbf{f}}(t) + \bar{q}(t)\,\overset{\circ}{\mathbf{E}},$$

$$\mathrm{div}_{\mathbf{x}}\boldsymbol{\Delta} = \bar{q}(t),$$

$$\mathrm{rot}_{\mathbf{x}}\mathbf{e}(t) = \mathbf{0}. \tag{6.65}$$

The associated boundary conditions are:

$$\overset{\circ}{\mathbf{n}} \cdot \boldsymbol{\Delta} = -\bar{w}(t), \qquad \overset{\circ}{\mathbf{n}} \times \mathbf{e}(t) = \mathbf{0},$$

$$\bar{\mathbf{t}}_{\mathbf{on}}(t) = \boldsymbol{\Sigma}\,\overset{\circ}{\mathbf{n}} \quad \text{on } \partial \overset{\circ}{V}. \tag{6.66}$$

6.2.4 Dynamic and Static Energy Balance. Incremental Initial and Boundary Value Problems

In next three sections we briefly describe the properties of the model derived previously.

First of all, we formulate dynamic and static energy balance laws as a consequence of incremental field equations derived in the preceding section. Here the differential operators are shown to be self-adjoint. This important property results from the form of constitutive equations (6.62) and from the symmetry properties (6.64).

To obtain the dynamic incremental energy balance law we multiply the equation $(6.65)_1$ with the velocity $\dot{\mathbf{u}}$. Integrating the resulting expression over $\overset{\circ}{V}$ and using the flux-divergence theorem, we derive:

$$\frac{\mathrm{d}}{\mathrm{d}t}\int_{\overset{\circ}{V}} \frac{1}{2}\overset{\circ}{\rho}\,\dot{\mathbf{u}}\cdot\dot{\mathbf{u}}\,\mathrm{d}v = \int_{\partial\overset{\circ}{V}} \dot{\mathbf{u}}\cdot(\boldsymbol{\Sigma}\,\overset{\circ}{\mathbf{n}})\mathrm{d}s + \int_{\overset{\circ}{V}}(\overset{\circ}{\rho}\,\bar{\mathbf{f}}\cdot\dot{\mathbf{u}} + \bar{q}\,\overset{\circ}{\mathbf{E}}\cdot\dot{\mathbf{u}})\mathrm{d}v$$

$$- \int_{\overset{\circ}{V}}(\nabla\dot{\mathbf{u}} : \boldsymbol{\Sigma})\mathrm{d}v. \tag{6.67}$$

Here ":" is the symbol for the inner product between second-order tensors.

Furthermore, from the balance equation $(6.65)_3$ it yields that the incremental electric field $\mathbf{e}(t)$ has a scalar electric potential $\phi(\mathbf{x}, t)$; that is, $\mathbf{e}(t) = -\nabla\phi$. Multiplying the time derivative of Equation $(6.65)_2$ by ϕ, integrating over $\overset{\circ}{V}$, and using the flux-divergence theorem, we obtain:

$$0 = -\int_{\partial\overset{\circ}{V}} \dot{\boldsymbol{\Delta}}\cdot(\phi\,\overset{\circ}{\mathbf{n}})\mathrm{d}s + \int_{\overset{\circ}{V}} \dot{\bar{q}}\phi\,\mathrm{d}v - \int_{\overset{\circ}{V}}(\dot{\boldsymbol{\Delta}}\cdot\mathbf{e})\mathrm{d}v. \tag{6.68}$$

Addition of the two preceding equations leads to

$$\frac{\mathrm{d}}{\mathrm{d}t}\int_{\overset{\circ}{V}}\frac{1}{2}\overset{\circ}{\rho}\,\dot{\mathbf{u}}\cdot\dot{\mathbf{u}}\mathrm{d}v = \int_{\partial\overset{\circ}{V}}\overset{\circ}{\mathbf{n}}\cdot(\boldsymbol{\Sigma}\dot{\mathbf{u}} - \phi\dot{\boldsymbol{\Delta}})\mathrm{d}s + \int_{\overset{\circ}{V}}(\overset{\circ}{\rho}\,\bar{\mathbf{f}}\cdot\dot{\mathbf{u}} + \bar{q}\,\overset{\circ}{\mathbf{E}}\cdot\dot{\mathbf{u}} + \phi\dot{\bar{q}})\mathrm{d}v$$

$$- \int_{\overset{\circ}{V}}(\nabla\dot{\mathbf{u}} : \boldsymbol{\Sigma} + \dot{\boldsymbol{\Delta}}\cdot\mathbf{e})\mathrm{d}v. \tag{6.69}$$

Using the constitutive laws (6.62) and the symmetry of tensors $\overset{\circ}{\boldsymbol{\Omega}}$ and $\overset{\circ}{\boldsymbol{\epsilon}}$ we see that the last term from (6.69) reduces to

$$-\int_{\overset{\circ}{V}}\frac{\mathrm{d}}{\mathrm{d}t}\left(\frac{1}{2}\nabla\mathbf{u}\cdot\overset{\circ}{\boldsymbol{\Omega}}\,\nabla\mathbf{u} + \frac{1}{2}\mathbf{e}\cdot\overset{\circ}{\boldsymbol{\epsilon}}\,\mathbf{e}\right)\mathrm{d}v.$$

Thus, we obtain the dynamic energy balance law in the form:

$$\frac{1}{2}\frac{\mathrm{d}}{\mathrm{d}t}\int_{\overset{\circ}{V}}(\overset{\circ}{\rho}\,\dot{\mathbf{u}}\cdot\dot{\mathbf{u}} + \nabla\mathbf{u}\cdot\overset{\circ}{\boldsymbol{\Omega}}\,\nabla\mathbf{u} + \mathbf{e}\cdot\overset{\circ}{\boldsymbol{\epsilon}}\,\mathbf{e})\mathrm{d}v$$

$$= \int_{\partial\overset{\circ}{V}}\overset{\circ}{\mathbf{n}}\cdot(\boldsymbol{\Sigma}\dot{\mathbf{u}} - \phi\dot{\boldsymbol{\Delta}})\mathrm{d}s + \int_{\overset{\circ}{V}}(\overset{\circ}{\rho}\,\bar{\mathbf{f}}\cdot\dot{\mathbf{u}} + \bar{q}\,\overset{\circ}{\mathbf{E}}\cdot\dot{\mathbf{u}} + \phi\dot{\bar{q}})\mathrm{d}v. \tag{6.70}$$

Analyzing the first integral, we remark that the first term represents the incremental kinetic energy of the body, the second term defines the incremental strain energy of the dielectric (including the effect of initial fields via the tensor $\overset{\circ}{\boldsymbol{\Omega}}$), and the last one is related to the energy of the incremental electric field. The surface integral appearing in Equation (6.70) introduces an important vector of electromechanics, that is, the incremental electromechanical energy-flux vector $\boldsymbol{\phi} = -\boldsymbol{\Sigma}\dot{\mathbf{u}} + \phi\dot{\boldsymbol{\Delta}}$. It generalizes the Poynting vector from the classical case, plays an important role in the reflection/refraction problems of the electromechanical waves, and characterizes the amount of electromechanical energy gained, or lost, by the boundary.

Let W be the sum of the incremental strain energy density and the energy density of the incremental electric field $W = \nabla\mathbf{u}\cdot\overset{\circ}{\boldsymbol{\Omega}}\,\nabla\mathbf{u} + \mathbf{e}\cdot\overset{\circ}{\boldsymbol{\epsilon}}\,\mathbf{e}$. In this way, the dynamic energy balance law (6.70) could be rewritten in the following final form.

$$\frac{1}{2}\frac{\mathrm{d}}{\mathrm{d}t}\int_{\overset{\circ}{V}}(\overset{\circ}{\rho}\,\dot{\mathbf{u}}\cdot\dot{\mathbf{u}} + W)\mathrm{d}v = \int_{\partial\overset{\circ}{V}}\overset{\circ}{\mathbf{n}}\cdot\boldsymbol{\Sigma}\dot{\mathbf{u}}\mathrm{d}s - \int_{\partial\overset{\circ}{V}}\phi\,\overset{\circ}{\mathbf{n}}\cdot\dot{\boldsymbol{\Delta}}\mathrm{d}s$$

$$+ \int_{\overset{\circ}{V}}(\overset{\circ}{\rho}\,\bar{\mathbf{f}}\cdot\dot{\mathbf{u}} + \bar{q}\,\overset{\circ}{\mathbf{E}}\cdot\dot{\mathbf{u}} + \phi\dot{\bar{q}})\mathrm{d}v. \tag{6.71}$$

If the energy density W is a positive definite quadratic form, we find that the initial and boundary incremental value problems, for this model, have unique solutions. This important property of W takes place if the tensors $\overset{\circ}{\boldsymbol{\Omega}}$ and $\overset{\circ}{\boldsymbol{\epsilon}}$ are positive definite. Note that, as in the classical linear theory of piezoelectric materials, without initial fields, the foregoing uniqueness condition does not impose any restriction on the instantaneous coupling tensor $\overset{\circ}{\boldsymbol{\Lambda}}$.

Using the same procedure for static problems, the static incremental energy balance law may be obtained in the form

$$\frac{1}{2}\int_{\overset{\circ}{V}} W\,dv = \int_{\partial\overset{\circ}{V}} \overset{\circ}{\mathbf{n}}\cdot\boldsymbol{\Sigma}\mathbf{u}\,ds - \int_{\partial\overset{\circ}{V}} \phi\,\overset{\circ}{\mathbf{n}}\cdot\boldsymbol{\Delta}\,ds + \int_{\overset{\circ}{V}}(\overset{\circ}{\rho}\,\bar{\mathbf{f}}\cdot\mathbf{u} + \bar{q}\,\overset{\circ}{\mathbf{E}}\cdot\mathbf{u} + \phi\bar{q})\,dv.$$

$$(6.72)$$

As in the dynamic case, the uniqueness of the solutions for initial and boundary incremental value problems is assured, up to a rigid-body motion, if W is positive definite, that is, if the tensors $\overset{\circ}{\boldsymbol{\Omega}}$ and $\overset{\circ}{\boldsymbol{\epsilon}}$ are positive definite.

In conclusion, the loss of uniqueness and the instability of solutions may occur if at least one of the tensors $\overset{\circ}{\boldsymbol{\Omega}}$ and $\overset{\circ}{\boldsymbol{\epsilon}}$ ceases to be positive definite. In this particular case, direct methods are used to obtain the solutions, as we show in Section 6.3 of this chapter.

6.2.5 Static and Dynamic Local Stability. Propagation of Plane Harmonic Waves

In this section we present the local stability criteria for dynamic (resp., static) problems concerning elastic dielectrics subject to initial electromechanical fields. In the final part we derive the conditions for propagation of plane harmonic waves in this electromechanical framework.

Under the previous conditions, we consider the following homogeneous static boundary value problem:

$$\mathrm{div}\,\boldsymbol{\Sigma} = \mathbf{0}, \qquad \mathrm{div}\,\boldsymbol{\Delta} = 0, \qquad \mathbf{e} = -\nabla\phi \quad \text{in } \overset{\circ}{V},$$

$$\boldsymbol{\Sigma}\,\overset{\circ}{\mathbf{n}} = \mathbf{0} \quad \text{on } \partial\,\overset{\circ}{V}_1, \qquad \mathbf{u} = \mathbf{0} \quad \text{on } \partial\,\overset{\circ}{V}_2,$$

$$\boldsymbol{\Delta}\cdot\overset{\circ}{\mathbf{n}} = 0 \quad \text{on } \partial\,\overset{\circ}{V}_3, \qquad \phi = 0 \quad \text{on } \partial\,\overset{\circ}{V}_4. \qquad (6.73)$$

Here $\partial\,\overset{\circ}{V}_1$ and $\partial\,\overset{\circ}{V}_2$ (resp., $\partial\,\overset{\circ}{V}_3$ and $\partial\,\overset{\circ}{V}_4$) represent partitions of the boundary $\partial\,\overset{\circ}{V}$.

Adapting here the concepts of stability from the linear stability theory of hyperelastic solids, developed in [13] and [8], we define the *static stability criterion*, as follows.

- The initially deformed and polarised equilibrium state $\overset{\circ}{B}$ is *locally stable* if and only if the homogeneous incremental boundary value problem (6.73) has only the null solution. If this condition is not fulfilled, we say that $\overset{\circ}{B}$ is *locally unstable*.

It is obvious that the static local stability criterion is linked to the uniqueness of the solution; that is, if, for a given configuration $\overset{\circ}{B}$, the uniqueness of solution stands, then $\overset{\circ}{B}$ is locally stable, and vice versa. In particular, we may infer that if the material tensors $\overset{\circ}{\boldsymbol{\Omega}}$ and $\overset{\circ}{\boldsymbol{\epsilon}}$ are positive definite, then the initially deformed and polarised equilibrium state $\overset{\circ}{B}$ is locally stable.

As regards the dynamic stability criterion, we say that the equilibrium configuration $\overset{\circ}{B}$ is locally stable if initial small time-dependent perturbations of $\overset{\circ}{B}$ remain small in time. In our special case, we assume the ordinary space–time decomposition for the incremental displacement $\mathbf{u}$ and the incremental electric potential ϕ

$$\mathbf{u}(\mathbf{x},t) = e^{i\omega t}\tilde{\mathbf{u}}(\mathbf{x}),$$
$$\phi(\mathbf{x},t) = e^{i\omega t}\tilde{\phi}(\mathbf{x}), \tag{6.74}$$

where ω is a complex number.

The homogeneous dynamic incremental boundary value problem to be considered is:

$$\overset{\circ}{\rho}\,\ddot{\mathbf{u}}(t) = \mathrm{div}\,\boldsymbol{\Sigma}, \qquad \mathrm{div}\,\boldsymbol{\Delta} = 0, \qquad \mathbf{e} = -\nabla\phi \quad \text{in} \quad \overset{\circ}{V},$$

$$\boldsymbol{\Sigma}\,\overset{\circ}{\mathbf{n}} = \mathbf{0} \quad \text{on} \quad \partial\overset{\circ}{V}_1, \qquad \mathbf{u} = \mathbf{0} \quad \text{on} \quad \partial\overset{\circ}{V}_2, \qquad \boldsymbol{\Delta}\cdot\overset{\circ}{\mathbf{n}} = 0 \quad \text{on} \quad \partial\overset{\circ}{V}_3,$$

$$\phi = 0 \quad \text{on} \quad \partial\overset{\circ}{V}_4. \tag{6.75}$$

Here $\partial\overset{\circ}{V}_1$ and $\partial\overset{\circ}{V}_2$ (resp., $\partial\overset{\circ}{V}_3$ and $\partial\overset{\circ}{V}_4$) represent partitions of the boundary $\partial\overset{\circ}{V}$.

From the incremental constitutive equations we obtain that $\boldsymbol{\Sigma}$, $\boldsymbol{\Delta}$, and $\mathbf{e}$ may be written in the form

$$\boldsymbol{\Sigma}(\mathbf{x},t) = e^{i\omega t}\tilde{\boldsymbol{\Sigma}}(\mathbf{x}), \qquad \boldsymbol{\Delta}(\mathbf{x},t) = e^{i\omega t}\tilde{\boldsymbol{\Delta}}(\mathbf{x}), \qquad \mathbf{e}(\mathbf{x},t) = e^{i\omega t}\tilde{\mathbf{e}}(\mathbf{x}). \tag{6.76}$$

Consequently, from relations (6.74)–(6.76) we derive the following eigenvalue problem for the parameter ω:

$$\mathrm{div}\,\tilde{\boldsymbol{\Sigma}} = \overset{\circ}{\rho}\,\omega^2\tilde{\mathbf{u}}, \qquad \mathrm{div}\,\tilde{\boldsymbol{\Delta}} = 0, \qquad \tilde{\mathbf{e}} = -\nabla\tilde{\phi} \quad \text{in} \quad \overset{\circ}{V}, \quad \tilde{\boldsymbol{\Sigma}}\,\overset{\circ}{\mathbf{n}} = \mathbf{0} \quad \text{on} \quad \partial\overset{\circ}{V}_1,$$

$$\tilde{\mathbf{u}} = \mathbf{0} \quad \text{on} \quad \partial\overset{\circ}{V}_2, \qquad \tilde{\boldsymbol{\Delta}}\cdot\overset{\circ}{\mathbf{n}} = 0 \quad \text{on} \quad \partial\overset{\circ}{V}_3, \qquad \tilde{\phi} = 0 \quad \text{on} \quad \partial\overset{\circ}{V}_4. \tag{6.77}$$

Now, we can define the *dynamic stability criterion* as follows.

- The initially deformed and polarised equilibrium state $\overset{\circ}{B}$ is *locally stable* if and only if the eigenvalues ω of the problem (6.77) have nonnegative imaginary part; that is, $\mathrm{Im}\,\omega \geq 0$.

It is evident that the eigenvalues ω depend on the initially applied deformation and electric fields. Hence, assuming that $\overset{\circ}{B}$ is a locally stable reference configuration, the critical loads producing dynamic instability correspond to those values for which the condition $\mathrm{Im}\,\omega = 0$ is fulfilled.

To express the relation between the critical electromechanical loads leading to instability, a standard calculation shows that all the eigenvalues ω^2 are real numbers; that is, $\mathrm{Im}\,\omega^2 = 0$ (see [2] for details). As we have seen, the dynamic stability criterion indicates the loss of stability at those critical values of the loading parameters for which $\mathrm{Im}\,\omega$ passes from positive to negative values. Consequently, because $\omega \neq 0$ is a real number, assuming a continuous dependency of ω on the loading parameters, we conclude that at the moment of the loss of stability ω must vanish; that is, $\omega = 0$.

Hence, according to the assumed dynamic stability criterion, the boundary of the stability domain is determined for those critical values of the loading parameters for which the equality $\omega = 0$ is satisfied for the first time on a given loading path. In other words, the equilibrium configuration $\overset{\circ}{B}$ becomes locally unstable when the eigenvalue problem (6.77) has a nonzero solution corresponding to a zero eigenvalue.

In this case we obtain an important conclusion.

- The problems (6.73) and (6.77) give the same critical value of the loading parameters, for a given loading path.

This result extends to hyperelastic dielectrics subject to electromechanical initial fields similar conclusions obtained in [8] for prestressed hyperelastic solids. It is interesting to note that this equivalence is a direct consequence of the self-adjointness of the differential operator describing the incremental behaviour of the body.

It is known that an important time-dependent perturbation used for dynamic stability analysis is the *plane harmonic (progressive) wave*. For our electromechanical problem this type of wave is defined by:

$$\mathbf{u}(\mathbf{x}, t) = \mathbf{a}e^{\mathrm{i}(\mathbf{p}\cdot\mathbf{x}-\omega t)},$$

$$\phi(\mathbf{x}, t) = ae^{\mathrm{i}(\mathbf{p}\cdot\mathbf{x}-\omega t)}, \tag{6.78}$$

where $\mathbf{a}$ and a are constant quantities characterizing the *amplitude* of the wave, $\mathbf{p} = p\mathbf{n}$ is a constant vector, p represents the *wave number*, and $\mathbf{n}$ is the *direction of propagation*. Here ω is the *frequency* of the wave. We suppose that this kind of wave propagates in an unbounded domain.

It is easy to show that the fields (6.78) satisfy the homogeneous balance equations (6.65) only if $\mathbf{a}$, ω, and $\mathbf{p}$ satisfy the following condition of propagation for harmonic waves:

$$\overset{\circ}{Q}(\mathbf{p})\,\boldsymbol{a} = \overset{\circ}{\rho}\,\omega^2\boldsymbol{a} \tag{6.79}$$

with

$$\overset{\circ}{Q}_{lm}(\mathbf{p}) = \overset{\circ}{A}_{lm}(\mathbf{p}) + \frac{\overset{\circ}{\Gamma}_l(\mathbf{p})\,\overset{\circ}{\Gamma}_m(\mathbf{p})}{\overset{\circ}{\Gamma}(\mathbf{p})},$$

$$\overset{\circ}{A}_{lm}(\mathbf{p}) = \overset{\circ}{\Omega}_{klmn}\,p_k p_n = p^2\,\overset{\circ}{\Omega}_{klmn}\,n_k n_n,$$

$$\overset{\circ}{\Gamma}_l(\mathbf{p}) = \overset{\circ}{\Lambda}_{mkl}\,p_m p_k = p^2\,\overset{\circ}{\Lambda}_{mkl}\,n_m n_k,$$

$$\overset{\circ}{\Gamma}(\mathbf{p}) = \overset{\circ}{\epsilon}_{kl}\,p_k p_l = p^2\,\overset{\circ}{\epsilon}_{kl}\,n_k n_l. \tag{6.80}$$

The fundamental propagation condition (6.79) is an eigenvector problem for the amplitude $\mathbf{a}$, linked in a natural way to the eigenvalue problem (*dispersion equation*):

$$\det\left(\overset{\circ}{\boldsymbol{Q}}(\mathbf{p}) - \overset{\circ}{\rho}\,\omega^2(\mathbf{p})\mathbf{1}\right) = 0, \tag{6.81}$$

which is satisfied by the frequency ω, by any given $\mathbf{p}$. The *velocity of propagation* (*phase velocity*) of the harmonic wave is defined by $v = \omega(\mathbf{p})/p$.

Relations (6.80) imply that $\overset{\circ}{\boldsymbol{Q}}(\mathbf{p}) = \overset{\circ}{\boldsymbol{Q}}^T(\mathbf{p})$, for any $\mathbf{p}$. It follows that the dispersion equation (6.81) has real eigenvalues $\overset{\circ}{\rho}\,\omega^2(\mathbf{p})$.

Moreover, because we assumed the positive definiteness of the instantaneous material tensors $\overset{\circ}{\Omega}$ and $\overset{\circ}{\epsilon}$, we obtain from (6.80) that the *instantaneous acoustic tensor* $\overset{\circ}{\boldsymbol{Q}}(\mathbf{p})$ is positive definite for any $\mathbf{p}$. Consequently, the eigenvalues $\overset{\circ}{\rho}\,\omega^2(\mathbf{p})$ are positive quantities. Therefore, we infer that there are three mutually orthogonal eigenvectors $\mathbf{a}_1$, $\mathbf{a}_2$, and $\mathbf{a}_3$, provided that the initial configuration is locally stable. Then, plane harmonic waves may propagate in any direction in a prestressed and prepolarised hyperelastic material.

In conclusion, we obtained that the incremental behaviour of the body in its initially deformed and polarised configuration $\overset{\circ}{B}$ is similar to classical linear piezoelectric theory (see [14, 28]).

6.2.6 Characteristic Surfaces

In the last part of this section we define and analyze the characteristic surfaces related to the problem of infinitesimal plane harmonic wave propagation in prestressed and prepolarised piezoelectric crystals (see [14] and [1]).

We recall here the main result of the previous part.

- If the instantaneous acoustic tensor $\overset{\circ}{\boldsymbol{Q}}(\mathbf{p})$ is positive definite, for any $\mathbf{p}$, then the eigenvalues $\overset{\circ}{\rho}\,\omega^2(\mathbf{p})$ are positive quantities. Therefore, there are three mutually orthogonal eigenvectors $\mathbf{a}_1$, $\mathbf{a}_2$, and $\mathbf{a}_3$, provided that the initial configuration is locally stable.

In physical terms this means that three harmonic plane waves may propagate with real positive frequencies, given by the dispersion equation (6.81). The amplitudes of the corresponding waves are the eigenvectors of the instantaneous acoustic tensor. Moreover, if the three frequencies are distinct, the corresponding amplitudes are orthogonal, and the waves are linearly polarised. In general, these waves are not pure (i.e., longitudinal, nor transverse). As in the classical case, we speak of incremental quasi-longitudinal and quasi-transverse waves. Finally, we define the *group velocity* vector by $\mathbf{v}_g(\mathbf{p}) = \partial \omega(\mathbf{p})/\partial \mathbf{p}$.

If we study the properties of the wave vector $\mathbf{p} = p\mathbf{n}$, we may define the *wave surface* as the locus of the tip of the vector $\mathbf{p}$, when the dispersion equation (6.81) is satisfied for a fixed, real, and positive frequency ω. Because ω is constant on the wave vector surface, the group velocity $\mathbf{v}_g(\mathbf{p})$ is aligned with the normal to this surface, the direction being that along which ω increases. This surface possesses three branches, corresponding to the distinct roots of the dispersion equation. It is an *equi-phase* surface, because all their points vibrate at the same time.

On the other hand, we define the *slowness surface* as the locus of the tip of the vector $\mathbf{m} = \mathbf{n}/v$, where $v = \omega/p$ is the phase velocity. The slowness surface is analogous to the index surface from optics, and plays an important role in reflection/refraction problems. As before, there are three such surfaces. The wave surface and the slowness surface are interconnected by the relationship $\mathbf{m} = \mathbf{p}/\omega$. Later in the chapter we analyse the properties of the slowness surfaces for various crystal classes.

Finally, we define the *velocity surface* as the locus of the tip of the vector $\mathbf{v} = v\mathbf{n}$. Generally, there are three such surfaces, one for quasi-longitudinal waves and two corresponding to the quasi-transverse waves. Because $\mathbf{v}$ and $\mathbf{m}$ are collinear and $|\mathbf{v}||\mathbf{m}| = 1$, the velocity and slowness surfaces are related by inversion through the origin.

6.3 Plane Harmonic Wave Propagation in Crystals Subject to Initial Electromechanical Fields

In this section we describe the propagation conditions of plane harmonic waves in various type of crystals subject to initial electromechanical fields.

6.3.1 Isotropic Solids. The Electrostrictive Effect

Following the paper [17], we study here the electrostrictive effect on plane harmonic wave propagation in an isotropic solid subject to initial mechanical and electric fields. In this case, even the piezoelectric effect is, obviously, absent, we find that the initial fields significantly influence the velocities of propagation and the polarisation of the waves, via the electrostrictive effect due to the initial electric field components. Considering the particular cases of longitudinal, transverse, and oblique initial electric fields, we find the velocities of propagation as closed form solutions, and we analyze the polarisation of the corresponding waves, showing the essential role of the initial electric field. Finally, we compute the velocities of propagation, for a longitudinal initial electric field, in quartz glass (amorphous SiO_2), applying initial strain fields of order 1%, respectively, 2%.

We have the homogenous field equations (6.65) (i.e., $\bar{\mathbf{f}} = \mathbf{0}$ and $\bar{q} = 0$) and the constitutive equations (6.62), with

$$\overset{\circ}{\Omega}_{klmn} = \overset{\circ}{\Omega}_{nmlk} = c_{klmn} + \overset{\circ}{S}_{kn}\,\delta_{lm} - \eta_{kn}\,\overset{\circ}{E}_l\overset{\circ}{E}_m,$$

$$\overset{\circ}{\Lambda}_{mkl} = \eta_{mk}\,\overset{\circ}{E}_l, \qquad \overset{\circ}{\epsilon}_{kl} = \overset{\circ}{\epsilon}_{lk} = \epsilon_{kl} = \delta_{kl} + \eta_{kl}, \tag{6.82}$$

where c_{klmn} are the components of the constant elasticity tensor, ϵ_{kl} are the components of the constant dielectric tensor, $\overset{\circ}{E}_i$ are the components of the initial applied electric field, and $\overset{\circ}{S}_{kn}$ are the components of the initial applied symmetric (Cauchy) stress tensor. From now on, we drop the superposed "$\circ$" for the components of constant material tensors.

We study the propagation of incremental progressive plane waves in an unbounded three-dimensional material described by the previous constitutive equations. Therefore, the displacement vector and the electric potential have the form (6.78). So, we obtain the condition of propagation (6.79) and (6.80) of plane harmonic waves. Because the acoustic tensor $\overset{\circ}{\mathbf{Q}}$ is symmetric, the eigenvalues $\overset{\circ}{\rho}\,\omega^2$ are real. Moreover, if we assume the positive definiteness of the instantaneous moduli tensor $\overset{\circ}{\boldsymbol{\Omega}}$ and $\boldsymbol{\epsilon}$ (i.e., if the initial configuration $\overset{\circ}{B}$ is locally stable), it follows from the definition of the acoustic tensor that it is positive definite. Consequently, the eigenvalues $\overset{\circ}{\rho}\,\omega^2$ are positive quantities for any $\mathbf{p}$. Thus, if $\boldsymbol{\Omega}$ and $\boldsymbol{\epsilon}$ satisfy the given conditions, in a prestressed and prepolarised electroelastic material, then incremental progressive waves can propagate in any direction, the direction of propagation $\mathbf{n}$, the wave number p, and the frequency ω being connected by the dispersion equation (6.81).

In the particular case of an isotropic material, the elasticity tensor contains two independent constants (see [14]). Using Voigt's convention we have:

$$\mathbf{c} = \begin{pmatrix} c_{11} & c_{12} & c_{12} & 0 & 0 & 0 \\ c_{12} & c_{11} & c_{12} & 0 & 0 & 0 \\ c_{12} & c_{12} & c_{11} & 0 & 0 & 0 \\ 0 & 0 & 0 & c_{66} & 0 & 0 \\ 0 & 0 & 0 & 0 & c_{66} & 0 \\ 0 & 0 & 0 & 0 & 0 & c_{66} \end{pmatrix}, \tag{6.83}$$

with $c_{11} = \lambda + 2\mu$, $c_{12} = \lambda$, and $c_{66} = (c_{11} - c_{12})/2 = \mu$. Here λ and μ are Lamé's constants.

The dielectric tensor has only one constant:

$$\boldsymbol{\eta} = \begin{pmatrix} \eta & 0 & 0 \\ 0 & \eta & 0 \\ 0 & 0 & \eta \end{pmatrix}, \tag{6.84}$$

and $\epsilon = 1 + \eta$.

Assuming that the direction of propagation coincides with the x_1-axis (i.e., $n_1 = 1$, $n_2 = n_3 = 0$), it yields that the tensor $\overset{\circ}{\boldsymbol{A}}$ has the following components (which must be multiplied by p^2).

$$\overset{\circ}{A}_{11} = c_{11} + \overset{\circ}{S}_{11} - \eta \, \overset{\circ}{E}_1^2, \qquad \overset{\circ}{A}_{12} = \overset{\circ}{A}_{21} = -\eta \, \overset{\circ}{E}_1 \overset{\circ}{E}_2,$$

$$\overset{\circ}{A}_{13} = \overset{\circ}{A}_{31} = -\eta \, \overset{\circ}{E}_1 \overset{\circ}{E}_3, \qquad \overset{\circ}{A}_{22} = c_{66} + \overset{\circ}{S}_{11} - \eta \, \overset{\circ}{E}_2^2,$$

$$\overset{\circ}{A}_{23} = \overset{\circ}{A}_{32} = -\eta \, \overset{\circ}{E}_2 \overset{\circ}{E}_3, \qquad \overset{\circ}{A}_{33} = c_{66} + \overset{\circ}{S}_{11} - \eta \, \overset{\circ}{E}_3^2 . \tag{6.85}$$

The components of $\overset{\circ}{\boldsymbol{\Gamma}}$ and $\overset{\circ}{\Gamma}$ become (the expressions must be multiplied by p^2):

$$\overset{\circ}{\Gamma}_1 = \eta \, \overset{\circ}{E}_1, \qquad \overset{\circ}{\Gamma}_2 = \eta \, \overset{\circ}{E}_2,$$

$$\overset{\circ}{\Gamma}_3 = \eta \, \overset{\circ}{E}_3, \qquad \overset{\circ}{\Gamma} = \epsilon = 1 + \eta. \tag{6.86}$$

Consequently, the dispersion equation (6.81) has in this case the following form.

$$F(V) = \begin{vmatrix} \overset{\circ}{A}_{11} + \dfrac{\overset{\circ}{\Gamma}_1^2}{\overset{\circ}{\Gamma}} - V & \overset{\circ}{A}_{12} + \dfrac{\overset{\circ}{\Gamma}_1 \overset{\circ}{\Gamma}_2}{\overset{\circ}{\Gamma}} & \overset{\circ}{A}_{13} + \dfrac{\overset{\circ}{\Gamma}_1 \overset{\circ}{\Gamma}_3}{\overset{\circ}{\Gamma}} \\[2.5ex] \overset{\circ}{A}_{12} + \dfrac{\overset{\circ}{\Gamma}_1 \overset{\circ}{\Gamma}_2}{\overset{\circ}{\Gamma}} & \overset{\circ}{A}_{22} + \dfrac{\overset{\circ}{\Gamma}_2^2}{\overset{\circ}{\Gamma}} - V & \overset{\circ}{A}_{23} + \dfrac{\overset{\circ}{\Gamma}_2 \overset{\circ}{\Gamma}_3}{\overset{\circ}{\Gamma}} \\[2.5ex] \overset{\circ}{A}_{13} + \dfrac{\overset{\circ}{\Gamma}_1 \overset{\circ}{\Gamma}_3}{\overset{\circ}{\Gamma}} & \overset{\circ}{A}_{23} + \dfrac{\overset{\circ}{\Gamma}_2 \overset{\circ}{\Gamma}_3}{\overset{\circ}{\Gamma}} & \overset{\circ}{A}_{33} + \dfrac{\overset{\circ}{\Gamma}_3^2}{\overset{\circ}{\Gamma}} - V \end{vmatrix} = 0, \tag{6.87}$$

where $V = \overset{\circ}{\rho} \, v^2$. In this way, we obtain the dispersion equation having the form:

$$F(V) = \begin{vmatrix} a - V & b & c \\ b & d - V & e \\ c & e & f - V \end{vmatrix} = 0, \tag{6.88}$$

where

$$a = c_{11} + \overset{\circ}{S}_{11} - \frac{\eta}{1+\eta}\, \overset{\circ}{E}_1^2, \qquad b = -\frac{\eta}{1+\eta}\, \overset{\circ}{E}_1 \overset{\circ}{E}_2,$$

$$c = -\frac{\eta}{1+\eta}\, \overset{\circ}{E}_1 \overset{\circ}{E}_3, \qquad d = c_{66} + \overset{\circ}{S}_{11} - \frac{\eta}{1+\eta}\, \overset{\circ}{E}_2^2,$$

$$e = -\frac{\eta}{1+\eta}\, \overset{\circ}{E}_2 \overset{\circ}{E}_3, \qquad f = c_{66} + \overset{\circ}{S}_{11} - \frac{\eta}{1+\eta}\, \overset{\circ}{E}_3^2 . \tag{6.89}$$

The condition of propagation of progressive waves (6.79) reduces here to the following eigenvector problem.

$$\begin{cases} a\, a_1 + b\, a_2 + c\, a_3 = V\, a_1 \\ b\, a_1 + d\, a_2 + e\, a_3 = V\, a_2 \\ c\, a_1 + e\, a_2 + f\, a_3 = V\, a_3 \end{cases} \tag{6.90}$$

with a, b, c, d, e, f given by (6.89) and a_i, $i = 1, 2, 3$, being the components of the unknown amplitude vector **a**.

In what follows, we present three important particular cases and a numerical example.

(a) *Longitudinal initial electric field* $(\overset{\circ}{E}_1 \neq 0,\ \overset{\circ}{E}_2 = \overset{\circ}{E}_3 = 0)$

In this case, the expressions b, c, e being zero, the dispersion equation (6.88) has the roots:

$$V_1 = c_{11} + \overset{\circ}{S}_{11} - \frac{\eta}{1+\eta}\, \overset{\circ}{E}_1^2,$$

$$V_2 = V_3 = c_{66} + \overset{\circ}{S}_{11}, \tag{6.91}$$

corresponding to the following wave velocities,

$$v_1 = \sqrt{\frac{\left(\lambda + 2\mu + \overset{\circ}{S}_{11} - \frac{\eta}{1+\eta}\, \overset{\circ}{E}_1^2\right)}{\overset{\circ}{\rho}}}\,,$$

$$v_2 = v_3 = \sqrt{\frac{\left(\mu + \overset{\circ}{S}_{11}\right)}{\overset{\circ}{\rho}}}. \tag{6.92}$$

As regards the polarisation of the obtained waves, using the condition of propagation (6.90) in this particular case, we can easily see that v_1 corresponds to a longitudinal wave, and $v_2 = v_3$ is the velocity of a transverse wave, arbitrarily polarised. On the other hand, we remark that both initial fields influence the wave velocities.

(b) *Transverse initial electric field* ($\overset{\circ}{E}_1 = 0$, $\overset{\circ}{E}_2 \neq 0$, $\overset{\circ}{E}_3 \neq 0$)

In this case, the coefficients b and c being zero, the dispersion equation (6.88) has the following three roots.

$$V_1 = c_{11} + \overset{\circ}{S}_{11}, \qquad V_2 = c_{66} + \overset{\circ}{S}_{11}, \qquad V_3 = c_{66} + \overset{\circ}{S}_{11} - \frac{\eta}{1+\eta}(\overset{\circ}{E}_2^2 + \overset{\circ}{E}_3^2).$$

$$(6.93)$$

These roots correspond to wave velocities:

$$v_1 = \sqrt{\frac{\left(\lambda + 2\mu + \overset{\circ}{S}_{11}\right)}{\overset{\circ}{\rho}}}, \qquad v_2 = \sqrt{\frac{\left(\mu + \overset{\circ}{S}_{11}\right)}{\overset{\circ}{\rho}}},$$

$$v_3 = \sqrt{\frac{\left(\mu + \overset{\circ}{S}_{11} - \frac{\eta}{1+\eta}(\overset{\circ}{E}_2^2 + \overset{\circ}{E}_3^2)\right)}{\overset{\circ}{\rho}}}.$$

$$(6.94)$$

If we analyze the polarisation of the waves having the previous velocities, we obtain from the propagation condition (6.90) that:

1. If $V = V_1 = \lambda + 2\mu + \overset{\circ}{S}_{11}$, it yields $a_2 = a_3 = 0$. Consequently, this wave is *longitudinal*.

2. If $V = V_2 = \mu + \overset{\circ}{S}_{11}$, it results in $a_1 = 0$ and $\overset{\circ}{E}_2\, a_2 + \overset{\circ}{E}_3\, a_3 = 0$. This wave is *transverse*, with a fixed polarisation direction, given by the initial electric field components.

3. If $V = V_3 = \mu + \overset{\circ}{S}_{11} - (\eta/(1+\eta))(\overset{\circ}{E}_2^2 + \overset{\circ}{E}_3^2)$, we obtain $a_1 = 0$ and $\overset{\circ}{E}_3\, a_2 - \overset{\circ}{E}_2\, a_3 = 0$. In this case the wave is *transverse*, too. It has a fixed polarisation direction, given by the initial electric field components, normal to the preceding one.

(c) *Oblique initial electric field* ($\overset{\circ}{E}_1 \neq 0$, $\overset{\circ}{E}_2 \neq 0$, $\overset{\circ}{E}_3 = 0$)

In this case, the coefficients c and e being zero, the dispersion equation (6.88) has the following roots.

$$V_{1,2} = \frac{\left(\lambda + 3\mu + 2\overset{\circ}{S}_{11} - \frac{\eta}{1+\eta}(\overset{\circ}{E}_1^2 + \overset{\circ}{E}_2^2) \pm \sqrt{\Delta}\right)}{2}, \qquad V_3 = \mu + \overset{\circ}{S}_{11},$$

$$(6.95)$$

where

$$\Delta = \left[\lambda + \mu - \frac{\eta}{1+\eta}(\overset{\circ}{E}_1^2 + \overset{\circ}{E}_2^2)\right]^2 + \frac{4\eta^2}{(1+\eta)^2}\,\overset{\circ}{E}_1^2\,\overset{\circ}{E}_2^2 > 0. \qquad (6.96)$$

One obtains the wave velocities in the form:

$$v_{1,2} = \sqrt{\frac{V_{1,2}}{\overset{\circ}{\rho}}}, \qquad v_3 = \sqrt{\frac{V_3}{\overset{\circ}{\rho}}}. \tag{6.97}$$

As regards the propagation condition (6.90), in this particular case we observe that:

1. If

$$V = V_1 = \frac{\left(\lambda + 3\mu + 2\ \overset{\circ}{S}_{11} - \dfrac{\eta}{1+\eta}(\overset{\circ}{E}_1^2 + \overset{\circ}{E}_2^2) + \sqrt{\Delta}\right)}{2},$$

we obtain $a_3 = 0$ and $(a - V_1)a_1 + ba_2 = 0$. This means that this is a *quasi-longitudinal* wave, having the polarisation direction restricted by the initial electric field components, only.

2. If

$$V = V_2 = \frac{\left(\lambda + 3\mu + 2\ \overset{\circ}{S}_{11} - \dfrac{\eta}{1+\eta}(\overset{\circ}{E}_1^2 + \overset{\circ}{E}_2^2) - \sqrt{\Delta}\right)}{2},$$

we obtain the polarisation direction given by $a_3 = 0$ and $(a - V_2)a_1 + ba_2 = 0$. This wave is quasi-longitudinal, its polarisation direction is restricted only by the initial electric field components, and is mutually perpendicular to the previous polarisation direction.

3. If $V = V_3 = \mu + \overset{\circ}{S}_{11}$ we easily find that $a_1 = a_2 = 0$. Consequently, it is a *transverse* wave, arbitrarily polarised.

(d) *A numerical example*

We analyze the influence of the initial mechanical and electric fields on wave velocities, corresponding to a longitudinal initial electric field, in the case of amorphous quartz SiO_2, that is, quartz glass. It has the following material constants (see [9, 14]): $c_{11} = 7.85 \cdot 10^{10}$ N/m^2, $c_{12} = 1.61 \cdot 10^{10}$ N/m^2, $\overset{\circ}{\rho} = 2203$ kg/m^3, and $\epsilon = 3.8$. Here the dielectric constant is reported to the vacuum dielectric constant.

In Table 6.1 we compute the wave velocities in the case without initial fields, and with initial strain fields of order 1%, respectively, 2%. The influence of the initial electric field on wave velocities is very weak, even if its intensity is important: $\overset{\circ}{E}_1 = 10^3 \sqrt{\text{Pa}} = 10^8$ V/m. The superior value corresponds to a traction stress $\overset{\circ}{S}_{11}$, and the inferior value is related to a compression stress $\overset{\circ}{S}_{11}$, that both generate the initial strain fields. One can observe quite important differences between these values, due to the initial strain fields.

6.3.2 Cubic Crystals. Generalised Anisotropy Factor, Slowness Surfaces

In this part we investigate, following the paper [18], the conditions of plane wave propagation in cubic crystals subject to initial deformations and electric fields. The analysis is extended to all symmetry classes belonging to the cubic system, exhibiting, or not, the piezoelectric effect. We show the influence of the electrostrictive and piezoelectric effects on wave propagation in such media. We derive the velocities of propagation as closed-form solutions, and we analyse the influence of the initial fields on the wave polarisation in two main cases: (i) propagation along a cube edge; and (ii) propagation along a cube face. In the second case we define a generalised anisotropy factor and we show the influence of the initial fields on it and on the shape of slowness surfaces.

In this problem we aim to satisfy the homogeneous field equations (6.65) and the constitutive equations (6.62) and (6.63). In order to study the conditions of propagation for incremental harmonic plane waves in an unbounded three-dimensional material described by the previous constitutive equations, we assume that the displacement vector and the electric potential have the form (6.78). So, we obtain the condition of propagation (6.79) with the dispersion equation (6.81). Because the acoustic tensor is symmetric, and supposed positive definite, then incremental progressive waves can propagate in any direction, with a velocity of propagation of the wave defined by $v = \omega/p$.

It is known that, in the particular case of a cubic crystal, the elasticity tensor contains three independent constants (see [14] or [28]). Using Voigt's convention we have:

$$\mathbf{c} = \begin{pmatrix} c_{11} & c_{12} & c_{12} & 0 & 0 & 0 \\ c_{12} & c_{11} & c_{12} & 0 & 0 & 0 \\ c_{12} & c_{12} & c_{11} & 0 & 0 & 0 \\ 0 & 0 & 0 & c_{44} & 0 & 0 \\ 0 & 0 & 0 & 0 & c_{44} & 0 \\ 0 & 0 & 0 & 0 & 0 & c_{44} \end{pmatrix}. \tag{6.98}$$

Among the five symmetry classes belonging to the cubic system, only $\bar{4}3m$ and 23 classes exhibit the piezoelectric effect, for the others (i.e., $m3m, 432, m3$) the piezoelectric effect being absent. In the first case, the

Table 6.1 The influence of initial strain fields on wave velocities, for longitudinal initial electric field

Wave Velocities	0% Initial Strain m/s	1% Initial Strain m/s	2% Initial Strain m/s
v_1	5969	5999/5939	6028/5909
$v_2 = v_3$	3763	3782/3744	3800/3725

piezoelectric tensor contains only one constant:

$$\mathbf{e} = \begin{pmatrix} 0 & 0 & 0 & e_{14} & 0 & 0 \\ 0 & 0 & 0 & 0 & e_{14} & 0 \\ 0 & 0 & 0 & 0 & 0 & e_{14} \end{pmatrix}, \tag{6.99}$$

and the dielectric tensor has one constant, for all five symmetry classes:

$$\boldsymbol{\eta} = \begin{pmatrix} \eta & 0 & 0 \\ 0 & \eta & 0 \\ 0 & 0 & \eta \end{pmatrix}. \tag{6.100}$$

In this case, to obtain the acoustic tensor $\overset{\circ}{\boldsymbol{Q}}$, we calculate the components of the tensor $\overset{\circ}{\boldsymbol{A}}$, the vector $\overset{\circ}{\boldsymbol{\Gamma}}$, and the coefficient $\overset{\circ}{\Gamma}$ in the following forms (which must be multiplied by p^2):

$$\overset{\circ}{A}_{11} = (c_{11} + \overset{\circ}{S}_{11})n_1^2 + (c_{44} + \overset{\circ}{S}_{22})n_2^2 + (c_{44} + \overset{\circ}{S}_{33})n_3^2 - \eta \, \overset{\circ}{E}_1^2$$
$$+ 2\,\overset{\circ}{S}_{12}\,n_1 n_2 + 2\,\overset{\circ}{S}_{13}\,n_1 n_3 + 2(\overset{\circ}{S}_{23} - 2e_{14}\,\overset{\circ}{E}_1)n_2 n_3$$
$$\overset{\circ}{A}_{12} = \overset{\circ}{A}_{21} = -\eta\,\overset{\circ}{E}_1\overset{\circ}{E}_2 + (c_{12} + c_{44})n_1 n_2 - 2e_{14}\,\overset{\circ}{E}_1\,n_1 n_3 - 2e_{14}\,\overset{\circ}{E}_2\,n_2 n_3$$
$$\overset{\circ}{A}_{13} = \overset{\circ}{A}_{31} = -\eta\,\overset{\circ}{E}_1\overset{\circ}{E}_3 + (c_{12} + c_{44})n_1 n_3 - 2e_{14}\,\overset{\circ}{E}_1\,n_1 n_2 - 2e_{14}\,\overset{\circ}{E}_3\,n_2 n_3$$

$$\overset{\circ}{A}_{22} = (c_{44} + \overset{\circ}{S}_{11})n_1^2 + (c_{11} + \overset{\circ}{S}_{22})n_2^2 + (c_{44} + \overset{\circ}{S}_{33})n_3^2 - \eta \, \overset{\circ}{E}_2^2$$
$$+ 2\,\overset{\circ}{S}_{12}\,n_1 n_2 + 2\,\overset{\circ}{S}_{23}\,n_2 n_3 + 2(\overset{\circ}{S}_{13} - 2e_{14}\,\overset{\circ}{E}_2)n_1 n_3$$
$$\overset{\circ}{A}_{23} = \overset{\circ}{A}_{32} = -\eta\,\overset{\circ}{E}_2\overset{\circ}{E}_3 + (c_{12} + c_{44})n_2 n_3 - 2e_{14}\,\overset{\circ}{E}_2\,n_1 n_2 - 2e_{14}\,\overset{\circ}{E}_3\,n_1 n_3$$
$$\overset{\circ}{A}_{33} = (c_{44} + \overset{\circ}{S}_{11})n_1^2 + (c_{44} + \overset{\circ}{S}_{22})n_2^2 + (c_{11} + \overset{\circ}{S}_{33})n_3^2 - \eta \, \overset{\circ}{E}_3^2$$
$$+ 2\,\overset{\circ}{S}_{13}\,n_1 n_3 + 2\,\overset{\circ}{S}_{23}\,n_2 n_3 + 2(\overset{\circ}{S}_{12} - 2e_{14}\,\overset{\circ}{E}_3)n_1 n_2, \tag{6.101}$$

respectively:

$$\overset{\circ}{\Gamma}_1 = \eta\,\overset{\circ}{E}_1 + 2e_{14}n_2 n_3,$$
$$\overset{\circ}{\Gamma}_2 = \eta\,\overset{\circ}{E}_2 + 2e_{14}n_1 n_3,$$
$$\overset{\circ}{\Gamma}_3 = \eta\,\overset{\circ}{E}_3 + 2e_{14}n_1 n_2, \qquad \overset{\circ}{\Gamma} = 1 + \eta. \tag{6.102}$$

Here n_i, $i = 1, 2, 3$, are the components of the direction of propagation vector $\boldsymbol{n}$ in an orthonormal basis oriented after the cube's edges. These coefficients generalise obviously, in the case of initial electromechanical fields, those presented in [14].

6.3.2.1 Propagation Along a Cube Edge

If we consider the problem of progressive wave propagation along the $[001]$ edge of a cubic crystal, we take $n_1 = n_2 = 0$ and $n_3 = 1$. In this case the dispersion equation (6.81) has the form:

$$f(V) = \begin{vmatrix} a_c - V & b_c & c_c \\ b_c & d_c - V & e_c \\ c_c & e_c & f_c - V \end{vmatrix} = 0, \tag{6.103}$$

where $V = \overset{\circ}{\rho}\, v^2$. From (6.101) we obtain the coefficients having the following form.

$$a_c = c_{44} + \overset{\circ}{S}_{33} - \frac{\eta}{1+\eta}\, \overset{\circ}{E}{}_1^2, \qquad b_c = -\frac{\eta}{1+\eta}\, \overset{\circ}{E}_1 \overset{\circ}{E}_2,$$

$$c_c = -\frac{\eta}{1+\eta}\, \overset{\circ}{E}_1 \overset{\circ}{E}_3, \qquad d_c = c_{44} + \overset{\circ}{S}_{33} - \frac{\eta}{1+\eta}\, \overset{\circ}{E}{}_2^2,$$

$$e_c = -\frac{\eta}{1+\eta}\, \overset{\circ}{E}_2 \overset{\circ}{E}_3, \qquad f_c = c_{11} + \overset{\circ}{S}_{33} - \frac{\eta}{1+\eta}\, \overset{\circ}{E}{}_3^2. \tag{6.104}$$

Even if the crystals possess piezoelectric behaviour, in this case the wave propagation is piezoelectrically inactive, depending on the electrostrictive effect, only.

The condition of propagation of progressive waves (6.79) reduces to the following eigenvector problem:

$$\begin{cases} a_c\, a_1 + b_c\, a_2 + c_c\, a_3 = V\, a_1 \\ b_c\, a_1 + d_c\, a_2 + e_c\, a_3 = V\, a_2 \\ c_c\, a_1 + e_c\, a_2 + f_c\, a_3 = V\, a_3 \end{cases} \tag{6.105}$$

with a_c, b_c, c_c, d_c, e_c, f_c given by (6.104) and a_i, $i = 1, 2, 3$, being the components of the unknown amplitude vector $\mathbf{a}$.

(a) *Longitudinal initial electric field* $(\overset{\circ}{E}_1 = \overset{\circ}{E}_2 = 0,\ \overset{\circ}{E}_3 \neq 0)$

In this particular case the previous dispersion equation (6.103) has three roots, corresponding to the following wave velocities,

$$v_1 = v_2 = \sqrt{\frac{\left(c_{44} + \overset{\circ}{S}_{33}\right)}{\overset{\circ}{\rho}}}, \qquad v_3 = \sqrt{\frac{\left(c_{11} + \overset{\circ}{S}_{33} - \dfrac{\eta}{1+\eta}\, \overset{\circ}{E}{}_3^2\right)}{\overset{\circ}{\rho}}}. \tag{6.106}$$

We can easily see from the condition of propagation (6.105) that the velocities $v_1 = v_2$ correspond to transverse waves, arbitrarly polarised. Moreover, v_3 is the velocity of a longitudinal wave.

(b) *Transverse initial electric field* ($\overset{\circ}{E}_1 \neq 0,\ \overset{\circ}{E}_2 \neq 0,\ \overset{\circ}{E}_3 = 0$)

In this case the dispersion equation (6.103) has three distinct roots, corresponding to the following wave velocities,

$$v_1 = \sqrt{\frac{\left(c_{11} + \overset{\circ}{S}_{33}\right)}{\overset{\circ}{\rho}}}, \qquad v_2 = \sqrt{\frac{\left(c_{44} + \overset{\circ}{S}_{33}\right)}{\overset{\circ}{\rho}}},$$

$$v_3 = \sqrt{\frac{\left(c_{44} + \overset{\circ}{S}_{33} - \frac{\eta}{1+\eta}(\overset{\circ}{E}_1^2 + \overset{\circ}{E}_2^2)\right)}{\overset{\circ}{\rho}}}. \tag{6.107}$$

As we analyse the polarisation of these waves, from Equations (6.105) we find that v_2 is the velocity of a transverse wave with linear polarisation, fixed by the initial electric field. Its amplitude components satisfy the relation $\overset{\circ}{E}_1\, a_1 + \overset{\circ}{E}_2\, a_2 = 0$. v_3 corresponds to a transverse wave with linear polarisation, fixed by the initial electric field, normal to the previous polarisation direction. The amplitude components satisfy the relation $\overset{\circ}{E}_2\, a_1 - \overset{\circ}{E}_1\, a_2 = 0$. Finally, we can easily see that v_1 is the velocity of a longitudinal wave.

6.3.2.2 Propagation Along a Cube Face

For the problem of plane wave propagation along the (001) plane of a cubic crystal, we choose $n_1 = \cos\varphi$, $n_2 = \sin\varphi$, and $n_3 = 0$, where φ is the angle between the propagation direction $\boldsymbol{n}$ and the [100] axis of the crystal.

Consequently, the dispersion equation (6.81) has, in this case, the form:

$$g(V) = \begin{vmatrix} a_c{}' - V & b_c{}' & c_c{}' \\ b_c{}' & d_c{}' - V & e_c{}' \\ c_c{}' & e_c{}' & f_c{}' - V \end{vmatrix} = 0, \tag{6.108}$$

where $V = \overset{\circ}{\rho}\, v^2$, and the coefficients have the following form,

$$a_c{}' = c_{11}\cos^2\varphi + c_{44}\sin^2\varphi + A(\varphi) - \frac{\eta}{1+\eta}\,\overset{\circ}{E}_1^2,$$

$$b_c{}' = \frac{c_{12} + c_{44}}{2}\sin 2\varphi - \frac{\eta}{1+\eta}\,\overset{\circ}{E}_1\overset{\circ}{E}_2,$$

$$c_c{}' = -\frac{e_{14}\,\overset{\circ}{E}_1}{1+\eta}\sin 2\varphi - \frac{\eta}{1+\eta}\,\overset{\circ}{E}_1\overset{\circ}{E}_3,$$

$$d_c{}' = c_{44}\cos^2\varphi + c_{11}\sin^2\varphi + A(\varphi) - \frac{\eta}{1+\eta}\,\overset{\circ}{E}_2^2,$$

$$e_c{}' = -\frac{e_{14}\,\overset{\circ}{E}_2}{1+\eta}\sin 2\varphi - \frac{\eta}{1+\eta}\,\overset{\circ}{E}_2\overset{\circ}{E}_3,$$

$$f_c{}' = c_{44} + A(\varphi) - \overset{\circ}{E}_3^2 + \frac{(e_{14}\sin 2\varphi - \overset{\circ}{E}_3)^2}{1+\eta}, \tag{6.109}$$

with

$$A(\varphi) = \overset{\circ}{S}_{11}\cos^2\varphi + \overset{\circ}{S}_{22}\sin^2\varphi + \overset{\circ}{S}_{12}\sin 2\varphi. \tag{6.110}$$

The condition of propagation of progressive waves (6.79) reduces in this case to the following eigenvector problem.

$$\begin{cases} a_c{}'\,a_1 + b_c{}'\,a_2 + c_c{}'\,a_3 = V\,a_1 \\ b_c{}'\,a_1 + d_c{}'\,a_2 + e_c{}'\,a_3 = V\,a_2 \\ c_c{}'\,a_1 + e_c{}'\,a_2 + f_c{}'\,a_3 = V\,a_3 \end{cases} \tag{6.111}$$

with $a_c{}'$, $b_c{}'$, $c_c{}'$, $d_c{}'$, $e_c{}'$, $f_c{}'$ given by (6.109) and a_i, $i = 1,2,3$, being the components of the unknown amplitude vector **a**.

(a) *Transverse initial electric field* ($\overset{\circ}{E}_1 = 0$, $\overset{\circ}{E}_2 = 0$, $\overset{\circ}{E}_3 \neq 0$)

In this particular case the dispersion equation (6.108) has three distinct roots, corresponding to the following wave velocities.

$$v_1 = \sqrt{\frac{\left(c_{11} + c_{44} + 2A(\varphi) + \sqrt{(c_{11} - c_{44})^2\cos^2 2\varphi + (c_{12} + c_{44})^2\sin^2 2\varphi}\right)}{2\,\overset{\circ}{\rho}}}, \tag{6.112}$$

$$v_2 = \sqrt{\frac{\left(c_{11} + c_{44} + 2A(\varphi) - \sqrt{(c_{11} - c_{44})^2\cos^2 2\varphi + (c_{12} + c_{44})^2\sin^2 2\varphi}\right)}{2\,\overset{\circ}{\rho}}}, \tag{6.113}$$

respectively:

$$v_3 = \sqrt{\frac{\left(c_{44} + A(\varphi) - \overset{\circ}{E}_3^2 + \frac{(e_{14}\sin 2\varphi - \overset{\circ}{E}_3)^2}{1+\eta}\right)}{\overset{\circ}{\rho}}}. \tag{6.114}$$

As regards the polarisation of these waves, we can see from the condition of propagation (6.111), that the velocity v_3, which is piezoelectrically and electrostrictively coupled, corresponds to a pure transverse wave. The remaining two waves are polarised in the plane (001) on mutually perpendicular directions, one being quasi-transverse, and the other quasi-longitudinal. It is interesting to note that the directions of polarisation of the last two waves are not influenced by the initial fields.

6.3.2.3 Anisotropy Factor. Slowness Surfaces

In the problem of plane wave propagation in the plane (001) of a cubic crystal, subject to initial transverse electric field and to an initial mechanical stress field, we consider three particular cases.

- For $\varphi = 0$ (i.e., $[100]$ axis), relations (6.112) and (6.113) yield the wave velocities:

$$v_1^L = \sqrt{\frac{\left(c_{11} + \overset{\circ}{S}_{11}\right)}{\overset{\circ}{\rho}}}, \qquad v_2^T = \sqrt{\frac{\left(c_{44} + \overset{\circ}{S}_{11}\right)}{\overset{\circ}{\rho}}}. \tag{6.115}$$

The first one corresponds to a pure longitudinal wave, whereas the second refers to a pure transverse one.

- For $\varphi = \pi/2$ (i.e., $[010]$ axis), from relations (6.112) and (6.113) we obtain the following wave velocities:

$$v_1^L = \sqrt{\frac{\left(c_{11} + \overset{\circ}{S}_{22}\right)}{\overset{\circ}{\rho}}}, \qquad v_2^T = \sqrt{\frac{\left(c_{44} + \overset{\circ}{S}_{22}\right)}{\overset{\circ}{\rho}}}, \tag{6.116}$$

the first one corresponding to a pure longitudinal wave, and the second to a pure transverse wave.

- For $\varphi = \pi/4$ (i.e., $[110]$ axis) we easily obtain that:

$$v_1^L = \sqrt{\frac{\left(c_{11} + c_{12} + 2c_{44} + \overset{\circ}{S}_{11} + \overset{\circ}{S}_{22} + 2\,\overset{\circ}{S}_{12}\right)}{2\,\overset{\circ}{\rho}}},$$

$$v_2^T = \sqrt{\frac{\left(c_{11} - c_{12} + \overset{\circ}{S}_{11} + \overset{\circ}{S}_{22} + 2\,\overset{\circ}{S}_{12}\right)}{2\,\overset{\circ}{\rho}}}. \tag{6.117}$$

These waves polarise longitudinal, respectively, transverse.

The ratio of the transverse wave velocities in the directions $[100]$ and $[110]$ defines the *generalised anisotropy factor* $\overset{\circ}{\mathcal{A}}$, in the following manner:

$$\overset{\circ}{\mathcal{A}} = \left(\frac{v_2^T[100]}{v_2^T[110]}\right)^2 = \frac{2(c_{44} + \overset{\circ}{S}_{11})}{c_{11} - c_{12} + \overset{\circ}{S}_{11} + \overset{\circ}{S}_{22} + 2\,\overset{\circ}{S}_{12}}. \tag{6.118}$$

It generalises the classical anisotropy factor $\mathcal{A}$, as defined in [14]:

$$\mathcal{A} = \frac{2c_{44}}{c_{11} - c_{12}}.$$

(6.119)

From relations (6.118) and (6.119) it is obvious that only the initial stress components influence the generalised anisotropy factor, and not the initial electric field. Most common, the anisotropy factors are greater than unity, but there are few crystals for which this factor is less than unity (e.g., $Bi_{12}GeO_{20}$ crystal from 23 symmetry class). It is evident that the anisotropy factor is unity for isotropic materials. This is not the case for the generalised anisotropy factor, which may be different from unity, even if the material is isotropic, due to the induced anisotropy generated by the initial stress components $\overset{\circ}{S}_{11}$, $\overset{\circ}{S}_{22}$, and $\overset{\circ}{S}_{12}$.

To study the influence of the initial fields on the generalised anisotropy factor $\overset{\circ}{\mathcal{A}}$, we consider two crystals: GaAs, from the symmetry class $\overline{4}3\,m$, and $Bi_{12}GeO_{20}$, from the symmetry class 23. In Table 6.2 we show the variation of the generalised anisotropy factor $\overset{\circ}{\mathcal{A}}$ with the initial mechanical field, for 1% initial hydrostatic deformations, respectively, 2% initial hydrostatic deformations. We remark that increasing the initial deformations leads $\overset{\circ}{\mathcal{A}}$ to approach unity.

As regards the angles β_1 and β_2 between the quasi-longitudinal and quasi-transverse amplitude vectors, related to velocities (6.112) and (6.113), and the Ox_1 axis (i.e., the angles between the direction of polarisation for QL and QT waves and the [100] axis of the crystal), they are the solutions of the following equation,

$$\tan 2\beta = -\frac{2b_c{}'}{d_c{}' - a_c{}'} = \frac{c_{12} + c_{44}}{c_{11} - c_{44}} \tan 2\varphi,$$

(6.120)

where the relations (6.109) are used for $\overset{\circ}{E}_1 = 0$, $\overset{\circ}{E}_2 = 0$, $\overset{\circ}{E}_3 \neq 0$. This equation, which is piezoelectrically inactive, has the same form as in the case without initial fields, expressing the fact that the initial fields do not modify the polarisation directions in the (001) plane.

Finally, we present the influence of initial fields on the shape of slowness surfaces. These surfaces are defined to be the locus of the end of the

Table 6.2 The dependency of the anisotropy factors on the initial deformations field

Anisotropy Factor	Initial Deformations Field (%)	GaAs	$Bi_{12}GeO_{20}$
$\mathcal{A}$	–	1.828	0.559
$\overset{\circ}{\mathcal{A}}$	1	1.798	0.570
	2	1.771	0.581

vector $\mathbf{n}/v$, where $\mathbf{n}$ is the direction of propagation and v is the velocity of propagation of the plane wave. The properties of these surfaces are derived in monographs [7, 14]. The slowness surfaces are analogous to index surfaces from optics, refraction surfaces from reflection/refraction problems. For a long time, it was believed that these surfaces were convex (see, e.g., [7]). Actually, one can find several crystals having concave parts on their slowness surfaces. This fact is linked to the anisotropy factor, which can be less than unity in particular directions of propagation.

In Figure 6.1 we represent sections of the slowness surfaces with the plane (001) for a $Bi_{12}GeO_{20}$ crystal subject to a 1% (resp., 2%) initial hydrostatic deformation field, and in Figure 6.2 are shown the sections of the slowness surfaces for the 1% (resp., 2%) initial shear deformation field. We remark that, even if the initial deformations are small, their influence on the shape of slowness surfaces is important. Here the continuous lines represent the slowness curves in the case without initial fields, and the broken lines show the corresponding curves in the case with initial fields.

6.3.3 6 mm-Type Crystals. Generalised Electromechanical Coupling Coefficient, Slowness Surfaces

In this section we study the conditions of propagation for plane harmonic waves in crystals from an hexagonal system, of 6 mm type, subject to initial electromechanical fields. Following the papers [27, 16, 15] we show that plane progressive waves propagate along the symmetry axis, in the plane normal to the symmetry axis (resp., in the meridian plane of the crystal). For the last two cases we study the influence of initial fields on slowness surfaces; we define and analyse the generalised electromechanical coupling coefficient.

In this problem we aim to satisfy the homogeneous field equations (6.65) and the constitutive equations (6.62) and (6.63). In order to study the conditions of propagation for incremental harmonic plane waves in an unbounded three-dimensional material described by the previous constitutive equations, we assume that the displacement vector and the electric potential have the form (6.78). Thus we obtain the condition of propagation (6.79) with the dispersion equation (6.81). Because the acoustic tensor is symmetric, and supposed positive definite, then incremental progressive waves can propagate in any direction, with a velocity of propagation of the wave defined by $v = \omega/p$.

It is known that, in the particular case of an hexagonal crystal, the elasticity tensor contains five independent constants (see [14] or [28]).

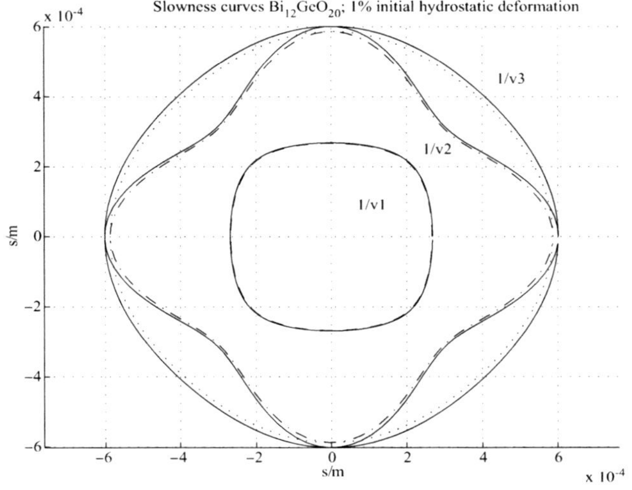

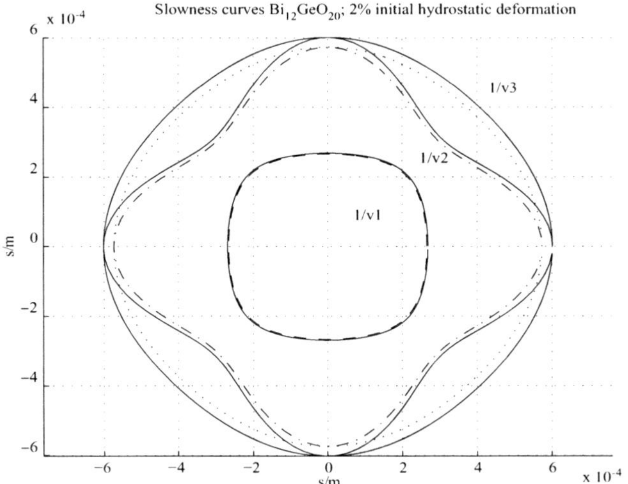

Fig. 6.1 Section of slowness surfaces with (001) plane for a Bi$_{12}$GeO$_{20}$ crystal: (a) 1% initial hydrostatic deformations; (b) 2% initial hydrostatic deformations.

Using Voigt's convention we have:

$$
\mathbf{c} = \begin{pmatrix}
c_{11} & c_{12} & c_{13} & 0 & 0 & 0 \\
c_{12} & c_{11} & c_{13} & 0 & 0 & 0 \\
c_{13} & c_{13} & c_{33} & 0 & 0 & 0 \\
0 & 0 & 0 & c_{44} & 0 & 0 \\
0 & 0 & 0 & 0 & c_{44} & 0 \\
0 & 0 & 0 & 0 & 0 & c_{66}
\end{pmatrix},
\tag{6.121}
$$

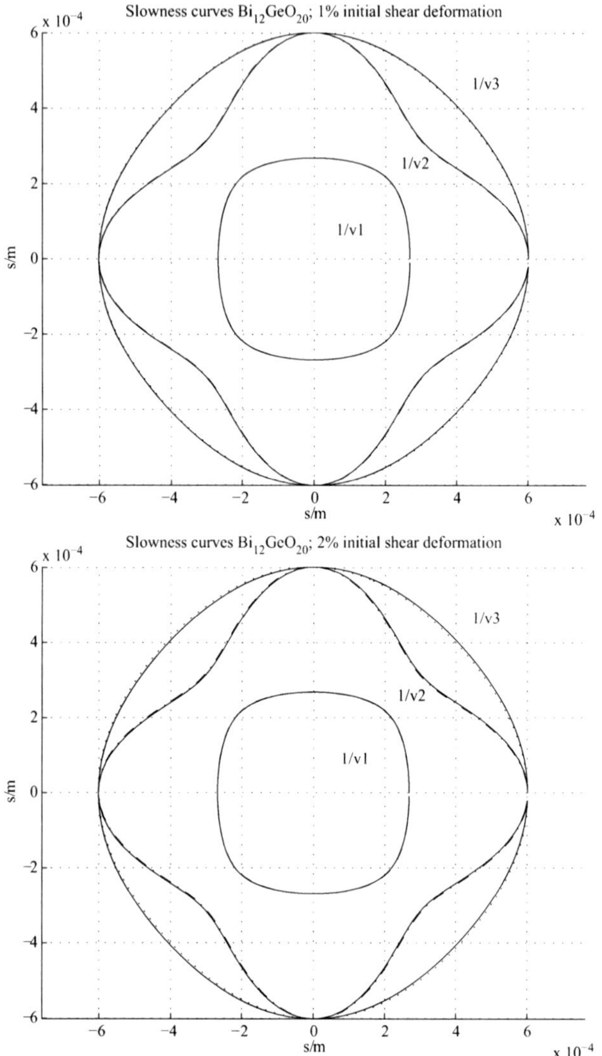

Fig. 6.2 Section of slowness surfaces with (001) plane for a Bi$_{12}$GeO$_{20}$ crystal: (a) 1% initial shear deformations; (b) 2% initial shear deformations.

with $c_{66} = (c_{11} - c_{12})/2$. The piezoelectric tensor contains three independent constants:

$$\mathbf{e} = \begin{pmatrix} 0 & 0 & 0 & 0 & e_{15} & 0 \\ 0 & 0 & 0 & e_{15} & 0 & 0 \\ e_{31} & e_{31} & e_{33} & 0 & 0 & 0 \end{pmatrix}, \tag{6.122}$$

and the dielectric tensor has two independent constants:

$$\boldsymbol{\eta} = \begin{pmatrix} \eta_{11} & 0 & 0 \\ 0 & \eta_{11} & 0 \\ 0 & 0 & \eta_{33} \end{pmatrix}. \tag{6.123}$$

6.3.3.1 Propagation Along the Symmetry Axis

Assuming that the symmetry axis of the crystal coincides with the x_3-axis, we study the propagation of progressive waves along this axis (i.e., $n_1 = n_2 = 0$ and $n_3 = 1$). Therefore, the tensor $\overset{\circ}{\boldsymbol{A}}$ has the following components (which must be multiplied by p^2),

$$\overset{\circ}{A}_{11} = c_{44} + \overset{\circ}{S}_{33} - \eta_{33}\,\overset{\circ}{E}_1^2, \qquad \overset{\circ}{A}_{12} = -\eta_{33}\,\overset{\circ}{E}_1\overset{\circ}{E}_2,$$

$$\overset{\circ}{A}_{13} = -e_{33}\,\overset{\circ}{E}_1 - \eta_{33}\,\overset{\circ}{E}_1\overset{\circ}{E}_3,$$

$$\overset{\circ}{A}_{21} = \overset{\circ}{A}_{12}, \qquad \overset{\circ}{A}_{22} = c_{44} + \overset{\circ}{S}_{33} - \eta_{33}\,\overset{\circ}{E}_2^2,$$

$$\overset{\circ}{A}_{23} = -e_{33}\,\overset{\circ}{E}_2 - \eta_{33}\,\overset{\circ}{E}_2\overset{\circ}{E}_3,$$

$$\overset{\circ}{A}_{31} = \overset{\circ}{A}_{13}, \qquad \overset{\circ}{A}_{32} = \overset{\circ}{A}_{23}, \qquad \overset{\circ}{A}_{33} = c_{33} + \overset{\circ}{S}_{33} - 2e_{33}\,\overset{\circ}{E}_3 - \eta_{33}\,\overset{\circ}{E}_3^2 \tag{6.124}$$

and the components of $\overset{\circ}{\boldsymbol{\Gamma}}$ and $\overset{\circ}{\Gamma}$ have the form (the components must be multiplied by p^2):

$$\overset{\circ}{\Gamma}_1 = \eta_{33}\,\overset{\circ}{E}_1, \qquad \overset{\circ}{\Gamma}_2 = \eta_{33}\,\overset{\circ}{E}_2,$$

$$\overset{\circ}{\Gamma}_3 = \eta_{33}\,\overset{\circ}{E}_3 + e_{33}, \qquad \overset{\circ}{\Gamma} = \epsilon_{33}. \tag{6.125}$$

Consequently, the dispersion equation (6.81) has in this case the following form,

$$f(V) = \begin{vmatrix} \overset{\circ}{A}_{11} + \dfrac{\overset{\circ}{\Gamma}_1^2}{\overset{\circ}{\Gamma}} - V & \overset{\circ}{A}_{12} + \dfrac{\overset{\circ}{\Gamma}_1\overset{\circ}{\Gamma}_2}{\overset{\circ}{\Gamma}} & \overset{\circ}{A}_{13} + \dfrac{\overset{\circ}{\Gamma}_1\overset{\circ}{\Gamma}_3}{\overset{\circ}{\Gamma}} \\[2ex] \overset{\circ}{A}_{12} + \dfrac{\overset{\circ}{\Gamma}_1\overset{\circ}{\Gamma}_2}{\overset{\circ}{\Gamma}} & \overset{\circ}{A}_{22} + \dfrac{\overset{\circ}{\Gamma}_2^2}{\overset{\circ}{\Gamma}} - V & \overset{\circ}{A}_{23} + \dfrac{\overset{\circ}{\Gamma}_2\overset{\circ}{\Gamma}_3}{\overset{\circ}{\Gamma}} \\[2ex] \overset{\circ}{A}_{13} + \dfrac{\overset{\circ}{\Gamma}_1\overset{\circ}{\Gamma}_3}{\overset{\circ}{\Gamma}} & \overset{\circ}{A}_{23} + \dfrac{\overset{\circ}{\Gamma}_2\overset{\circ}{\Gamma}_3}{\overset{\circ}{\Gamma}} & \overset{\circ}{A}_{33} + \dfrac{\overset{\circ}{\Gamma}_3^2}{\overset{\circ}{\Gamma}} - V \end{vmatrix} = 0, \tag{6.126}$$

where $V = \overset{\circ}{\rho}\, v^2$.

To obtain closed-form solutions of this equation, we suppose that the initial electric field has small intensity. Consequently, we neglect all the products of the initial electric field components. Thus, we derive:

$$f(V) = \begin{vmatrix} a_h - V & 0 & -b_h \\ 0 & a_h - V & -c_h \\ -b_h & -c_h & d_h - V \end{vmatrix} = 0 \qquad (6.127)$$

with
$$a_h = c_{44} + \overset{\circ}{S}_{33}, \qquad b_h = \frac{e_{33}}{\epsilon_{33}} \overset{\circ}{E}_1,$$

$$c_h = \frac{e_{33}}{\epsilon_{33}} \overset{\circ}{E}_2, \qquad d_h = c_{33} + \overset{\circ}{S}_{33} + \frac{e_{33}^2 - 2e_{33} \overset{\circ}{E}_3}{\epsilon_{33}}. \qquad (6.128)$$

It is easy to see that this equation reduces to:

$$f(V) = (a_h - V)[V^2 - (a_h + d_h)V + a_h d_h - b_h^2 - c_h^2] = 0. \qquad (6.129)$$

Using realistic values of the initially applied fields and of the material constants (see [14]) we can assume that b_h^2 and c_h^2 may be neglected, compared to the product $a_h d_h$, in Equation(6.129). Indeed, for piezoelectric coefficients on the order of $10\ \mathrm{C/m}^2 = 10^6\ \mathrm{Pa}^{1/2}$, for dielectric coefficients on the order of $10^{-8}\ \mathrm{F/m} = 10^2$ and for elastic constants of order $10^{11}\ \mathrm{Pa}$ we obtain, for an initial electric field of order $10^6\ \mathrm{V/m} = 10\ \mathrm{Pa}^{1/2}$, that b_h^2 and c_h^2 are of order $10^{10}\ \mathrm{Pa}^2$, and $a_h d_h$ is of order $10^{14}\ \mathrm{Pa}^2$. Moreover, even if the initial electric field is of order $10^2\ \mathrm{Pa}^{1/2}$, b_h^2 and c_h^2 are of order $10^{12}\ \mathrm{Pa}^2$, and $a_h d_h$ is of order $10^{15}\ \mathrm{Pa}^2$.

Consequently, in the case of wave propagation along the x_3-axis, the dispersion relation (6.129) has three real roots $V_1 = V_2 = c_{44} + \overset{\circ}{S}_{33}$ and $V_3 = c_{33} + \overset{\circ}{S}_{33} + ((e_{33}^2 - 2e_{33} \overset{\circ}{E}_3)/\epsilon_{33})$. Thus, using the previous assumptions, we obtain the velocities of propagation as closed-form solutions:

$$v_{QL} = \sqrt{\frac{\left(c_{33} + \overset{\circ}{S}_{33} + \dfrac{e_{33}^2 - 2e_{33} \overset{\circ}{E}_3}{\epsilon_{33}} \right)}{\overset{\circ}{\rho}}}, \qquad v_T = \sqrt{\frac{(c_{44} + \overset{\circ}{S}_{33})}{\overset{\circ}{\rho}}}. \qquad (6.130)$$

Regarding the condition of propagation of progressive waves (6.79), it reduces in our case to the following eigenvector problem,

$$\begin{cases} a_h\, a_1 - b_h\, a_3 = \overset{\circ}{\rho}\, v^2\, a_1 \\ a_h\, a_2 - c_h\, a_3 = \overset{\circ}{\rho}\, v^2\, a_2 \\ -b_h\, a_1 - c_h\, a_2 + d_h\, a_3 = \overset{\circ}{\rho}\, v^2\, a_3, \end{cases} \qquad (6.131)$$

with a_h, b_h, c_h, d_h given by (6.128) and a_i, $i = 1, 2, 3$ are the components of the unknown amplitude vector.

- If we put into the last system $v = v_T$, it reduces to $a_3 = 0$ and $b_h\, a_1 + c_h\, a_2 = 0$. It follows that v_T is the velocity of propagation of a transverse wave T, the amplitude vector having an arbitrary module a and a fixed direction in the $x_1 x_2$ plane, determined by equation $\overset{\circ}{E}_1\, x_1 + \overset{\circ}{E}_2\, x_2 = 0$.
- If we consider in (6.131) that $v = v_{QL}$, we obtain the amplitude vector:

$$\boldsymbol{a} = \left(\frac{b_h}{a_h - d_h} \boldsymbol{i}_1 + \frac{c_h}{a_h - d_h} \boldsymbol{i}_2 + \boldsymbol{i}_3 \right) a_3 \qquad (6.132)$$

with a_3 an arbitrary real constant and $\boldsymbol{i}_j$, $j = 1, 2, 3$, the unitary vectors of the x_j-axes. The last of equations (6.131) is satisfied, because we neglected b_h^2 and c_h^2. Consequently, from the form (6.132) of the amplitude vector we conclude that v_{QL} is the velocity of propagation of a quasi-longitudinal wave QL with an arbitrary module.

We define the *generalized coupling coefficient* $\overset{\circ}{K}$, concerning the QL wave propagation along the x_3-axis in the case of presence of initial fields, by the formula:

$$\overset{\circ}{K} = \frac{|e_{33}|}{\sqrt{\epsilon_{33}(c_{33} + \overset{\circ}{S}_{33}) + e_{33}^2 - 2e_{33}\,\overset{\circ}{E}_3}}, \qquad (6.133)$$

which generalises the coupling coefficient K, for the L wave, from the case without initial fields, as defined in [14]:

$$K = \frac{|e_{33}|}{\sqrt{\epsilon_{33} c_{33} + e_{33}^2}}. \qquad (6.134)$$

Analysing the influence of the initial fields on the coupling coefficient $\overset{\circ}{K}$, for three crystals of 6 mm-type (PZT-4, ZnO, and CdS) and for the proposed restrictions on the initial electric field ($10\ \mathrm{Pa}^{1/2}$, or $10^2\ \mathrm{Pa}^{1/2}$), we find that the initial electric field has small influence on the coupling coefficient. On the other hand, an important influence is due to the initial deformation field, via the $\overset{\circ}{S}_{33}$ stress component. In what follows, all the numerical calculations are made using the material constants given in [14].

In Table 6.3 we show the dependency of the generalised coupling coefficient for an estimated initial deformation of order 1%, (resp., 2%). The T wave, with the velocity of propagation v_T, is not piezoelectrically coupled, which is also the case without initial fields.

Table 6.3 The dependency of the coupling coefficient on the initial strain field. Propagation along x_3-axis

Coupling Coefficient	Initial Strain Field (%)	PZT-4	ZnO	CdS
K	–	0.511	0.270	0.154
$\overset{\circ}{K}$	1	0.5094	0.2688	0.1539
	2	0.5076	0.2676	0.1531

6.3.3.2 Propagation in the $x_1 x_2$ Plane

We consider now the problem of progressive wave propagation in the $x_1 x_2$ plane. In this case, we must take $n_1 = \cos\theta$, $n_2 = \sin\theta$, and $n_3 = 0$, where θ is the polar angle in this plane. Therefore, the tensor $\overset{\circ}{\boldsymbol{A}}$ has the following components (which must be multiplied by p^2):

$$\overset{\circ}{A}_{11} = c_{11}\cos^2\theta + c_{66}\sin^2\theta + A(\theta) - \eta_{11}\,\overset{\circ}{E}_1^2,$$

$$\overset{\circ}{A}_{12} = \overset{\circ}{A}_{21} = (c_{66} + c_{12})\frac{\sin 2\theta}{2} - \eta_{11}\,\overset{\circ}{E}_1\overset{\circ}{E}_2,$$

$$\overset{\circ}{A}_{13} = \overset{\circ}{A}_{31} = -e_{15}\,\overset{\circ}{E}_1 - \eta_{11}\,\overset{\circ}{E}_1\overset{\circ}{E}_3,$$

$$\overset{\circ}{A}_{22} = c_{66}\cos^2\theta + c_{11}\sin^2\theta + A(\theta) - \eta_{11}\,\overset{\circ}{E}_2^2,$$

$$\overset{\circ}{A}_{23} = \overset{\circ}{A}_{32} = -e_{15}\,\overset{\circ}{E}_2 - \eta_{11}\,\overset{\circ}{E}_2\overset{\circ}{E}_3,$$

$$\overset{\circ}{A}_{33} = c_{44} + A(\theta) - 2e_{15}\,\overset{\circ}{E}_3 - \eta_{11}\,\overset{\circ}{E}_3^2, \tag{6.135}$$

where

$$A(\theta) = \overset{\circ}{S}_{11}\cos^2\theta + \overset{\circ}{S}_{22}\sin^2\theta + \overset{\circ}{S}_{12}\sin 2\theta. \tag{6.136}$$

The components of $\overset{\circ}{\boldsymbol{\Gamma}}$ and $\overset{\circ}{\Gamma}$ become (the components must be multiplied by p^2):

$$\overset{\circ}{\Gamma}_1 = \eta_{11}\,\overset{\circ}{E}_1, \qquad \overset{\circ}{\Gamma}_2 = \eta_{11}\,\overset{\circ}{E}_2, \qquad \overset{\circ}{\Gamma}_3 = \eta_{11}\,\overset{\circ}{E}_3 + e_{15}, \qquad \overset{\circ}{\Gamma} = \epsilon_{11}. \tag{6.137}$$

In this case the dispersion equation has the form (6.126), with the new coefficients obtained previously. Using the approximations discussed in the last paragraph we obtain:

$$g(V) = \begin{vmatrix} a'_h + A(\theta) - V & b'_h & c'_h \\ b'_h & d'_h + A(\theta) - V & e'_h \\ c'_h & e'_h & f'_h + A(\theta) - V \end{vmatrix} = 0, \tag{6.138}$$

where

$$a'_h = c_{11}\cos^2\theta + c_{66}\sin^2\theta, \qquad b'_h = (c_{66} + c_{12})\frac{\sin2\theta}{2}, \qquad c'_h = -\frac{e_{15}\,\overset{\circ}{E}_1}{\epsilon_{11}},$$

$$d'_h = c_{66}\cos^2\theta + c_{11}\sin^2\theta, \qquad e'_h = -\frac{e_{15}\,\overset{\circ}{E}_2}{\epsilon_{11}}, \qquad f'_h = c_{44} + \frac{e_{15}^2 - 2e_{15}\,\overset{\circ}{E}_3}{\epsilon_{11}}.$$

$$(6.139)$$

We can easily see (neglecting ${c'_h}^2$, ${e'_h}^2$, and $c'_h e'_h$) that (6.138) possesses three distinct roots, corresponding to the wave velocities:

$$v_1 = \sqrt{\frac{\left(c_{44} + \dfrac{e_{15}^2 - 2e_{15}\,\overset{\circ}{E}_3}{\epsilon_{11}} + \overset{\circ}{S}_{11}\cos^2\theta + \overset{\circ}{S}_{22}\sin^2\theta + \overset{\circ}{S}_{12}\sin2\theta\right)}{\overset{\circ}{\rho}}},$$

$$(6.140)$$

$$v_2 = \sqrt{\frac{\left(c_{11} + \overset{\circ}{S}_{11}\cos^2\theta + \overset{\circ}{S}_{22}\sin^2\theta + \overset{\circ}{S}_{12}\sin2\theta\right)}{\overset{\circ}{\rho}}}, \qquad (6.141)$$

respectively:

$$v_3 = \sqrt{\frac{\left(c_{66} + \overset{\circ}{S}_{11}\cos^2\theta + \overset{\circ}{S}_{22}\sin^2\theta + \overset{\circ}{S}_{12}\sin2\theta\right)}{\overset{\circ}{\rho}}}. \qquad (6.142)$$

The condition of propagation of progressive waves in the $x_1 x_2$ plane reduces to the following eigenvector problem,

$$\begin{cases} [a'_h + A(\theta)]\,a_1 + b'_h\,a_2 + c'_h\,a_3 = \overset{\circ}{\rho}\,v^2\,a_1 \\ b'_h\,a_1 + [d'_h + A(\theta)]\,a_2 + e'_h\,a_3 = \overset{\circ}{\rho}\,v^2\,a_2 \\ c'_h\,a_1 + e'_h\,a_2 + [f'_h + A(\theta)]\,a_3 = \overset{\circ}{\rho}\,v^2\,a_3, \end{cases} \qquad (6.143)$$

with a'_h, b'_h, c'_h, d'_h, e'_h, f'_h given by (6.139) and a_i, $i = 1, 2, 3$, the components of the unknown amplitude vector.

- If $v = v_1$, the previous propagation condition has the solution:

$$a_1 = \frac{e'_h(a'_h - f'_h) - b'_h c'_h}{{b'_h}^2 - (a'_h - f'_h)(d'_h - f'_h)}\,a_3, \qquad a_2 = \frac{c'_h(d'_h - f'_h) - b'_h e'_h}{{b'_h}^2 - (a'_h - f'_h)(d'_h - f'_h)}\,a_3,$$

$$(6.144)$$

with a_3 arbitrary, because the determinant of the corresponding system vanishes considering the approximations made. This is a quasi-transverse wave (QT_1 wave).

- If $v = v_2$ the system (6.143) has the following solution,

$$a_1 = \frac{f'_h - c_{11}}{c'_h(d'_h - c_{11}) - b'_h e'_h} a_3, \qquad a_2 = \frac{b'_h(f'_h - c_{11})}{c'_h(d'_h - c_{11}) - b'_h e'_h} a_3, \quad (6.145)$$

with a_3 arbitrary, characterizing a quasi-longitudinal wave (QL wave).

- Finally, if $v = v_3$, we obtain the solution:

$$a_1 = \frac{f'_h - c_{66}}{c'_h(d'_h - c_{66}) - b'_h e'_h} a_3, \qquad a_2 = \frac{b'_h(f'_h - c_{66})}{c'_h(d'_h - c_{66}) - b'_h e'_h} a_3, \quad (6.146)$$

with a_3 arbitrary, which is a quasi-transverse wave (QT_2 wave).

We remark that in all these cases the polarisation of the waves is restricted only by the initial electric field components, and not by the components of the initial stress field. If there is not an initial applied electric field, the polarisation of the waves is arbitrary.

In this case, we define the *generalised coupling coefficient* by the following formula:

$$\overset{\circ}{K^*}(\theta) = \frac{|e_{15}|}{\sqrt{\epsilon_{11} c_{44} + e_{15}^2 - 2 e_{15} \overset{\circ}{E}_3 + A(\theta)}}. \quad (6.147)$$

It generalises the coupling coefficient from the case without initial fields (see [14]):

$$K^*(\theta) = \frac{|e_{15}|}{\sqrt{\epsilon_{11} c_{44} + e_{15}^2}}. \quad (6.148)$$

The previous coupling coefficient $\overset{\circ}{K^*}$ is linked to the QT_1 wave, which has the velocity v_1, whereas the other two waves, QL and QT_2, propagating with velocities v_2 (resp., v_3) are not piezoelectrically active, similar to the case without initial fields. The influence of the initial electric field on the coupling coefficient $\overset{\circ}{K^*}$ is weak, which is not the case of initial deformation field, as we can see in Table 6.4. There we compute the values of $\overset{\circ}{K^*}$ for three 6 mm-type

Table 6.4 The dependency of the coupling coefficient on the initial strain field. Propagation in $x_1 x_2$-plane

Coupling Coefficient	Initial Strain Field (%)	PZT-4	ZnO	CdS
K^*	–	0.7015	0.3160	0.1890
$\overset{\circ}{K^*}$	1	0.7014	0.3069	0.1827
	2	0.7012	0.2986	0.1770

crystals and for two initial strain fields of order 1% (resp., 2%) generated by the initial hydrostatic stress fields $\overset{\circ}{S}_{11} = \overset{\circ}{S}_{22}$. Even if for the PZT-4 crystal this influence remains small, for ZnO and CdS crystals it is important.

In Figure 6.3 we plot the sections of the slowness surfaces with the x_1x_2 plane, for a PZT-4 crystal and for previously described initial hydrostatic strain fields. We remark that for an initial deformation field of order 2% the differences between slowness curves without initial fields (continuous lines) and those with initial fields (broken lines) are important.

Finally, in Figure 6.4 we show the sections of the slowness surfaces with the x_1x_2 plane, for a PZT-4 crystal and for initial shear strain fields of order 1% (resp., 2%). We remark that for an initial deformation field of order 2% the

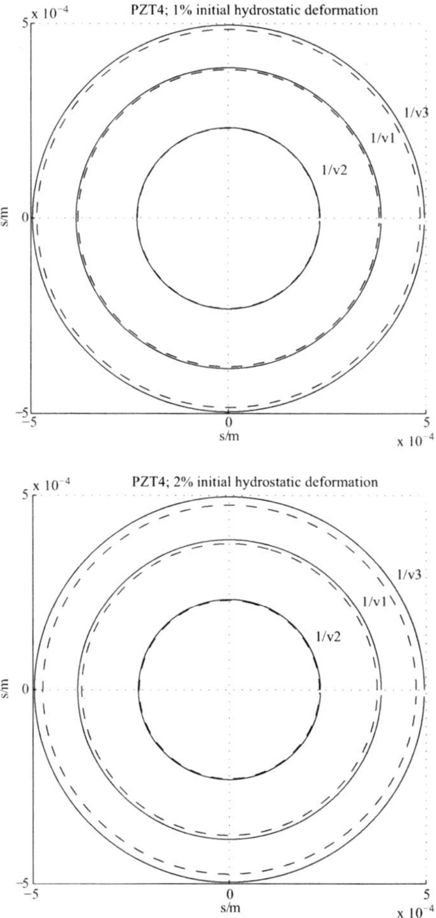

Fig. 6.3 Section of slowness surfaces with x_1x_2 plane for PZT-4 crystal: (a) 1% initial hydrostatic deformations; (b) 2% initial hydrostatic deformations.

differences between slowness curves without initial fields (continuous lines) and those with initial fields (broken lines) become important, changing the shape of the curves and attaining the maximum value at $\theta = 45°$, $135°$, $225°$, and $315°$. This may have great importance if we study the reflection–refraction problem for this new situation.

6.3.3.3 Propagation in the Meridian Plane

If we take into account the problem of progressive wave propagation in the meridian plane of a 6 mm-type crystal, we choose $n_1 = 0$, $n_2 = \sin \varphi$, and

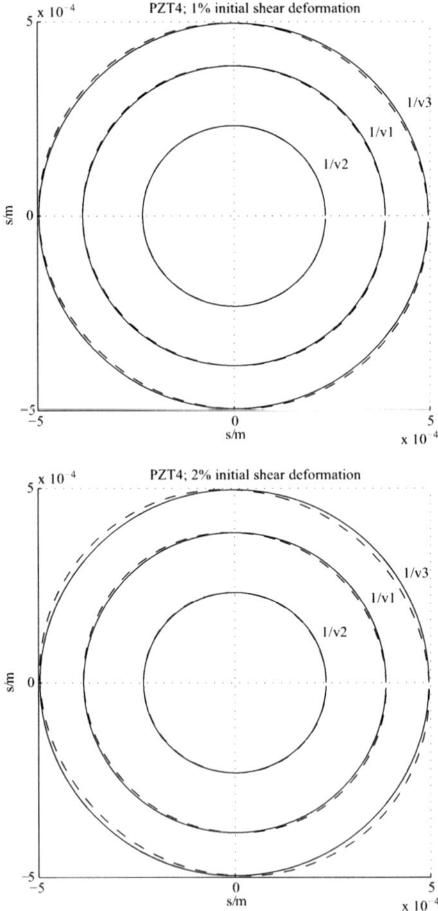

Fig. 6.4 Section of slowness surfaces with x_1x_2 plane for PZT-4 crystal: (a) 1% initial shear deformations; (b) 2% initial shear deformations.

$n_3 = \cos \varphi$, where φ is the angle between the propagation direction and the A_6 symmetry axis of the crystal. Therefore, the symmetric tensor $\overset{\circ}{A}$ has the following components (which must be multiplied by p^2).

$$\overset{\circ}{A}_{11} = (c_{66} - \eta_{11} \overset{\circ}{E}_1^2)\sin^2\varphi + (c_{44} - \eta_{33} \overset{\circ}{E}_1^2)\cos^2\varphi + B(\varphi),$$

$$\overset{\circ}{A}_{12} = -\eta_{11} \overset{\circ}{E}_1 \overset{\circ}{E}_2 \sin^2\varphi - \eta_{33} \overset{\circ}{E}_1 \overset{\circ}{E}_2 \cos^2\varphi - (e_{15} + e_{31}) \overset{\circ}{E}_1 \frac{\sin2\varphi}{2},$$

$$\overset{\circ}{A}_{13} = (-e_{15} \overset{\circ}{E}_1 - \eta_{11} \overset{\circ}{E}_1 \overset{\circ}{E}_3)\sin^2\varphi + (-e_{33} \overset{\circ}{E}_1 - \eta_{33} \overset{\circ}{E}_1 \overset{\circ}{E}_3)\cos^2\varphi,$$

$$\overset{\circ}{A}_{22} = (c_{11} - \eta_{11} \overset{\circ}{E}_2^2)\sin^2\varphi + (c_{44} - \eta_{33} \overset{\circ}{E}_2^2)\cos^2\varphi$$
$$\qquad - (e_{15} + e_{31}) \overset{\circ}{E}_2 \sin2\varphi + B(\varphi),$$

$$\overset{\circ}{A}_{23} = (-e_{15} \overset{\circ}{E}_2 - \eta_{11} \overset{\circ}{E}_2\overset{\circ}{E}_3)\sin^2\varphi + (-e_{33} \overset{\circ}{E}_2 - \eta_{33} \overset{\circ}{E}_2\overset{\circ}{E}_3)\cos^2\varphi$$
$$\qquad + [c_{13} + c_{44} - (e_{15} + e_{31}) \overset{\circ}{E}_3]\frac{\sin2\varphi}{2},$$

$$\overset{\circ}{A}_{33} = (c_{44} - 2e_{15} \overset{\circ}{E}_3 - \eta_{11} \overset{\circ}{E}_3^2)\sin^2\varphi + (c_{33} - 2e_{33} \overset{\circ}{E}_3$$
$$\qquad - \eta_{33} \overset{\circ}{E}_3^2)\cos^2\varphi + B(\varphi), \tag{6.149}$$

where

$$B(\varphi) = \overset{\circ}{S}_{33}\cos^2\varphi + \overset{\circ}{S}_{22}\sin^2\varphi + \overset{\circ}{S}_{23}\sin2\varphi. \tag{6.150}$$

The components of $\overset{\circ}{\boldsymbol{\Gamma}}$ and $\overset{\circ}{\Gamma}$ become (the expressions must be multiplied by p^2):

$$\overset{\circ}{\Gamma}_1 = \eta_{11} \overset{\circ}{E}_1 \sin^2\varphi + \eta_{33} \overset{\circ}{E}_1 \cos^2\varphi,$$

$$\overset{\circ}{\Gamma}_2 = \eta_{11} \overset{\circ}{E}_2 \sin^2\varphi + \eta_{33} \overset{\circ}{E}_2 \cos^2\varphi + (e_{15} + e_{31})\frac{\sin2\varphi}{2},$$

$$\overset{\circ}{\Gamma}_3 = (e_{15} + \eta_{11} \overset{\circ}{E}_3)\sin^2\varphi + (e_{33} + \eta_{33} \overset{\circ}{E}_3)\cos^2\varphi,$$

$$\overset{\circ}{\Gamma} = 1 + \eta_{11}\sin^2\varphi + \eta_{33}\cos^2\varphi. \tag{6.151}$$

Consequently, the dispersion equation has in this case the form (6.126), with the previous coefficients. Neglecting all the products of the initial electric field components, we obtain:

$$h(V) = \begin{vmatrix} a_h{}'' + B(\varphi) - V & b_h{}'' & c_h{}'' \\ b_h{}'' & d_h{}'' + B(\varphi) - V & e_h{}'' \\ c_h{}'' & e_h{}'' & f_h{}'' + B(\varphi) - V \end{vmatrix} = 0, \tag{6.152}$$

where $V = \overset{\circ}{\rho}\, v^2$ and

$$a_h{}'' = c_{66}\sin^2\varphi + c_{44}\cos^2\varphi,$$

$$b_h{}'' = -\frac{(e_{15} + e_{31})\sin2\varphi\, \overset{\circ}{E}_1}{2(1 + \eta_{11}\sin^2\varphi + \eta_{33}\cos^2\varphi)},$$

$$c_h{}'' = -\frac{(e_{15}\sin^2\varphi + e_{33}\cos^2\varphi)\, \overset{\circ}{E}_1}{1 + \eta_{11}\sin^2\varphi + \eta_{33}\cos^2\varphi},$$

$$d_h{}'' = c_{11}\sin^2\varphi + c_{44}\cos^2\varphi$$
$$+ \frac{(e_{15} + e_{31})\sin2\varphi[(e_{15} + e_{31})\sin2\varphi/4 - \overset{\circ}{E}_2]}{1 + \eta_{11}\sin^2\varphi + \eta_{33}\cos^2\varphi},$$

$$e_h{}'' = (c_{13} + c_{44})\frac{\sin2\varphi}{2} - \frac{(e_{15}\sin^2\varphi + e_{33}\cos^2\varphi)\, \overset{\circ}{E}_2}{1 + \eta_{11}\sin^2\varphi + \eta_{33}\cos^2\varphi}$$
$$+ \frac{(e_{15} + e_{31})\sin2\varphi(e_{15}\sin^2\varphi + e_{33}\cos^2\varphi - \overset{\circ}{E}_3)}{2(1 + \eta_{11}\sin^2\varphi + \eta_{33}\cos^2\varphi)},$$

$$f_h{}'' = c_{44}\sin^2\varphi + c_{33}\cos^2\varphi - \frac{2(e_{15}\sin^2\varphi + e_{33}\cos^2\varphi)\, \overset{\circ}{E}_3}{1 + \eta_{11}\sin^2\varphi + \eta_{33}\cos^2\varphi}$$
$$+ \frac{(e_{15}\sin^2\varphi + e_{33}\cos^2\varphi)^2}{1 + \eta_{11}\sin^2\varphi + \eta_{33}\cos^2\varphi}. \tag{6.153}$$

As we can see from these relations, the coefficients involved in the dispersion equation (6.152) have more complicated forms than those from the cases analysed previously (i.e., for wave propagation along the symmetry axis and in the plane normal to this axis). In the absence of the initial fields one obtains the same coefficients as in [14]. Even so, the dispersion equation (6.152) has three distinct real roots, corresponding to three wave velocities:

$$v_1 = \sqrt{\frac{(a_h{}'' + B(\varphi))}{\overset{\circ}{\rho}}}, \tag{6.154}$$

$$v_2 = \sqrt{\frac{\left(d_h{}'' + f_h{}'' + 2B(\varphi) + \sqrt{(d_h{}'' - f_h{}'')^2 + 4e_h{}''^2}\right)}{2\overset{\circ}{\rho}}}, \tag{6.155}$$

respectively,

$$v_3 = \sqrt{\frac{\left(d_h{}'' + f_h{}'' + 2B(\varphi) - \sqrt{(d_h{}'' - f_h{}'')^2 + 4e_h{}''^2}\right)}{2\overset{\circ}{\rho}}}. \tag{6.156}$$

The condition of propagation of progressive waves (6.79) reduces, in this particular case, to the following eigenvector problem.

$$\begin{cases} \left[a_h{}'' + B(\varphi)\right] a_1 + b_h{}'' a_2 + c_h{}'' a_3 = \overset{\circ}{\rho}\, v^2\, a_1 \\ b_h{}'' a_1 + \left[d_h{}'' + B(\varphi)\right] a_2 + e_h{}'' a_3 = \overset{\circ}{\rho}\, v^2\, a_2 \\ c_h{}'' a_1 + e_h{}'' a_2 + \left[f_h{}'' + B(\varphi)\right] a_3 = \overset{\circ}{\rho}\, v^2\, a_3 \end{cases} \tag{6.157}$$

with $a_h{}''$, $b_h{}''$, $c_h{}''$, $d_h{}''$, $e_h{}''$, $f_h{}''$ given by (6.153) and a_i, $i = 1, 2, 3$, being the components of the unknown amplitude vector $\mathbf{a}$.

- If $v = v_1$ the previous propagation condition has the solution:

$$\begin{aligned} a_1 &= \frac{e_h{}''^2 + a_h{}'' f_h{}'' - a_h{}''^2 - d_h{}'' f_h{}'' + a_h{}'' d_h{}''}{c_h{}'' d_h{}'' - b_h{}'' e_h{}'' - a_h{}'' c_h{}''}\, a_3, \\ a_2 &= \frac{b_h{}'' f_h{}'' - a_h{}'' b_h{}'' - c_h{}'' e_h{}''}{c_h{}'' d_h{}'' - b_h{}'' e_h{}'' - a_h{}'' c_h{}''}\, a_3 \end{aligned} \tag{6.158}$$

with a_3 arbitrary, because the determinant of the corresponding system (6.157) vanishes considering the approximations made (i.e., $b''^2 = c''^2 = b'' c'' = 0$, neglecting all the products of the initial electric field components).

If we analyse this solution, we can see that it corresponds to the transverse wave from the case of absence of initial fields. In our case, this wave is a quasi-transverse one, its polarisation being affected by the initial electric field only. Moreover, it preserves a velocity piezoelectrically inactive, without influence of the initial electric field, but determined by the initial stress field.

- If $v = v_{2,3}$ the system (6.157) has the following form.

$$\begin{cases} \left[2\,a_h'' - d_h{}'' - f_h{}'' \pm \sqrt{(d_h{}'' - f_h{}'')^2 + 4e_h{}''^2}\right] a_1 + 2b_h{}'' a_2 + 2c_h{}'' a_3 = 0 \\ 2b_h{}'' a_1 + \left[d_h{}'' - f_h{}'' \pm \sqrt{(d_h{}'' - f_h{}'')^2 + 4e_h{}''^2}\right] a_2 + 2e_h{}'' a_3 = 0 \\ 2c_h{}'' a_1 + 2e_h{}'' a_2 + \left[-d_h{}'' + f_h{}'' \pm \sqrt{(d_h{}'' - f_h{}'')^2 + 4e_h{}''^2}\right] a_3 = 0. \end{cases} \tag{6.159}$$

One obtains easily that its determinant is zero, neglecting the products of the initial electric field components. Consequently, we can express a_1, a_2 as functions of a_3. On the other hand, as the coefficients of the system (6.159) depend on the initial electric field components only, we can conclude that the waves with velocities (6.155) and (6.156) are quasi-longitudinal, (resp., quasi-transverse), their polarisation depending only on the initial electric field.

In Figures 6.5 and 6.6 we plot the sections of slowness surfaces with the meridian plane $x_2 x_3$, in the case of initial stress fields that generate initial strain fields of order 1% (resp., 2%), and for initial electric field components $\overset{\circ}{E}_1 = \overset{\circ}{E}_2 = \overset{\circ}{E}_3 = 10^2$ Pa, concerning PZT-4 and ZnO crystals. In our numerical computations all the material constants are taken from the book [14].

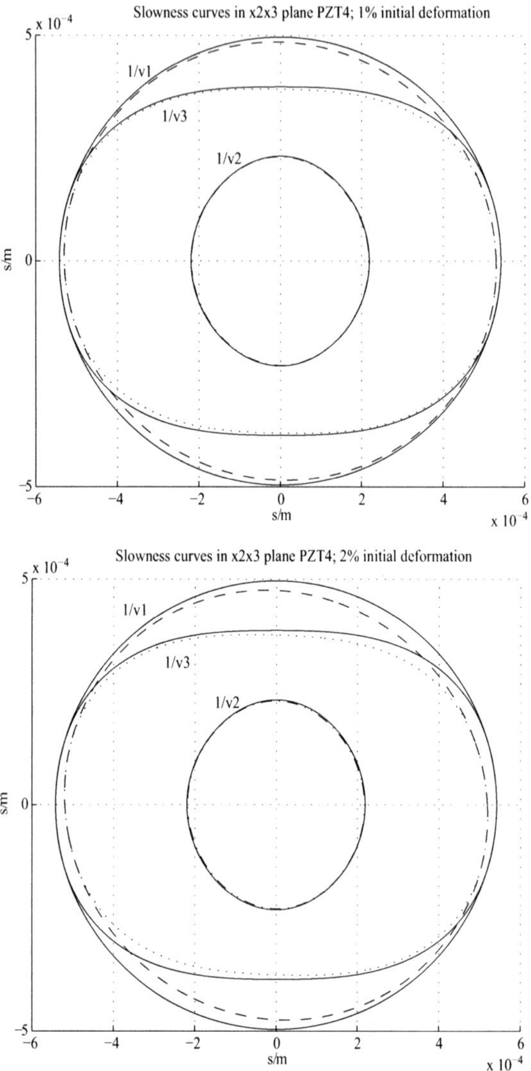

Fig. 6.5 Section of slowness surfaces with the meridian plane for PZT-4 crystal: (a) 1% initial deformation; (b) 2% initial deformation.

We note the differences between slowness curves ignoring initial fields (continuous lines) and those including the previous initial fields' influence (broken lines), which grow with the initial deformation field magnitude, and the fact that the initial fields may change the convexity of these curves, which could be important in reflection–refraction problems. It is interesting to note the asymmetry of the gap between the continuous and the broken lines, for angles φ and $\varphi + 90°$, due to the shear term from the initial stress field.

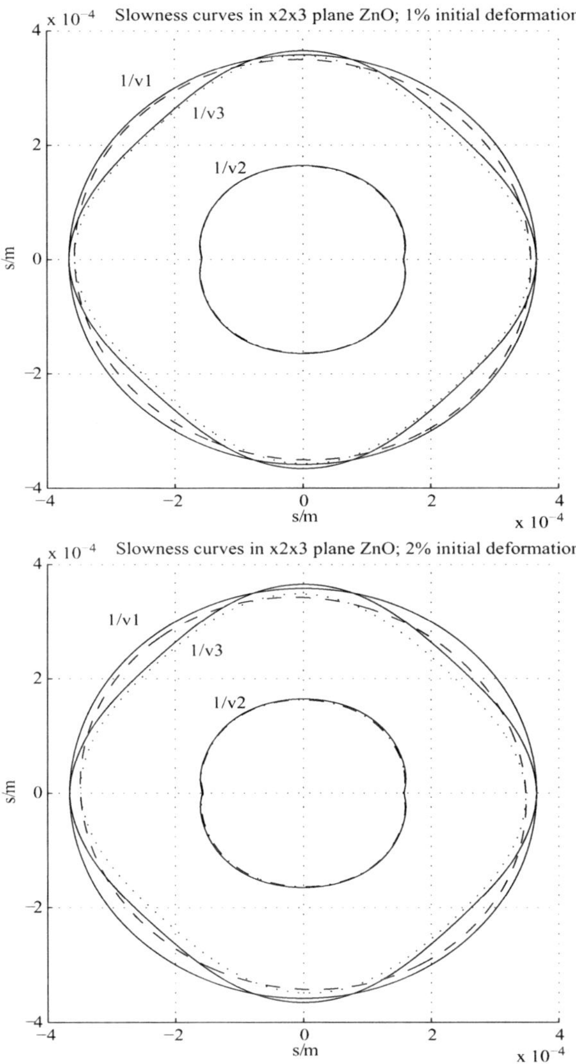

Fig. 6.6 Section of slowness surfaces with the meridian plane for ZnO crystal: (a) 1% initial deformation; (b) 2% initial deformation.

6.4 Propagation of Attenuated Waves in Crystals Subject to Initial Electromechanical Fields

In this part, we deal with the study of the conditions of propagation of attenuated waves in isotropic solids and in cubic crystals, subject to initial deformation and electric fields (see papers [19–24]). We show the influence of the electrostrictive and piezoelectric effects on attenuated wave propagation in such media. We derive the velocities of propagation and the attenuation coefficients in closed-form, and we analyse the influence of the initial fields on the wave polarisation. For a particular choice of the initial electric field, we obtain and analyse the generalised anisotropy factor. We derive approximate expressions of the obtained solutions in order to compare them with the classical solutions.

6.4.1 Fundamental Equations. Acoustic Tensor

We assume that we are in the framework described in Section 6.2.3. In this case, we recall the form of the homogeneous field equations:

$$\overset{\circ}{\rho}\,\ddot{\boldsymbol{u}} = \operatorname{div}\boldsymbol{\Sigma}, \qquad \operatorname{div}\boldsymbol{\Delta} = 0, \qquad \operatorname{rot}\boldsymbol{e} = 0 \;\Leftrightarrow\; \boldsymbol{e} = -\operatorname{grad}\phi. \tag{6.160}$$

Furthermore, we generalise the incremental constitutive equations (6.62), in the form:

$$\Sigma_{kl} = \overset{\circ}{\Omega}_{klmn} u_{m,n} + \overset{\circ}{\Lambda}_{mkl}\phi,_{m} + d_{klmn}\dot{u}_{m,n}$$

$$\Delta_k = \overset{\circ}{\Lambda}_{kmn} u_{n,m} + \overset{\circ}{\epsilon}_{kl}e_l = \overset{\circ}{\Lambda}_{kmn}\,u_{n,m} - \overset{\circ}{\epsilon}_{kl}\,\phi,_{l}. \tag{6.161}$$

In these equations $\overset{\circ}{\Omega}_{klmn}$ are the components of the instantaneous elasticity tensor, d_{klmn} are the components of attenuation (damping) tensor, $\overset{\circ}{\Lambda}_{kmn}$ are the components of the instantaneous coupling tensor, and $\overset{\circ}{\epsilon}_{kl}$ are the components of the instantaneous dielectric tensor.

The instantaneous coefficients can be expressed in terms of the classical moduli of the material and on the initial applied fields as follows.

$$\overset{\circ}{\Omega}_{klmn} = \overset{\circ}{\Omega}_{nmlk} = c_{klmn} + \overset{\circ}{S}_{kn}\delta_{lm} - e_{kmn}\overset{\circ}{E}_l - e_{nkl}\overset{\circ}{E}_m - \eta_{kn}\overset{\circ}{E}_l\overset{\circ}{E}_m,$$

$$\overset{\circ}{\Lambda}_{mkl} = e_{mkl} + \eta_{mk}\overset{\circ}{E}_l, \qquad \overset{\circ}{\epsilon}_{kl} = \overset{\circ}{\epsilon}_{lk} = \epsilon_{kl} = \delta_{kl} + \eta_{kl}, \tag{6.162}$$

where c_{klmn} are the components of the constant elasticity tensor, e_{kmn} are the components of the constant piezoelectric tensor, ϵ_{kl} are the components of the constant dielectric tensor, $\overset{\circ}{E}_i$ are the components of the initial applied electric field, and $\overset{\circ}{S}_{kn}$ are the components of the initial applied symmetric (Cauchy) stress tensor.

We remark that the attenuation tensor $\mathbf{d}$ has the same symmetry properties as the elasticity tensor $\mathbf{c}$. Hence, in general there are 21 independent elastic coefficients c_{klmn}, as well as 21 independent attenuation components d_{klmn}, 18 independent piezoelectric coefficients e_{klm}, and 6 independent dielectric coefficients ϵ_{kl}.

We study here the propagation of incremental mechanical attenuated plane waves in an unbounded three-dimensional material described by the previous constitutive equations. Therefore, the displacement vector and the electric potential have the following form.

$$
\begin{aligned}
\boldsymbol{u}(\boldsymbol{x},t) &= \boldsymbol{a}\,\exp(-\boldsymbol{\alpha}\cdot\boldsymbol{x})\exp[\mathrm{i}(\omega t - \boldsymbol{p}\cdot\boldsymbol{x})], \\
\phi(\boldsymbol{x},t) &= \bar{a}\,\exp[\mathrm{i}(\omega t - \boldsymbol{p}\cdot\boldsymbol{x})].
\end{aligned}
\tag{6.163}
$$

Here $\boldsymbol{a}$ and $\bar{a}$ are constants, characterizing the amplitude of the wave, $\boldsymbol{p} = p\,\boldsymbol{n}$ (with $\boldsymbol{n}^2 = 1$) is a constant vector, p representing the wave number and $\boldsymbol{n}$ denoting the direction of propagation of the wave, and $\boldsymbol{\alpha} = \alpha\,\boldsymbol{n}$ (with α defining the attenuation coefficient). Here ω is the frequency of the wave. The velocity of propagation of the wave is defined by $v = \omega/p$. The validity of the hypothesis saying that the direction of propagation coincides with the direction of attenuation was analysed in monograph [11].

Introducing these forms of $\boldsymbol{u}$ and ϕ into the field equations and taking into account the constitutive equations, we obtain the condition of propagation of attenuated waves:

$$
\overset{\circ}{\boldsymbol{Q}}\,\boldsymbol{a} = \overset{\circ}{\rho}\,\omega^2 \boldsymbol{a},
\tag{6.164}
$$

where the components of the instantaneous acoustic tensor $\overset{\circ}{\boldsymbol{Q}}$ have the following form.

$$
\overset{\circ}{Q}_{lm} = \overset{\circ}{\Omega}_{klmn}\,(p_n - \mathrm{i}\alpha_n)(p_k - \mathrm{i}\alpha_k) + \frac{(\overset{\circ}{\Lambda}_{uvl}\,p_u p_v)[\overset{\circ}{\Lambda}_{rsm}\,(p_r - \mathrm{i}\alpha_r)(p_s - \mathrm{i}\alpha_s)]}{\overset{\circ}{\epsilon}_{ij}\,p_i p_j}
$$
$$
+\, \mathrm{i}\omega d_{klmn}(p_k - \mathrm{i}\alpha_k)(p_n - \mathrm{i}\alpha_n).
\tag{6.165}
$$

It is evident, in this case, that these components are complex numbers and that the tensor $\overset{\circ}{\boldsymbol{Q}}$ is not symmetric. Consequently, the arguments used in Section 6.2.5 to derive the condition of propagation, supposing the symmetry and the positive definiteness of the acoustic tensor, are no longer valid here, for the general formulation. However, for isotropic solids and for cubic crystals, assuming a particular form of the initial electric fields, we obtain the solutions of attenuated waves propagation problem.

6.4.2 Isotropic Solids. Approximate Solutions

In the particular case of an isotropic material, the elasticity tensor contains two independent components, namely $c_{11} = \lambda + 2\mu$ and $c_{12} = \lambda$, with $c_{66} = (c_{11} - c_{12})/2 = \mu$ (see (6.83)). Here λ and μ are Lamé's coefficients. Similarly, the attenuation tensor has only two independent components, namely d_{11} and d_{12}. The component $d_{66} = (d_{11} - d_{12})/2$. The dielectric tensor has only one component η, with $\epsilon = 1 + \eta$ (see (6.84)).

When we assume that the direction of propagation coincides with the x_1-axis (i.e., $n_1 = 1$, $n_2 = n_3 = 0$), it yields that the acoustic tensor $\overset{\circ}{\boldsymbol{Q}}$ has the following components,

$$\overset{\circ}{Q}_{lm} = \left[\overset{\circ}{\Omega}_{1lm1} + \frac{\overset{\circ}{\Lambda}_{11l}\overset{\circ}{\Lambda}_{11m}}{\overset{\circ}{\epsilon}_{11}} + i\omega d_{1lm1} \right] (p - i\alpha)^2 = \overset{\circ}{Q}'_{lm}\,(p - i\alpha)^2. \quad (6.166)$$

If we denote by $V' = \overset{\circ}{\rho}\,\omega^2/(p - i\alpha)^2$, the condition of propagation (6.164) will take the form of the following eigenvector problem,

$$\overset{\circ}{\boldsymbol{Q}}'\,\boldsymbol{a} = V'\boldsymbol{a}, \quad \text{or} \quad (\overset{\circ}{Q}'_{lm} - V'\delta_{lm})a_m = 0, \qquad l = \overline{1,3}. \qquad (6.167)$$

This problem is usually associated with the following eigenvalue problem (the characteristic or dispersion equation),

$$\det(\overset{\circ}{Q}'_{lm} - V'\delta_{lm}) = 0. \qquad (6.168)$$

In our case, the components of the tensor $\overset{\circ}{\boldsymbol{Q}}'$ are:

$$\overset{\circ}{Q}'_{11} = a' = c_{11} + i\omega d_{11} + \overset{\circ}{S}_{11} - \frac{\eta}{1+\eta}\,\overset{\circ}{E}_1^2, \qquad \overset{\circ}{Q}'_{12} = b' = -\frac{\eta}{1+\eta}\,\overset{\circ}{E}_1\overset{\circ}{E}_2,$$

$$\overset{\circ}{Q}'_{13} = c' = -\frac{\eta}{1+\eta}\,\overset{\circ}{E}_1\overset{\circ}{E}_3, \qquad \overset{\circ}{Q}'_{22} = d' = c_{66} + i\omega d_{66} + \overset{\circ}{S}_{11} - \frac{\eta}{1+\eta}\,\overset{\circ}{E}_2^2,$$

$$\overset{\circ}{Q}'_{23} = e' = -\frac{\eta}{1+\eta}\overset{\circ}{E}_2\overset{\circ}{E}_3, \qquad \overset{\circ}{Q}'_{33} = f' = c_{66} + i\omega d_{66} + \overset{\circ}{S}_{11} - \frac{\eta}{1+\eta}\,\overset{\circ}{E}_3^2\,.$$

$$(6.169)$$

With these notations, the condition of propagation (6.167) becomes

$$\begin{cases} a'\,a_1 + b'\,a_2 + c'\,a_3 = V'\,a_1 \\ b'\,a_1 + d'\,a_2 + e'\,a_3 = V'\,a_2 \\ c'\,a_1 + e'\,a_2 + f'\,a_3 = V'\,a_3 \end{cases} \qquad (6.170)$$

whereas, the characteristic equation (6.168) has the form

$$F(V') = \begin{vmatrix} a' - V' & b' & c' \\ b' & d' - V' & e' \\ c' & e' & f' - V' \end{vmatrix} = 0. \qquad (6.171)$$

6.4.2.1 Analysis of Particular Cases

In order to obtain the phase velocities and the attenuation coefficients as closed-form solutions, in what follows we present two important particular cases.

(a) *Longitudinal initial electric field* $(\overset{\circ}{E}_1 \neq 0,\ \overset{\circ}{E}_2 = \overset{\circ}{E}_3 = 0)$

In this case, the expressions b', c', e' being zero, the characteristic equation (6.171) has the roots:

$$V_1' = c_{11} + i\omega d_{11} + \overset{\circ}{S}_{11} - \frac{\eta}{1+\eta}\,\overset{\circ}{E}_1^2, \qquad V_2' = V_3' = c_{66} + i\omega d_{66} + \overset{\circ}{S}_{11}.$$

$$(6.172)$$

As regards the polarisation of the obtained waves, using the condition of propagation (6.170) in this particular case, we can easily see that V_1' corresponds to a longitudinal wave, whereas $V_2' = V_3'$ are linked to a transverse wave, arbitrarily polarised.

To find the phase velocities and the attenuation coefficients related to the previous roots, we denote by:

$$V_1' = A_L + iB_L = \frac{\overset{\circ}{\rho}\,\omega^2}{(p_L - i\alpha_L)^2},$$

$$A_L = c_{11} + \overset{\circ}{S}_{11} - \frac{\eta}{1+\eta}\,\overset{\circ}{E}_1^2, \qquad B_L = \omega d_{11}. \qquad (6.173)$$

It yields a phase velocity v_L, in the form:

$$v_L^2 = \frac{\omega^2}{p_L^2} = \frac{2(A_L^2 + B_L^2)}{\overset{\circ}{\rho}\left(\sqrt{A_L^2 + B_L^2} + A_L\right)}, \qquad (6.174)$$

and an attenuation coefficient α_L, done by the relation:

$$\alpha_L^2 = \frac{\overset{\circ}{\rho}\,\omega^2}{2} \cdot \frac{\sqrt{A_L^2 + B_L^2} - A_L}{A_L^2 + B_L^2}. \qquad (6.175)$$

We can conclude that the displacement vector, in this particular case, has the form $\boldsymbol{u}_L = (u_1^L, 0, 0)$, with:

$$u_1^L(x_1, t) = a_1 \exp(-\alpha_L x_1)\exp\left[i\omega\left(t - \frac{x_1}{v_L}\right)\right]. \qquad (6.176)$$

We easily observe that the attenuation affects the phase velocity v_L (by ω), and the amplitude of the longitudinal wave (by α_L). Moreover, the electrostrictive effect is generated by the term $-\eta/(1+\eta)\,\overset{\circ}{E}_1^2$.

To obtain an approximate solution of this problem, we denote by $\bar{\epsilon} = \omega d_{11}/c_{11}$ a nondimensional parameter. Supposing that $\bar{\epsilon} \ll 1$, which is the case for silica SiO_2, where $\bar{\epsilon} = 2.3 \cdot 10^{-3}$, we develop the expression (6.174) and (6.175) of phase velocity, respectively, of attenuation coefficient after $\bar{\epsilon}$. Neglecting the terms containing powers of order greater than, or equal to 1 of $\bar{\epsilon}$, we derive the approximate form of the phase velocity:

$$v_L \simeq v_L^* \sqrt{1 + \psi}, \qquad v_L^* = \sqrt{\frac{c_{11}}{\overset{\circ}{\rho}}},$$

$$\psi = \frac{\overset{\circ}{S}_{11} - \frac{\eta}{1+\eta} \overset{\circ}{E}_1^2}{c_{11}}. \tag{6.177}$$

Here v_L^* is the longitudinal velocity in the classical case, without initial fields, and ψ is a nondimensional parameter describing the influence of the initial fields.

In a similar way, we can derive an approximate form of the attenuation coefficient:

$$\alpha_L \simeq \alpha_L^* \cdot \frac{1}{(1 + \psi)^{3/2}}, \qquad \alpha_L^* = \frac{\tau \omega^2}{2 v_L^*}, \qquad \tau = \frac{d_{11}}{c_{11}}. \tag{6.178}$$

Here α_L^* is the attenuation coefficient in the case without initial fields, defined in [14].

A numerical application for silica (amorphous quartz SiO_2) shows the report of these formulae with previous results. In Table 6.5 we compute the ratios between longitudinal wave velocities in the case without initial fields, and with initial strain fields of order 1%, 2%, and 5%, respectively, of attenuation coefficients (see formulae (6.177) and (6.178)). The influence of the initial electric field on wave velocities and attenuation coefficients is very weak, even if its intensity is important: $\overset{\circ}{E}_1 = 10^3 \sqrt{Pa} = 10^8$ V/m. The superior value corresponds to a traction stress $\overset{\circ}{S}_{11}$, and the inferior value is related to a compression stress $\overset{\circ}{S}_{11}$, that both generate the initial strain fields. One can observe important differences between these values, due to the initial strain fields.

Applying the same procedure as in the case of longitudinal waves, we find the phase velocity and attenuation coefficient for the transverse waves.

Table 6.5 The influence of initial strain fields on wave velocities and attenuation coefficients for longitudinal initial electric field

	Without Initial Strain	1%	2%	5%
v_L/v_L^*	1	1.005/0.995	1.010/0.990	1.025/0.975
α_L/α_L^*	1	0.985/1.015	0.971/1.031	0.929/1.080

So, denoting by

$$V_2' = V_3' = A_T + iB_T = \frac{\overset{\circ}{\rho}\,\omega^2}{(p_T - i\alpha_T)^2},$$

$$A_T = c_{66} + \overset{\circ}{S}_{11}, \qquad B_T = \omega d_{66}, \tag{6.179}$$

we obtain the phase velocity v_T, in the form

$$v_T^2 = \frac{\omega^2}{p_T^2} = \frac{2(A_T^2 + B_T^2)}{\overset{\circ}{\rho}\left(\sqrt{A_T^2 + B_T^2} + A_T\right)}, \tag{6.180}$$

and an attenuation coefficient α_T, done by the relation

$$\alpha_T^2 = \frac{\overset{\circ}{\rho}\,\omega^2}{2} \cdot \frac{\sqrt{A_T^2 + B_T^2} - A_T}{A_T^2 + B_T^2}. \tag{6.181}$$

We can conclude that the displacement vector, in this case, has the form $\boldsymbol{u}_T = (0, u_2^T, u_3^T)$, with:

$$u_k^T(x_1, t) = a_k \exp(-\alpha_T x_1)\exp\left[i\omega\left(t - \frac{x_1}{v_T}\right)\right], \qquad k = 2; 3. \tag{6.182}$$

We observe that the phase velocity v_T depends on ω and the amplitude is affected by α_T. Similar approximate forms for the phase velocity and attenuation coefficient can be obtained in this case.

(b) *Transverse initial electric field* $(\overset{\circ}{E}_1 = 0,\ \overset{\circ}{E}_2 \neq 0,\ \overset{\circ}{E}_3 \neq 0)$

In this case, the coefficients b' and c' being zero, the characteristic equation (6.171) has the following three roots,

$$V_1' = c_{11} + \overset{\circ}{S}_{11} + i\omega d_{11}, \qquad V_2' = c_{66} + \overset{\circ}{S}_{11} + i\omega d_{66},$$

$$V_3' = c_{66} + \overset{\circ}{S}_{11} + i\omega d_{66} - \frac{\eta}{1+\eta}(\overset{\circ}{E}_2^2 + \overset{\circ}{E}_3^2). \tag{6.183}$$

As regards the polarisation of the obtained waves, using the condition of propagation (6.170) in this particular case, we can easily see that V_1' corresponds to a longitudinal wave.

V_2' is linked to a transverse wave, whose polarisation direction is fixed by the initial electric field. Indeed, in this case, the system (6.170) reduces to the equation $\overset{\circ}{E}_2\, a_2 + \overset{\circ}{E}_3\, a_3 = 0$.

V_3' corresponds to a transverse wave, with a direction of polarisation fixed by the initial electric field, done by the equation $\overset{\circ}{E}_3\, a_2 - \overset{\circ}{E}_2\, a_3 = 0$, normal to the preceding direction.

To find the phase velocities and attenuation coefficients related to the previous roots, we proceed as in the case with longitudinal initial electric field. So, we denote by:

$$V_1' = A_L' + iB_L' = \frac{\overset{\circ}{\rho}\,\omega^2}{(p_L - i\alpha_L)^2}, \qquad A_L' = c_{11} + \overset{\circ}{S}_{11}, \qquad B_L' = \omega d_{11}.$$

$$(6.184)$$

It yields a phase velocity v_L, in the form:

$$v_L^2 = \frac{\omega^2}{p_L^2} = \frac{2(A_L'^{\,2} + B_L'^{\,2})}{\overset{\circ}{\rho}\left(\sqrt{A_L'^{\,2} + B_L'^{\,2}} + A_L'\right)} \tag{6.185}$$

and an attenuation coefficient α_L, done by the relation:

$$\alpha_L^2 = \frac{\overset{\circ}{\rho}\,\omega^2}{2} \cdot \frac{\sqrt{A_L'^{\,2} + B_L'^{\,2}} - A_L'}{A_L'^{\,2} + B_L'^{\,2}}. \tag{6.186}$$

We can conclude that the displacement vector, in this case, has the form $\boldsymbol{u}_L = (u_1^L, 0, 0)$, with:

$$u_1^L(x_1, t) = a_1 \exp(-\alpha_L x_1)\exp\left[i\omega\left(t - \frac{x_1}{v_L}\right)\right]. \tag{6.187}$$

We can observe that the attenuation affects the phase velocity v_L (by ω), and the amplitude of the longitudinal wave (by α_L). In this case, the electrostrictive effect is absent.

Applying the same procedure, we find the phase velocity and attenuation coefficient for the transverse waves. So, denoting by

$$V_2' = A_{T_1} + iB_{T_1} = \frac{\overset{\circ}{\rho}\,\omega^2}{(p_{T_1} - i\alpha_{T_1})^2},$$

$$A_{T_1} = c_{66} + \overset{\circ}{S}_{11}, \qquad B_{T_1} = \omega d_{66}, \tag{6.188}$$

we obtain the phase velocity v_{T_1}, in the form:

$$v_{T_1}^2 = \frac{\omega^2}{p_{T_1}^2} = \frac{2(A_{T_1}^2 + B_{T_1}^2)}{\overset{\circ}{\rho}\left(\sqrt{A_{T_1}^2 + B_{T_1}^2} + A_{T_1}\right)}, \tag{6.189}$$

and an attenuation coefficient α_{T_1}, done by the relation:

$$\alpha_{T_1}^2 = \frac{\overset{\circ}{\rho}\,\omega^2}{2} \cdot \frac{\sqrt{A_{T_1}^2 + B_{T_1}^2} - A_{T_1}}{A_{T_1}^2 + B_{T_1}^2}. \tag{6.190}$$

We can see that the displacement vector in this case has the form $\boldsymbol{u}_{T_1} = (0, u_2^{T_1}, u_3^{T_1})$, with:

$$u_k^{T_1}(x_1, t) = a_k \exp(-\alpha_{T_1} x_1) \exp\left[i\omega\left(t - \frac{x_1}{v_{T_1}}\right)\right], \qquad k = 2; 3. \tag{6.191}$$

Similar approximate forms for the phase velocity and attenuation coefficient can be obtained in this case. In conclusion, this transverse wave is attenuated, has the polarisation fixed by the initial electric field, and is not affected by the electrostrictive effect.

Finally, denoting by

$$V_3' = A_{T_2} + iB_{T_2} = \frac{\overset{\circ}{\rho}\,\omega^2}{(p_{T_2} - i\alpha_{T_2})^2},$$

$$A_{T_2} = c_{66} + \overset{\circ}{S}_{11} - \frac{\eta}{1+\eta}(\overset{\circ}{E}_2^2 + \overset{\circ}{E}_3^2), \qquad B_{T_2} = \omega d_{66}, \tag{6.192}$$

we obtain the phase velocity v_{T_2}, in the form:

$$v_{T_2}^2 = \frac{\omega^2}{p_{T_2}^2} = \frac{2(A_{T_2}^2 + B_{T_2}^2)}{\overset{\circ}{\rho}\left(\sqrt{A_{T_2}^2 + B_{T_2}^2} + A_{T_2}\right)} \tag{6.193}$$

and an attenuation coefficient α_{T_2}, done by the relation:

$$\alpha_{T_2}^2 = \frac{\overset{\circ}{\rho}\,\omega^2}{2} \cdot \frac{\sqrt{A_{T_2}^2 + B_{T_2}^2} - A_{T_2}}{A_{T_2}^2 + B_{T_2}^2}. \tag{6.194}$$

We derive that the displacement vector in this case has the form $\boldsymbol{u}_{T_2} = (0, u_2^{T_2}, u_3^{T_2})$, with

$$u_k^{T_2}(x_1, t) = a_k \exp(-\alpha_{T_2} x_1) \exp\left[i\omega\left(t - \frac{x_1}{v_{T_2}}\right)\right], \qquad k = 2; 3. \tag{6.195}$$

Similar approximate forms for the phase velocity and attenuation coefficient can be obtained in this case. In conclusion, this transverse wave is attenuated, has the polarisation fixed by the initial electric field, and is affected by the electrostrictive effect.

6.4.3 Cubic Crystals. Approximate Solutions. Generalised Anisotropy Factor

In this part we analyse the problem of attenuated wave propagation in a cubic crystal subject to initial electromechanical fields. For particular directions of propagation and attenuation we derive the phase velocities, the attenuation coefficients, and we study the corresponding polarisation. For a particular choice of the initial electric field, we obtain and analyse the generalised anisotropy factor.

To study the conditions for propagation of incremental mechanical attenuated waves in an unbounded three-dimensional material, we suppose the homogeneous field equations (6.160) and the incremental constitutive equations (6.161) and (6.162). Supposing that the displacement vector and the electric potential have the form (6.163), we obtain the condition of propagation (6.164).

It is known that, in the case of a cubic crystal, the elasticity tensor contains three independent constants c_{11}, c_{12}, and c_{44} (see (6.98)). Similarly, the attenuation tensor possesses three independent coefficients d_{11}, d_{12}, and d_{44}. Among the five symmetry classes belonging to the cubic system, only $\overline{4}3\mathrm{m}$ and 23 classes exhibit the piezoelectric effect, for the others (i.e., $\mathrm{m}3\mathrm{m}, 432, \mathrm{m}3$) the piezoelectric effect is absent. In case of symmetry classes $\overline{4}3\,\mathrm{m}$ and 23, the piezoelectric tensor contains one constant e_{14}, whereas the dielectric tensor has one constant η, for all five symmetry classes (see (6.99)and (6.100)).

6.4.3.1 Attenuated Wave Propagation Along an Edge of a Cubic Crystal Subject to Initial Fields

To study the attenuated wave propagation along the $[001]$ axis, we assume that the direction of propagation coincides with the x_3-axis (i.e., $n_3 = 1$, $n_1 = n_2 = 0$). It follows that the acoustic tensor $\overset{\circ}{\boldsymbol{Q}}$ has the following components.

$$\overset{\circ}{Q}_{lm} = \left[\overset{\circ}{\Omega}_{3lm3} + \frac{\overset{\circ}{\Lambda}_{33l}\overset{\circ}{\Lambda}_{33m}}{\overset{\circ}{\epsilon}_{33}} + i\omega d_{3lm3} \right] (p - i\alpha)^2 = \overset{\circ}{Q}'_{lm} (p - i\alpha)^2. \quad (6.196)$$

If we denote by $V = \overset{\circ}{\rho}\,\omega^2/(p - i\alpha)^2$, the condition of propagation (6.164) will take the form of an eigenvector problem:

$$\overset{\circ}{\boldsymbol{Q}}'\,\boldsymbol{a} = V\boldsymbol{a}, \quad \text{or} \quad (\overset{\circ}{Q}'_{lm} - V\delta_{lm})a_m = 0, \qquad l = \overline{1,3}. \quad (6.197)$$

This problem is usually associated with the following eigenvalue problem (characteristic equation),

$$\det(\overset{\circ}{Q}'_{lm} - V\delta_{lm}) = 0. \quad (6.198)$$

Note that, for this particular direction of propagation the acoustic tensor becomes symmetric $\overset{\circ}{Q}'_{lm} = \overset{\circ}{Q}'_{ml}$.

Thus, we obtain the components of the tensor $\overset{\circ}{\boldsymbol{Q}}{}'$ in the form:

$$\overset{\circ}{Q}{}'_{11} = a_c = c_{44} + i\omega d_{44} + \overset{\circ}{S}_{33} - \frac{\eta}{1+\eta}\,\overset{\circ}{E}_1^2,$$

$$\overset{\circ}{Q}{}'_{12} = \overset{\circ}{Q}{}'_{21} = b_c = -\frac{\eta}{1+\eta}\,\overset{\circ}{E}_1\overset{\circ}{E}_2,$$

$$\overset{\circ}{Q}{}'_{13} = \overset{\circ}{Q}{}'_{31} = c_c = -\frac{\eta}{1+\eta}\,\overset{\circ}{E}_1\overset{\circ}{E}_3,$$

$$\overset{\circ}{Q}{}'_{22} = d_c = c_{44} + i\omega d_{44} + \overset{\circ}{S}_{33} - \frac{\eta}{1+\eta}\,\overset{\circ}{E}_2^2,$$

$$\overset{\circ}{Q}{}'_{23} = \overset{\circ}{Q}{}'_{32} = e_c = -\frac{\eta}{1+\eta}\,\overset{\circ}{E}_2\overset{\circ}{E}_3,$$

$$\overset{\circ}{Q}{}'_{33} = f_c = c_{11} + i\omega d_{11} + \overset{\circ}{S}_{33} - \frac{\eta}{1+\eta}\,\overset{\circ}{E}_3^2 . \tag{6.199}$$

From the analysis of the form of the previous coefficients, we can easily observe that the piezoelectric effect is absent for this direction of propagation, even if the crystal is piezoelectric active.

With this notation, the condition of propagation (6.197) becomes

$$\begin{cases} a_c\, a_1 + b_c\, a_2 + c_c\, a_3 = V\, a_1 \\ b_c\, a_1 + d_c\, a_2 + e_c\, a_3 = V\, a_2, \\ c_c\, a_1 + e_c\, a_2 + f_c\, a_3 = V\, a_3 \end{cases} \tag{6.200}$$

and the characteristic equation (6.198) has the form

$$G(V) = \begin{vmatrix} a_c - V & b_c & c_c \\ b_c & d_c - V & e_c \\ c_c & e_c & f_c - V \end{vmatrix} = 0. \tag{6.201}$$

In order to obtain the phase velocities and the attenuation coefficients in closed-form, in what follows we present two important particular cases.

(a) *Longitudinal initial electric field* ($\overset{\circ}{E}_1 = \overset{\circ}{E}_2 = 0,\ \overset{\circ}{E}_3 \neq 0$)

This case can be defined as an *electroacoustic Pockels effect* (see [6] for the analogous electro-optical effect). Here, the expressions b_c, c_c, e_c being zero, the characteristic equation (6.201) has the following roots,

$$V_1 = V_2 = c_{44} + i\omega d_{44} + \overset{\circ}{S}_{33},$$

$$V_3 = c_{11} + i\omega d_{11} + \overset{\circ}{S}_{33} - \frac{\eta}{1+\eta}\,\overset{\circ}{E}_3^2 . \tag{6.202}$$

As regards the polarisation of the obtained waves, using the condition of propagation (6.200) in this particular case, we can easily see that V_3 corresponds to a longitudinal wave with electrostrictive effect, and $V_1 = V_2$ are linked to transverse waves, arbitrarily polarised.

To find the phase velocities and attenuation coefficients related to the previous roots, we denote by

$$V_3 = A_L + iB_L = \frac{\overset{\circ}{\rho}\,\omega^2}{(p_L - i\alpha_L)^2},$$

$$A_L = c_{11} + \overset{\circ}{S}_{33} - \frac{\eta}{1+\eta}\,\overset{\circ}{E}_3^2, \qquad B_L = \omega d_{11}. \tag{6.203}$$

It yields a phase velocity v_L, in the form:

$$v_L^2 = \frac{\omega^2}{p_L^2} = \frac{2(A_L^2 + B_L^2)}{\overset{\circ}{\rho}\left(\sqrt{A_L^2 + B_L^2} + A_L\right)}, \tag{6.204}$$

and an attenuation coefficient α_L, given by the relation:

$$\alpha_L^2 = \frac{\overset{\circ}{\rho}\,\omega^2}{2} \cdot \frac{\sqrt{A_L^2 + B_L^2} - A_L}{A_L^2 + B_L^2}. \tag{6.205}$$

We can conclude that the displacement vector in this particular case has the form $\boldsymbol{u}_L = (0, 0, u_3^L)$, with:

$$u_3^L(x_3, t) = a_3 \exp(-\alpha_L x_3)\exp\left[i\omega\left(t - \frac{x_3}{v_L}\right)\right]. \tag{6.206}$$

We easily observe that the attenuation affects the phase velocity v_L (by ω), and the amplitude of the longitudinal wave (by α_L). Moreover, the electrostrictive effect is represented by the term $-\eta/(1 + \eta)\,\overset{\circ}{E}_3^2$.

To obtain an approximate solution of this problem, we denote by $\bar{\epsilon} = \omega d_{11}/c_{11}$ a nondimensional parameter. Supposing that $\bar{\epsilon} \ll 1$, we approximate the expression (6.204) and (6.205) for phase velocity and attenuation coefficient, to first order in $\bar{\epsilon}$.

Neglecting the terms containing powers of order greater than $\bar{\epsilon}$, we derive the approximate form of the phase velocity:

$$v_L \simeq v_L^* \sqrt{1 + \psi}, \qquad v_L^* = \sqrt{\frac{c_{11}}{\overset{\circ}{\rho}}},$$

$$\psi = \frac{\overset{\circ}{S}_{33} - \frac{\eta}{1+\eta}\,\overset{\circ}{E}_3^2}{c_{11}}. \tag{6.207}$$

Here v_L^* is the longitudinal velocity in the classical case, without initial fields, and ψ is a nondimensional parameter describing the influence of the initial fields.

In a similar way, we can derive an approximate form of the attenuation coefficient:

$$\alpha_L \simeq \alpha_L^* \cdot \frac{1}{(1+\psi)^{3/2}}, \qquad \alpha_L^* = \frac{\tau \omega^2}{2 v_L^*}, \qquad \tau = \frac{d_{11}}{c_{11}}. \tag{6.208}$$

Here α_L^* is the attenuation coefficient in the case without initial fields, as defined in [14].

Applying the same procedure as in the case of the longitudinal wave, we find the phase velocity and attenuation coefficients for the transverse waves. So, using the notation:

$$V_1 = V_2 = A_T + iB_T = \frac{\overset{\circ}{\rho}\,\omega^2}{(p_T - i\alpha_T)^2},$$

$$A_T = c_{44} + \overset{\circ}{S}_{33}, \qquad B_T = \omega d_{44}, \tag{6.209}$$

we obtain the phase velocity v_T in the form:

$$v_T^2 = \frac{\omega^2}{p_T^2} = \frac{2(A_T^2 + B_T^2)}{\overset{\circ}{\rho}\left(\sqrt{A_T^2 + B_T^2} + A_T\right)}, \tag{6.210}$$

and the attenuation coefficient α_T, as

$$\alpha_T^2 = \frac{\overset{\circ}{\rho}\,\omega^2}{2} \cdot \frac{\sqrt{A_T^2 + B_T^2} - A_T}{A_T^2 + B_T^2}. \tag{6.211}$$

We can conclude that the displacement vector in this case has the form $\boldsymbol{u}_T = (u_1^T, u_2^T, 0)$, with

$$u_k^T(x_3, t) = a_k \exp(-\alpha_T x_3) \exp\left[i\omega\left(t - \frac{x_3}{v_T}\right)\right], \qquad k = 1; 2. \tag{6.212}$$

We observe that the phase velocity v_T depends on ω and the amplitude is affected by α_T. Similar approximate forms for the phase velocity and attenuation coefficient can be obtained in this case, too.

(b) *Transverse initial electric field* ($\overset{\circ}{E}_1 \neq 0$, $\overset{\circ}{E}_2 \neq 0$, $\overset{\circ}{E}_3 = 0$)

This case can be defined as an *electroacoustic Kerr effect* (see [6] for the analogous electrooptical effect). Here the coefficients c_c and e_c being zero, the characteristic equation (6.201) has the following three roots,

$$V_1' = c_{44} + \overset{\circ}{S}_{33} + i\omega d_{44}, \quad V_2' = c_{44} + \overset{\circ}{S}_{33} + i\omega d_{44} - \frac{\eta}{1+\eta}(\overset{\circ}{E}_1^2 + \overset{\circ}{E}_2^2),$$

$$V_3' = c_{11} + \overset{\circ}{S}_{33} + i\omega d_{11}. \tag{6.213}$$

As regards the polarisation of the obtained waves, using the condition of propagation (6.200) in this particular case, we can easily see that V_3' corresponds to a longitudinal wave, and V_1' is linked to a transverse wave, whose polarisation direction is fixed by the initial electric field. Indeed, in this case, the system (6.200) reduces to the equation $\overset{\circ}{E}_1 a_1 + \overset{\circ}{E}_2 a_2 = 0$.

V_2' corresponds to a transverse wave, with a direction of polarisation fixed by the initial electric field, given by the equation $\overset{\circ}{E}_2 a_1 - \overset{\circ}{E}_1 a_2 = 0$, normal to the preceding direction.

To find the phase velocities and attenuation coefficients related to the previous roots, we proceed as in the case with a longitudinal initial electric field. So, using the notation:

$$V_3' = A_L' + iB_L' = \frac{\overset{\circ}{\rho}\,\omega^2}{(p_L - i\alpha_L)^2},$$

$$A_L' = c_{11} + \overset{\circ}{S}_{33}, \qquad B_L' = \omega d_{11}, \tag{6.214}$$

we obtain a phase velocity v_L in the form,

$$v_L^2 = \frac{\omega^2}{p_L^2} = \frac{2(A_L'^2 + B_L'^2)}{\overset{\circ}{\rho}\,(\sqrt{A_L'^2 + B_L'^2} + A_L')}, \tag{6.215}$$

and an attenuation coefficient α_L,

$$\alpha_L^2 = \frac{\overset{\circ}{\rho}\,\omega^2}{2} \cdot \frac{\sqrt{A_L'^2 + B_L'^2} - A_L'}{A_L'^2 + B_L'^2}. \tag{6.216}$$

We conclude that the displacement vector in this case has the form $\boldsymbol{u}_L = (0, 0, u_3^L)$, with

$$u_3^L(x_3, t) = a_3 \exp(-\alpha_L x_3) \exp\left[i\omega\left(t - \frac{x_3}{v_L}\right)\right]. \tag{6.217}$$

We observe that the attenuation affects the phase velocity v_L (by ω), and the amplitude of the longitudinal wave (by α_L). In this case, the electrostrictive effect is absent.

Applying the same procedure, we find the phase velocity and attenuation coefficient for the transverse waves. So, letting

$$V_1' = A_{T_1} + iB_{T_1} = \frac{\overset{\circ}{\rho}\,\omega^2}{(p_{T_1} - i\alpha_{T_1})^2}\,, \qquad A_{T_1} = c_{44} + \overset{\circ}{S}_{33}, \qquad B_{T_1} = \omega d_{44},$$

(6.218)

we obtain the phase velocity v_{T_1} in the form

$$v_{T_1}^2 = \frac{\omega^2}{p_{T_1}^2} = \frac{2(A_{T_1}^2 + B_{T_1}^2)}{\overset{\circ}{\rho}\left(\sqrt{A_{T_1}^2 + B_{T_1}^2} + A_{T_1}\right)}\,,$$

(6.219)

and attenuation coefficient α_{T_1},

$$\alpha_{T_1}^2 = \frac{\overset{\circ}{\rho}\,\omega^2}{2} \cdot \frac{\sqrt{A_{T_1}^2 + B_{T_1}^2} - A_{T_1}}{A_{T_1}^2 + B_{T_1}^2}\,.$$

(6.220)

We can see that the displacement vector in this case has the form $\boldsymbol{u}_{T_1} = (u_1^{T_1}, u_2^{T_1}, 0)$, where

$$u_k^{T_1}(x_3, t) = a_k \exp(-\alpha_{T_1} x_3)\exp\left[i\omega\left(t - \frac{x_3}{v_{T_1}}\right)\right], \qquad k = 1; 2. \quad (6.221)$$

Similar approximate forms for the phase velocity and attenuation coefficient can be obtained in this case. This transverse wave is attenuated, has the polarisation fixed by the initial electric field, and is not affected by the electrostrictive effect.

Finally, on letting:

$$V_2' = A_{T_2} + iB_{T_2} = \frac{\overset{\circ}{\rho}\,\omega^2}{(p_{T_2} - i\alpha_{T_2})^2}\,,$$

$$A_{T_2} = c_{44} + \overset{\circ}{S}_{33} - \frac{\eta}{1 + \eta}(\overset{\circ}{E}_1^2 + \overset{\circ}{E}_2^2), \qquad B_{T_2} = \omega d_{44}, \qquad (6.222)$$

we obtain the phase velocity v_{T_2}, in the form

$$v_{T_2}^2 = \frac{\omega^2}{p_{T_2}^2} = \frac{2(A_{T_2}^2 + B_{T_2}^2)}{\overset{\circ}{\rho}\left(\sqrt{A_{T_2}^2 + B_{T_2}^2} + A_{T_2}\right)}\,,$$

(6.223)

and an attenuation coefficient α_{T_2},

$$\alpha_{T_2}^2 = \frac{\overset{\circ}{\rho}\,\omega^2}{2} \cdot \frac{\sqrt{A_{T_2}^2 + B_{T_2}^2} - A_{T_2}}{A_{T_2}^2 + B_{T_2}^2}\,.$$

(6.224)

We conclude that the displacement vector in this case has the form $\boldsymbol{u}_{T_2} = (u_1^{T_2}, u_2^{T_2}, 0)$, with:

$$u_k^{T_2}(x_3, t) = a_k \exp(-\alpha_{T_2} x_3)\exp\left[i\omega\left(t - \frac{x_3}{v_{T_2}}\right)\right], \qquad k = 1; 2. \qquad (6.225)$$

Similar approximate forms for the phase velocity and attenuation coefficient can be obtained in this case. This transverse wave is attenuated, has the polarisation fixed by the initial electric field, and is affected by the electrostrictive effect.

6.4.3.2 Attenuated Wave Propagation on a Face of a Cubic Crystal Subject to Initial Fields

In the problem of attenuated wave propagation along the (001) plane of a cubic crystal, we choose $n_1 = \cos\varphi$, $n_2 = \sin\varphi$, and $n_3 = 0$, where φ is the angle between the propagation direction $\boldsymbol{n}$ and the [100] axis of the cubic crystal.

It follows in this case that the acoustic tensor $\overset{\circ}{\boldsymbol{Q}}$ has the components,

$$\overset{\circ}{Q}_{lm} = \overset{\circ}{Q'}_{lm}(p - i\alpha)^2, \qquad l, m = \overline{1, 3}. \qquad (6.226)$$

If we denote by $V = \overset{\circ}{\rho}\,\omega^2/(p - i\alpha)^2$, the condition of propagation (6.164) takes the form of an eigenvector problem,

$$\overset{\circ}{\boldsymbol{Q'}}\,\boldsymbol{a} = V\boldsymbol{a}, \quad \text{or} \quad (\overset{\circ}{Q'}_{lm} - V\delta_{lm})a_m = 0, \qquad l, m = \overline{1, 3}. \qquad (6.227)$$

This problem is usually associated with the following eigenvalue problem (characteristic equation),

$$\det(\overset{\circ}{Q'}_{lm} - V\delta_{lm}) = 0. \qquad (6.228)$$

Note that, for the problem of attenuated wave propagation along the face (001) of a cubic crystal, the acoustic tensor becomes symmetric $\overset{\circ}{Q'}_{lm} = \overset{\circ}{Q'}_{ml}$. Indeed, we obtain the components of the tensor $\overset{\circ}{\boldsymbol{Q'}}$ in the following form.

$$\overset{\circ}{Q'}_{11} = a'_c = (c_{11} + i\omega d_{11})\cos^2\varphi + (c_{44} + i\omega d_{44})\sin^2\varphi + A(\varphi) - \frac{\eta}{1+\eta}\overset{\circ}{E}_1^2,$$

$$\overset{\circ}{Q'}_{12} = \overset{\circ}{Q'}_{21} = b'_c = \frac{c_{12} + c_{44} + i\omega(d_{12} + d_{44})}{2}\sin2\varphi - \frac{\eta}{1+\eta}\overset{\circ}{E}_1\overset{\circ}{E}_2,$$

$$\overset{\circ}{Q'}_{13} = \overset{\circ}{Q'}_{31} = c'_c = -\frac{e_{14}\overset{\circ}{E}_1}{1+\eta}\sin2\varphi - \frac{\eta}{1+\eta}\overset{\circ}{E}_1\overset{\circ}{E}_3,$$

$$\overset{\circ}{Q}'_{22} = d'_c = (c_{44} + i\omega d_{44})\cos^2\varphi + (c_{11} + i\omega d_{11})\sin^2\varphi + A(\varphi) - \frac{\eta}{1+\eta}\,\overset{\circ}{E}_2^2,$$

$$\overset{\circ}{Q}'_{23} = \overset{\circ}{Q}'_{32} = e'_c = -\frac{e_{14}\,\overset{\circ}{E}_2}{1+\eta}\sin2\varphi - \frac{\eta}{1+\eta}\,\overset{\circ}{E}_2\overset{\circ}{E}_3,$$

$$\overset{\circ}{Q}'_{33} = f'_c = c_{44} + i\omega d_{44} + A(\varphi) - \overset{\circ}{E}_3^2 + \frac{(e_{14}\sin2\varphi - \overset{\circ}{E}_3)^2}{1+\eta}, \tag{6.229}$$

with

$$A(\varphi) = \overset{\circ}{S}_{11}\cos^2\varphi + \overset{\circ}{S}_{22}\sin^2\varphi + \overset{\circ}{S}_{12}\sin2\varphi. \tag{6.230}$$

After a short inspection of the form of the previous components, we can observe that they generalise the components of the acoustic tensor, obtained in the problem of plane harmonic wave propagation along a face of a cubic crystal (see (6.109)). The present form exhibits the attenuation effect, the piezoelectric effect, and the electrostrictive effect.

With this notation, the condition of propagation (6.227) becomes

$$\begin{cases} a'_c\, a_1 + b'_c\, a_2 + c'_c\, a_3 = V\, a_1 \\ b'_c\, a_1 + d'_c\, a_2 + e'_c\, a_3 = V\, a_2 \\ c'_c\, a_1 + e'_c\, a_2 + f'_c\, a_3 = V\, a_3 \end{cases} \tag{6.231}$$

and the characteristic equation (6.228) has the form

$$H(V) = \begin{vmatrix} a'_c - V & b'_c & c'_c \\ b'_c & d'_c - V & e'_c \\ c'_c & e'_c & f'_c - V \end{vmatrix} = 0. \tag{6.232}$$

In order to derive the phase velocities and attenuation coefficients in closed-form, in what follows we analyse an important particular case: the transverse initial electric field. Studying the wave propagation for special directions of propagation in the plane (001) we obtain the generalised anisotropy factor.

(a) *Transverse initial electric field* ($\overset{\circ}{E}_1 = 0$, $\overset{\circ}{E}_2 = 0$, $\overset{\circ}{E}_3 \neq 0$)

In this particular case the components c'_c, e'_c vanish. The other components of the tensor $\overset{\circ}{Q}'$ become

$$a'_c = (c_{11} + i\omega d_{11})\cos^2\varphi + (c_{44} + i\omega d_{44})\sin^2\varphi + A(\varphi),$$

$$b'_c = \frac{c_{12} + c_{44} + i\omega(d_{12} + d_{44})}{2}\,\sin2\varphi,$$

$$d'_c = (c_{44} + i\omega d_{44})\cos^2\varphi + (c_{11} + i\omega d_{11})\sin^2\varphi + A(\varphi),$$

$$f_c' = c_{44} + \mathrm{i}\omega d_{44} + A(\varphi) - \overset{\circ}{E}_3^2 + \frac{(e_{14}\sin 2\varphi - \overset{\circ}{E}_3)^2}{1 + \eta}, \qquad (6.233)$$

with $A(\varphi)$ done by (6.230).

Consequently, the characteristic equation (6.232) has the following three roots,

$$V_{1,2} = \frac{a_c' + d_c' \pm \sqrt{(a_c' - d_c')^2 + 4b_c'^2}}{2}, \qquad V_3 = f_c'. \qquad (6.234)$$

As regards the polarisation of the corresponding waves, we can see from the condition of propagation (6.231), that the wave related to the root V_3 is an attenuated transverse wave, arbitrarily polarised, which is piezoelectrically and electrostrictively coupled. The remaining two waves are polarised in the plane (001) in mutually perpendicular directions, one being quasi-transverse, and the other quasi-longitudinal. It is interesting to note that the directions of polarisation of the last two waves are not influenced by the initial electric field.

To find the phase velocities and the attenuation coefficients related to the previous roots, we denote by

$$V_3 = A_T + \mathrm{i}B_T = \frac{\overset{\circ}{\rho}\,\omega^2}{(p_T - \mathrm{i}\alpha_T)^2},$$

$$A_T = c_{44} + A(\varphi) - \overset{\circ}{E}_3^2 + \frac{(e_{14}\sin 2\varphi - \overset{\circ}{E}_3)^2}{1 + \eta}, \qquad B_T = \omega d_{44}. \qquad (6.235)$$

It yields a phase velocity v_T, in the form,

$$v_T^2 = \frac{\omega^2}{p_T^2} = \frac{2(A_T^2 + B_T^2)}{\overset{\circ}{\rho}\left(\sqrt{A_T^2 + B_T^2} + A_T\right)}, \qquad (6.236)$$

and an attenuation coefficient α_T, given by the relation,

$$\alpha_T^2 = \frac{\overset{\circ}{\rho}\,\omega^2}{2} \cdot \frac{\sqrt{A_T^2 + B_T^2} - A_T}{A_T^2 + B_T^2}. \qquad (6.237)$$

We can conclude that the displacement vector, in this case, has the form $\boldsymbol{u}_T = (0, 0, u_3^T)$, with

$$u_3^T(x_1, x_2, t) = a_3\exp[-\alpha_T(x_1\cos\varphi + x_2\sin\varphi)]\exp\left[\mathrm{i}\omega\left(t - \frac{x_1\cos\varphi + x_2\sin\varphi}{v_T}\right)\right].$$
$$(6.238)$$

We easily observe that the attenuation affects the phase velocity v_T (by ω), and the amplitude of the longitudinal wave (by α_T). Moreover, the piezoelectric and electrostrictive effects are represented by the term

$$- \overset{\circ}{E_3^2} + \frac{(e_{14}\sin 2\varphi - \overset{\circ}{E_3})^2}{1 + \eta}.$$

Applying the same procedure as in the case of the transverse wave, we can find the phase velocity and attenuation coefficients for the quasi-longitudinal, respectively, quasi-transverse waves. A detailed analysis is made for three particular directions of propagation in the plane (001).

(b) *Special directions of propagation. Generalized anisotropy factor*

In the problem of attenuated wave propagation in the plane (001) of a cubic crystal, subject to initial transverse electric field and to an initial mechanical stress field, we consider the following particular directions of propagation:

• For $\varphi = 0$ (i.e., [100] axis), the roots $V_{1,2}$, done by (6.234), have the following form.

$$V_1 = c_{11} + i\omega d_{11} + \overset{\circ}{S}_{11},$$

$$V_2 = c_{44} + i\omega d_{44} + \overset{\circ}{S}_{11}. \tag{6.239}$$

The first root refers to a pure longitudinal wave, and the second to a pure transverse one. The corresponding phase velocities and attenuation coefficients have the form.

$$v_L^2 = \frac{2(A_L^2 + B_L^2)}{\overset{\circ}{\rho}\left(\sqrt{A_L^2 + B_L^2} + A_L\right)}, \qquad \alpha_L^2 = \frac{\overset{\circ}{\rho}\,\omega^2}{2} \cdot \frac{\sqrt{A_L^2 + B_L^2} - A_L}{A_L^2 + B_L^2}, \tag{6.240}$$

where $A_L = c_{11} + \overset{\circ}{S}_{11}$ and $B_L = \omega d_{11}$, respectively:

$$v_{T_1}^2 = \frac{2(A_{T_1}^2 + B_{T_1}^2)}{\overset{\circ}{\rho}\left(\sqrt{A_{T_1}^2 + B_{T_1}^2} + A_{T_1}\right)}, \qquad \alpha_{T_1}^2 = \frac{\overset{\circ}{\rho}\,\omega^2}{2} \cdot \frac{\sqrt{A_{T_1}^2 + B_{T_1}^2} - A_{T_1}}{A_{T_1}^2 + B_{T_1}^2},$$

$$\tag{6.241}$$

with $A_{T_1} = c_{44} + \overset{\circ}{S}_{11}$ and $B_{T_1} = \omega d_{44}$.

• For $\varphi = \pi/2$ (i.e., [010] axis) the roots $V_{1,2}$ are done by the expressions:

$$V_1 = c_{11} + i\omega d_{11} + \overset{\circ}{S}_{22}, \qquad V_2 = c_{44} + i\omega d_{44} + \overset{\circ}{S}_{22}. \tag{6.242}$$

The first root corresponds to a pure longitudinal wave, and the second one to a pure transverse wave. The phase velocities and attenuation coefficients will have, in this case, the form

$$v_L^2 = \frac{2(A_L^2 + B_L^2)}{\overset{\circ}{\rho}\left(\sqrt{A_L^2 + B_L^2} + A_L\right)}, \qquad \alpha_L^2 = \frac{\overset{\circ}{\rho}\,\omega^2}{2} \cdot \frac{\sqrt{A_L^2 + B_L^2} - A_L}{A_L^2 + B_L^2}, \qquad (6.243)$$

where $A_L = c_{11} + \overset{\circ}{S}_{22}$ and $B_L = \omega d_{11}$, respectively:

$$v_{T_1}^2 = \frac{2(A_{T_1}^2 + B_{T_1}^2)}{\overset{\circ}{\rho}\left(\sqrt{A_{T_1}^2 + B_{T_1}^2} + A_{T_1}\right)}, \qquad \alpha_{T_1}^2 = \frac{\overset{\circ}{\rho}\,\omega^2}{2} \cdot \frac{\sqrt{A_{T_1}^2 + B_{T_1}^2} - A_{T_1}}{A_{T_1}^2 + B_{T_1}^2},$$

$$(6.244)$$

with $A_{T_1} = c_{44} + \overset{\circ}{S}_{22}$ and $B_{T_1} = \omega d_{44}$.

- For $\varphi = \pi/4$ (i.e., [110] axis) the roots $V_{1,2}$ have the form

$$V_1 = \frac{c_{11} + c_{12} + 2c_{44} + i\omega(d_{11} + d_{12} + d_{44}) + \overset{\circ}{S}_{11} + \overset{\circ}{S}_{22} + 2\overset{\circ}{S}_{12}}{2},$$

$$V_2 = \frac{c_{11} - c_{12} + i\omega(d_{11} - d_{12}) + \overset{\circ}{S}_{11} + \overset{\circ}{S}_{22} + 2\overset{\circ}{S}_{12}}{2}. \qquad (6.245)$$

The first root is related to a pure longitudinal wave, and the second corresponds to a pure transverse wave. The phase velocities and attenuation coefficients have, in this case, the following expressions

$$v_L^2 = \frac{2(A_L^2 + B_L^2)}{\overset{\circ}{\rho}\left(\sqrt{A_L^2 + B_L^2} + A_L\right)}, \qquad \alpha_L^2 = \frac{\overset{\circ}{\rho}\,\omega^2}{2} \cdot \frac{\sqrt{A_L^2 + B_L^2} - A_L}{A_L^2 + B_L^2}, \qquad (6.246)$$

where $A_L = (c_{11} + c_{12} + 2c_{44} + \overset{\circ}{S}_{11} + \overset{\circ}{S}_{22} + 2\overset{\circ}{S}_{12})/2$ and $B_L = \omega(d_{11} + d_{12} + d_{44})/2$, respectively:

$$v_{T_1}^2 = \frac{2(A_{T_1}^2 + B_{T_1}^2)}{\overset{\circ}{\rho}\left(\sqrt{A_{T_1}^2 + B_{T_1}^2} + A_{T_1}\right)}, \qquad \alpha_{T_1}^2 = \frac{\overset{\circ}{\rho}\,\omega^2}{2} \cdot \frac{\sqrt{A_{T_1}^2 + B_{T_1}^2} - A_{T_1}}{A_{T_1}^2 + B_{T_1}^2}$$

$$(6.247)$$

with $A_{T_1} = (c_{11} - c_{12} + \overset{\circ}{S}_{11} + \overset{\circ}{S}_{22} + 2\overset{\circ}{S}_{12})/2$ and $B_{T_1} = \omega(d_{11} - d_{12})/2$.

The ratio of the transverse wave velocities in the directions [100] and [110] defines the generalised anisotropy *factor* $\mathcal{A}^*$ for the problem of attenuated wave propagation in cubic crystals subject to initial fields, in the following manner,

$$\mathcal{A}^* = \left(\frac{v_{T_1}[100]}{v_{T_1}[110]}\right)^2, \qquad (6.248)$$

where the velocities are done by $(6.241)_1$ and $(6.247)_1$. It generalises the classical anisotropy factor $\mathcal{A}$, as defined in [14] in the case without initial fields, respectively, the anisotropy factor $\overset{\circ}{\mathcal{A}}$, defined and analysed in Section 6.3.2.3 for the problem of plane harmonic wave propagation in cubic crystals subject to initial fields. It is obvious that only the initial stress components influence the generalised anisotropy factor, and not the initial electric field.

It is easy to see that the generalised anisotropy factor $\mathcal{A}^*$ can be expressed, via the classical anisotropy factor $\mathcal{A}$, as follows.

$$\mathcal{A}^* = \mathcal{A} \cdot \frac{\alpha^2 + \psi^2}{\beta^2 + \chi^2} \cdot \frac{\sqrt{\beta^2 + \chi^2} + \beta}{\sqrt{\alpha^2 + \psi^2} + \alpha}. \tag{6.249}$$

Here $\mathcal{A} = 2c_{44}/(c_{11} - c_{12})$ is the classical anisotropy factor, without initial fields. The nondimensional parameters $\psi = \omega d_{44}/c_{44}$ and $\chi = \omega(d_{11} - d_{12})/(c_{11} - c_{12})$ express the influence of attenuation, and the nondimensional parameters $\alpha = 1 + \overset{\circ}{S}_{11}/c_{44}$ and $\beta = 1 + (\overset{\circ}{S}_{11} + \overset{\circ}{S}_{22} + 2\,\overset{\circ}{S}_{12})/(c_{11} - c_{12})$ describe the influence of the initial deformation field.

It is obvious that if the initial fields are absent and the wave is unattenuated, we have $\mathcal{A}^* = \mathcal{A}$ (in this particular case $\alpha = \beta = 1$; resp., $\psi = \chi = 0$). Moreover, if we develop the expression of $\mathcal{A}^*$ after small parameters ψ and χ, and we retain only the linear terms, we obtain that

$$\mathcal{A}^* \simeq \overset{\circ}{\mathcal{A}} = \frac{2(c_{44} + \overset{\circ}{S}_{11})}{c_{11} - c_{12} + \overset{\circ}{S}_{11} + \overset{\circ}{S}_{22} + 2\,\overset{\circ}{S}_{12}}. \tag{6.250}$$

Thus, in this particular case, the generalised anisotropy factor reduces to the anisotropy factor from the problem of plane harmonic wave propagation on a face of a cubic crystal.

6.5 The Coupling of Guided Plane Waves in Piezoelectric Crystals Subject to Initial Electromechanical Fields

In this part we study the coupling conditions for propagation of planar guided waves in a piezoelectric semi-infinite plane subject to initial electromechanical fields (see papers [25, 26]). If the sagittal plane is normal to a direct (resp., inverse) dyad axis, we show that the fundamental system of equations decomposes for particular choices of the initial electric field. Moreover, we obtain a similar decomposition of mechanical and electrical boundary conditions, which enables us to characterize the obtained guided waves.

6.5.1 Coupling Conditions for Waveguide Propagation

In this part, we assume the hypotheses described in Section 6.2.3. In this case, we have the homogeneous field equations (6.65), the associated boundary conditions (6.66), and the incremental constitutive equations (6.62) and (6.63).

From the previous field and constitutive equations we obtain the following fundamental system of equations,

$$\overset{\circ}{\rho}\,\ddot{u}_l = \overset{\circ}{\Omega}_{klmn}\,u_{m,nk} + \overset{\circ}{\Lambda}_{mkl}\,\phi_{,mk},$$

$$\overset{\circ}{\Lambda}_{kmn}\,u_{n,mk} - \overset{\circ}{\epsilon}_{kn}\,\phi_{,nk} = 0, \qquad l = \overline{1,3}. \tag{6.251}$$

In what follows we describe the geometric hypotheses for our problem. The crystal is assumed to be semi-infinite, occupying the region $x_2 > 0$, and the waves are supposed to propagate along the x_1-axis. The plane x_1x_2 containing the surface normal and the propagation direction is called the *sagittal plane*. Furthermore, we suppose that the waveguide has the properties invariant with time t and with x_1 variable. In these conditions, if the material behaves linearly and without attenuation, the normal modes have the form

$$u_j(\boldsymbol{x},t) = u_j^0(x_2,x_3)\exp[\mathrm{i}(\omega t - px_1)], \qquad j = \overline{1,4}. \tag{6.252}$$

Here u_1, u_2, u_3 are the mechanical displacements, and u_4 stands for the electric potential ϕ. In the previous relations p represents the wave number, ω defines the frequency of the wave, and $\mathrm{i}^2 = -1$. Using these hypotheses, the equations (6.251) become:

$$\overset{\circ}{\Omega}_{klmn}\,u_{m,nk} + \overset{\circ}{\Lambda}_{mkl}\,\phi_{,mk} = -\overset{\circ}{\rho}\,\omega^2 u_l,$$

$$\overset{\circ}{\Lambda}_{kmn}\,u_{n,mk} = \overset{\circ}{\epsilon}_{kn}\,\phi_{,nk}, \qquad l = \overline{1,3}. \tag{6.253}$$

We define the nondimensional variable $X_2 = px_2$ and we neglect the effects of diffraction in the x_3-direction, so that $\partial/\partial x_3 = 0$. The other hypotheses yield the derivation rules $\partial/\partial x_1 = -\mathrm{i}p$ and $\partial/\partial x_2 = p\partial/\partial X_2$. Finally, we introduce the phase velocity of the guided wave as $V = \omega/p$.

To analyse the coupling conditions of the plane waveguide, under the previous hypotheses, we introduce the differential operators with complex coefficients, as follows.

$$\overset{\circ}{\Gamma}_{il} = \overset{\circ}{\Omega}_{1il1} - \overset{\circ}{\Omega}_{2il2}\,\frac{\partial^2}{\partial X_2^2} + \mathrm{i}(\overset{\circ}{\Omega}_{1il2} + \overset{\circ}{\Omega}_{1li2})\frac{\partial}{\partial X_2},$$

$$\overset{\circ}{\gamma}_l = \overset{\circ}{\Lambda}_{11l} - \overset{\circ}{\Lambda}_{22l}\,\frac{\partial^2}{\partial X_2^2} + \mathrm{i}(\overset{\circ}{\Lambda}_{12l} + \overset{\circ}{\Lambda}_{21l})\frac{\partial}{\partial X_2},$$

$$\overset{\circ}{\epsilon} = \overset{\circ}{\epsilon}_{11} - \overset{\circ}{\epsilon}_{22}\,\frac{\partial^2}{\partial X_2^2} + 2\mathrm{i}\,\overset{\circ}{\epsilon}_{12}\,\frac{\partial}{\partial X_2}. \tag{6.254}$$

In these conditions, after a lengthy but elementary calculus we obtain the differential system (6.253) in the form

$$
\begin{pmatrix}
\overset{\circ}{\Gamma}_{11} - \overset{\circ}{\rho}\, V^2 & \overset{\circ}{\Gamma}_{12} & \overset{\circ}{\Gamma}_{13} & \overset{\circ}{\gamma}_1 \\[4pt]
\overset{\circ}{\Gamma}_{12} & \overset{\circ}{\Gamma}_{22} - \overset{\circ}{\rho}\, V^2 & \overset{\circ}{\Gamma}_{23} & \overset{\circ}{\gamma}_2 \\[4pt]
\overset{\circ}{\Gamma}_{13} & \overset{\circ}{\Gamma}_{23} & \overset{\circ}{\Gamma}_{33} - \overset{\circ}{\rho}\, V^2 & \overset{\circ}{\gamma}_3 \\[4pt]
\overset{\circ}{\gamma}_1 & \overset{\circ}{\gamma}_2 & \overset{\circ}{\gamma}_3 & -\overset{\circ}{\epsilon}
\end{pmatrix}
\begin{pmatrix}
u_1 \\ u_2 \\ u_3 \\ u_4
\end{pmatrix} = 0.
\tag{6.255}
$$

Here the coefficients are defined by relations (6.254). The system (6.255) is a homogeneous differential system of four equations with four unknowns, that is, the components of the mechanical displacement and the electric potential, having as coefficients complex differential operators in nondimensional variable X_2. It generalises the similar system from the case without initial fields, derived in [14].

In what follows we analyse the coupling conditions of the guided plane wave propagation in two particular cases.

6.5.1.1 Sagittal Plane Normal to a Direct Axis of Order Two

In this case, we suppose that the sagittal plane $x_1 x_2$ is normal to a dyad axis (x_3 in our case). Then, the elastic constants with one index equal to 3 are zero. After a short inspection of the coefficients of the system (6.255), using the Voigt convention, we find:

$$
\begin{aligned}
\overset{\circ}{\Gamma}_{13} &= -\left[e_{15} + \mathrm{i}(e_{14} + e_{25})\frac{\partial}{\partial X_2} - e_{24}\frac{\partial^2}{\partial X_2^2} \right] \overset{\circ}{E}_1 \\[6pt]
&\quad -\left[\eta_{11} + 2\mathrm{i}\eta_{12}\frac{\partial}{\partial X_2} - \eta_{22}\frac{\partial^2}{\partial X_2^2} \right] \overset{\circ}{E}_1 \overset{\circ}{E}_3, \\[10pt]
\overset{\circ}{\Gamma}_{23} &= -\left[e_{15} + \mathrm{i}(e_{14} + e_{25})\frac{\partial}{\partial X_2} - e_{24}\frac{\partial^2}{\partial X_2^2} \right] \overset{\circ}{E}_2 \\[6pt]
&\quad -\left[\eta_{11} + 2\mathrm{i}\eta_{12}\frac{\partial}{\partial X_2} - \eta_{22}\frac{\partial^2}{\partial X_2^2} \right] \overset{\circ}{E}_2 \overset{\circ}{E}_3 .
\end{aligned}
\tag{6.256}
$$

We can easily observe that $\overset{\circ}{\Gamma}_{13}$ and $\overset{\circ}{\Gamma}_{23}$ do not depend on the initial stress field components, but on the initial electric field components, only. Thus, $\overset{\circ}{\Gamma}_{13} = \overset{\circ}{\Gamma}_{23} = 0$ if $\overset{\circ}{E}_1 = \overset{\circ}{E}_2 = 0$.

Moreover, if we suppose that the dyad axis is direct (this means that the sagittal plane is normal to a direct axis of order two), it follows that the crystal belongs to the class 2 of the monoclinic system ($A_2 \parallel x_3$). In this particular case the piezoelectric constants with no index equal to 3 are zero.

Therefore, we obtain:

$$
\overset{\circ}{\gamma}_1 = \left(\eta_{11} + 2\mathrm{i}\eta_{12}\frac{\partial}{\partial X_2} - \eta_{22}\frac{\partial^2}{\partial X_2^2} \right) \overset{\circ}{E}_1,
$$

$$
\overset{\circ}{\gamma}_2 = \left(\eta_{11} + 2\mathrm{i}\eta_{12}\frac{\partial}{\partial X_2} - \eta_{22}\frac{\partial^2}{\partial X_2^2} \right) \overset{\circ}{E}_2 . \tag{6.257}
$$

So, we obtain that $\overset{\circ}{\gamma}_1 = \overset{\circ}{\gamma}_2 = 0$ if $\overset{\circ}{E}_1 = \overset{\circ}{E}_2 = 0$.

In conclusion, we derive the following result concerning the decomposition of the fundamental system (6.255).

If the axis x_3 is a direct dyad axis and if $\overset{\circ}{E}_1 = \overset{\circ}{E}_2 = 0$, the system (6.255) *reduces to two independent subsystems*, as follows.

- The first subsystem

$$
\begin{pmatrix} \overset{\circ}{\Gamma}_{11} - \overset{\circ}{\rho}\, V^2 & \overset{\circ}{\Gamma}_{12} \\ \overset{\circ}{\Gamma}_{12} & \overset{\circ}{\Gamma}_{22} - \overset{\circ}{\rho}\, V^2 \end{pmatrix} \begin{pmatrix} u_1 \\ u_2 \end{pmatrix} = 0 \tag{6.258}
$$

defines a nonpiezoelectric guided wave, polarised in the sagittal plane $x_1 x_2$, which depends on the initial stress field, only. We denote it by $\overset{\circ}{P}_2$. These characteristics are due to the form of the involved coefficients:

$$
\overset{\circ}{\Gamma}_{11} = c_{11} + \overset{\circ}{S}_{11} + 2\mathrm{i}(c_{16} + \overset{\circ}{S}_{12})\frac{\partial}{\partial X_2} - (c_{66} + \overset{\circ}{S}_{22})\frac{\partial^2}{\partial X_2^2},
$$

$$
\overset{\circ}{\Gamma}_{12} = c_{16} + \mathrm{i}(c_{12} + c_{66})\frac{\partial}{\partial X_2} - c_{26}\frac{\partial^2}{\partial X_2^2},
$$

$$
\overset{\circ}{\Gamma}_{22} = c_{66} + \overset{\circ}{S}_{11} + 2\mathrm{i}(c_{26} + \overset{\circ}{S}_{12})\frac{\partial}{\partial X_2} - (c_{22} + \overset{\circ}{S}_{22})\frac{\partial^2}{\partial X_2^2}. \tag{6.259}
$$

- The second subsystem:

$$
\begin{pmatrix} \overset{\circ}{\Gamma}_{33} - \overset{\circ}{\rho}\, V^2 & \overset{\circ}{\gamma}_3 \\ \overset{\circ}{\gamma}_3 & -\overset{\circ}{\epsilon} \end{pmatrix} \begin{pmatrix} u_3 \\ u_4 \end{pmatrix} = 0 \tag{6.260}
$$

has as solution a transverse-horizontal wave, with polarisation after the axis x_3, which is piezoelectric and electrostrictive active, and depends on the initial mechanical and electrical fields. It is denoted by $\overline{\overset{\circ}{TH}}$ and generalises the famous Bleustein–Gulyaev wave (see [14], to compare). The components involved in this equation have the form:

$$
\overset{\circ}{\Gamma}_{33} = c_{55} + \overset{\circ}{S}_{11} + 2\mathrm{i}(c_{45} + \overset{\circ}{S}_{12})\frac{\partial}{\partial X_2} - (c_{44} + \overset{\circ}{S}_{22})\frac{\partial^2}{\partial X_2^2}
$$

$$- 2 \left[e_{15} + i(e_{14} + e_{25})\frac{\partial}{\partial X_2} - e_{24}\frac{\partial^2}{\partial X_2^2} \right] \overset{\circ}{E}_3$$

$$- \left[\eta_{11} + 2i\eta_{12}\frac{\partial}{\partial X_2} - \eta_{22}\frac{\partial^2}{\partial X_2^2} \right] \overset{\circ}{E}_3^{\,2},$$

$$\overset{\circ}{\gamma}_3 = e_{15} + i(e_{14} + e_{25})\frac{\partial}{\partial X_2} - e_{24}\frac{\partial^2}{\partial X_2^2}$$

$$+ \left[\eta_{11} + 2i\eta_{12}\frac{\partial}{\partial X_2} - \eta_{22}\frac{\partial^2}{\partial X_2^2} \right] \overset{\circ}{E}_3,$$

$$\overset{\circ}{\epsilon} = \overset{\circ}{\epsilon}_{11} + 2i\,\overset{\circ}{\epsilon}_{12}\,\frac{\partial}{\partial X_2} - \overset{\circ}{\epsilon}_{22}\,\frac{\partial^2}{\partial X_2^2} = 1 + \eta_{11} + 2i\eta_{12}\frac{\partial}{\partial X_2}$$

$$- (1 + \eta_{22})\frac{\partial^2}{\partial X_2^2}. \tag{6.261}$$

6.5.1.2 Sagittal Plane Parallel to a Mirror Plane

We suppose now that the sagittal plane $x_1 x_2$ is normal to an inverse dyad axis (x_3 in our case) or, equivalently, if the sagittal plane is parallel to a mirror plane M. It follows that the crystal belongs to the class m of the monoclinic system ($M \perp x_3$). In this particular case the elastic constants with one index equal to 3 are zero, as well as the piezoelectric constants with one index equal to 3, which vanish.

Analysing the coefficients of the system (6.255) in this case, we find:

$$\overset{\circ}{\Gamma}_{13} = - \left[e_{11} + i(e_{21} + e_{16})\frac{\partial}{\partial X_2} - e_{26}\frac{\partial^2}{\partial X_2^2} \right] \overset{\circ}{E}_3$$

$$- \left[\eta_{11} + 2i\eta_{12}\frac{\partial}{\partial X_2} - \eta_{22}\frac{\partial^2}{\partial X_2^2} \right] \overset{\circ}{E}_1\overset{\circ}{E}_3,$$

$$\overset{\circ}{\Gamma}_{23} = - \left[e_{16} + i(e_{26} + e_{12})\frac{\partial}{\partial X_2} - e_{22}\frac{\partial^2}{\partial X_2^2} \right] \overset{\circ}{E}_3$$

$$- \left[\eta_{11} + 2i\eta_{12}\frac{\partial}{\partial X_2} - \eta_{22}\frac{\partial^2}{\partial X_2^2} \right] \overset{\circ}{E}_2\overset{\circ}{E}_3,$$

$$\overset{\circ}{\gamma}_3 = \left(\eta_{11} + 2i\eta_{12}\frac{\partial}{\partial X_2} - \eta_{22}\frac{\partial^2}{\partial X_2^2} \right) \overset{\circ}{E}_3. \tag{6.262}$$

It yields that $\overset{\circ}{\Gamma}_{13} = \overset{\circ}{\Gamma}_{23} = 0$ and $\overset{\circ}{\gamma}_3 = 0$ if $\overset{\circ}{E}_3 = 0$.

Thus, *if the axis x_3 is an inverse dyad axis and if $\overset{\circ}{E}_3 = 0$, the fundamental system (6.255) splits into two parts*, as follows.

- The first subsystem has the form:

$$\begin{pmatrix} \overset{\circ}{\Gamma}_{11} - \overset{\circ}{\rho}\,V^2 & \overset{\circ}{\Gamma}_{12} & \overset{\circ}{\gamma}_1 \\ \overset{\circ}{\Gamma}_{12} & \overset{\circ}{\Gamma}_{22} - \overset{\circ}{\rho}\,V^2 & \overset{\circ}{\gamma}_2 \\ \overset{\circ}{\gamma}_1 & \overset{\circ}{\gamma}_2 & -\overset{\circ}{\epsilon} \end{pmatrix} \begin{pmatrix} u_1 \\ u_2 \\ u_4 \end{pmatrix} = 0. \tag{6.263}$$

It has as solution a guided wave with sagittal plane polarisation, associated with the electric field (via the electric potential $u_4 = \phi$), providing piezoelectric and electrostrictive effects, and depending on the initial stress and electric fields. It is denoted by $\overline{\overset{\circ}{P}}_2$. The electric field associated with this wave is contained in the sagittal plane because $E_3 = -\partial\phi/\partial x_3 = 0$. This fact is consistent with the hypothesis $\overset{\circ}{E}_3 = 0$. These features of the $\overline{\overset{\circ}{P}}_2$ wave are obtained from the analysis of the corresponding coefficients:

$$\overset{\circ}{\Gamma}_{11} = c_{11} + \overset{\circ}{S}_{11} - 2e_{11}\overset{\circ}{E}_1 - \eta_{11}\overset{\circ}{E}_1^2 + 2\mathrm{i}[c_{16} + \overset{\circ}{S}_{12} - (e_{16} + e_{21})\overset{\circ}{E}_1$$
$$- \eta_{12}\overset{\circ}{E}_1^2]\frac{\partial}{\partial X_2} - (c_{66} + \overset{\circ}{S}_{22} - 2e_{26}\overset{\circ}{E}_1 - \eta_{22}\overset{\circ}{E}_1^2)\frac{\partial^2}{\partial X_2^2},$$

$$\overset{\circ}{\Gamma}_{12} = c_{16} - e_{16}\overset{\circ}{E}_1 - e_{11}\overset{\circ}{E}_2 - \eta_{11}\overset{\circ}{E}_1\overset{\circ}{E}_2 + \mathrm{i}[c_{12} + c_{66} - (e_{12} + e_{26})\overset{\circ}{E}_1$$
$$- (e_{21} + e_{16})\overset{\circ}{E}_2 - 2\eta_{12}\overset{\circ}{E}_1\overset{\circ}{E}_2]\frac{\partial}{\partial X_2}$$
$$- (c_{26} - e_{22}\overset{\circ}{E}_1 - e_{26}\overset{\circ}{E}_2 - \eta_{22}\overset{\circ}{E}_1\overset{\circ}{E}_2)\frac{\partial^2}{\partial X_2^2},$$

$$\overset{\circ}{\Gamma}_{22} = c_{66} + \overset{\circ}{S}_{11} - 2e_{16}\overset{\circ}{E}_2 - \eta_{11}\overset{\circ}{E}_2^2 + 2\mathrm{i}[c_{26} + \overset{\circ}{S}_{12} - (e_{26} + e_{12})\overset{\circ}{E}_2$$
$$- \eta_{12}\overset{\circ}{E}_2^2]\frac{\partial}{\partial X_2} - (c_{22} + \overset{\circ}{S}_{22} - 2e_{22}\overset{\circ}{E}_2 - \eta_{22}\overset{\circ}{E}_2^2)\frac{\partial^2}{\partial X_2^2}, \tag{6.264}$$

respectively:

$$\overset{\circ}{\gamma}_1 = e_{11} + \eta_{11}\overset{\circ}{E}_1 + \mathrm{i}(e_{16} + e_{21} + 2\eta_{12}\overset{\circ}{E}_1)\frac{\partial}{\partial X_2} - (e_{26} + \eta_{22}\overset{\circ}{E}_1)\frac{\partial^2}{\partial X_2^2},$$

$$\overset{\circ}{\gamma}_2 = e_{16} + \eta_{11}\overset{\circ}{E}_2 + \mathrm{i}(e_{12} + e_{26} + 2\eta_{12}\overset{\circ}{E}_2)\frac{\partial}{\partial X_2} - (e_{22} + \eta_{22}\overset{\circ}{E}_2)\frac{\partial^2}{\partial X_2^2},$$

$$\overset{\circ}{\epsilon} = \overset{\circ}{\epsilon}_{11} + 2\mathrm{i}\,\overset{\circ}{\epsilon}_{12}\frac{\partial}{\partial X_2} - \overset{\circ}{\epsilon}_{22}\frac{\partial^2}{\partial X_2^2} = 1 + \eta_{11} + 2\mathrm{i}\eta_{12}\frac{\partial}{\partial X_2} - (1 + \eta_{22})\frac{\partial^2}{\partial X_2^2}. \tag{6.265}$$

- The second subsystem reduces to a single equation, as follows,

$$(\overset{\circ}{\Gamma}_{33} - \overset{\circ}{\rho}\,V^2)u_3 = 0. \tag{6.266}$$

Its root corresponds to a transverse-horizontal wave, nonpiezoelectric, and influenced by the initial stress field only. It is called the $\overset{\circ}{T}H$ wave. In this equation:

$$\overset{\circ}{\Gamma}_{33} = c_{55} + \overset{\circ}{S}_{11} + 2\mathrm{i}(c_{45} + \overset{\circ}{S}_{12})\frac{\partial}{\partial X_2} - (c_{44} + \overset{\circ}{S}_{22})\frac{\partial^2}{\partial X_2^2}. \tag{6.267}$$

6.5.2 The Decoupling of Mechanical and Electrical Boundary Conditions

In this section we analyse the decomposition of the mechanical (resp., electrical) boundary conditions on the surface $x_2 = 0$.

6.5.2.1 Mechanical Boundary Conditions

On the boundary surface $x_2 = 0$ the mechanical conditions are assumed to concern the surface stresses Σ_{2i} with $i = \overline{1,3}$. Following the incremental constitutive equations (6.62), we find in this case that:

$$\begin{aligned}
\Sigma_{21} &= \overset{\circ}{\Omega}_{2111}u_{1,1} + \overset{\circ}{\Omega}_{2121}u_{2,1} + \overset{\circ}{\Omega}_{2131}u_{3,1} + \overset{\circ}{\Omega}_{2112}u_{1,2} + \overset{\circ}{\Omega}_{2122}u_{2,2} \\
&\quad + \overset{\circ}{\Omega}_{2132}u_{3,2} + \overset{\circ}{\Lambda}_{121}u_{4,1} + \overset{\circ}{\Lambda}_{221}u_{4,2},
\end{aligned}$$

$$\begin{aligned}
\Sigma_{22} &= \overset{\circ}{\Omega}_{2211}u_{1,1} + \overset{\circ}{\Omega}_{2221}u_{2,1} + \overset{\circ}{\Omega}_{2231}u_{3,1} + \overset{\circ}{\Omega}_{2212}u_{1,2} + \overset{\circ}{\Omega}_{2222}u_{2,2} \\
&\quad + \overset{\circ}{\Omega}_{2232}u_{3,2} + \overset{\circ}{\Lambda}_{122}u_{4,1} + \overset{\circ}{\Lambda}_{222}u_{4,2},
\end{aligned} \tag{6.268}$$

$$\begin{aligned}
\Sigma_{23} &= \overset{\circ}{\Omega}_{2311}u_{1,1} + \overset{\circ}{\Omega}_{2321}u_{2,1} + \overset{\circ}{\Omega}_{2331}u_{3,1} + \overset{\circ}{\Omega}_{2312}u_{1,2} + \overset{\circ}{\Omega}_{2322}u_{2,2} \\
&\quad + \overset{\circ}{\Omega}_{2332}u_{3,2} + \overset{\circ}{\Lambda}_{123}u_{4,1} + \overset{\circ}{\Lambda}_{223}u_{4,2}.
\end{aligned}$$

(a) For a sagittal plane normal to a direct axis of order two and if $\overset{\circ}{E}_1 = \overset{\circ}{E}_2 = 0$ we obtain:

$$\Sigma_{21} = -k\mathrm{i}[(c_{16} + \overset{\circ}{S}_{12})u_1 + c_{66}u_2] + k\frac{\partial}{\partial X_2}[(c_{66} + \overset{\circ}{S}_{22})u_1 + c_{26}u_2],$$

$$\Sigma_{22} = -k\mathrm{i}[c_{12}u_1 + (c_{26} + \overset{\circ}{S}_{12})u_2] + k\frac{\partial}{\partial X_2}[c_{26}u_1 + (c_{22} + \overset{\circ}{S}_{22})u_2],$$

$$\begin{aligned}
\Sigma_{23} &= -k\mathrm{i}[(\overset{\circ}{S}_{12} + c_{45} - e_{25}\overset{\circ}{E}_3 - e_{14}\overset{\circ}{E}_3 - \eta_{12}\overset{\circ}{E}_3^2)u_3 + (e_{14} + \eta_{12}\overset{\circ}{E}_3)u_4] \\
&\quad + k\frac{\partial}{\partial X_2}[(c_{44} + \overset{\circ}{S}_{22} - 2e_{24}\overset{\circ}{E}_3 - \eta_{22}\overset{\circ}{E}_3^2)u_3 + (e_{24} + \eta_{22}\overset{\circ}{E}_3)u_4].
\end{aligned}$$

$$\tag{6.269}$$

Consequently, the mechanical boundary conditions on the plane $x_2 = 0$, under the previous conditions, reduce to the equalities (6.269), for given stresses Σ_{2i} with $i = \overline{1,3}$.

As regards the boundary conditions associated with the waves previously derived, for the $\overset{\circ}{P}_2$ wave we have relations (6.269) with $u_3 = u_4 = 0$ (it yields that $\Sigma_{23} = 0$ for this wave), whereas for $\overline{TH}$ we have the same relations with $u_1 = u_2 = 0$ (it results that $\Sigma_{12} = \Sigma_{22} = 0$ in this case).

(b) For a sagittal plane parallel to a mirror plane and if $\overset{\circ}{E}_3 = 0$ we derive:

$$
\Sigma_{21} = k\left[(-i)(c_{16} + \overset{\circ}{S}_{12} - e_{16}\overset{\circ}{E}_1 - e_{21}\overset{\circ}{E}_1 - \eta_{12}\overset{\circ}{E}_1{}^2)\right.
$$
$$
\left. + (c_{66} + \overset{\circ}{S}_{22} - 2e_{26}\overset{\circ}{E}_1 - \eta_{22}\overset{\circ}{E}_1{}^2)\frac{\partial}{\partial X_2}\right]u_1
$$
$$
+ k\left[(-i)(c_{66} - e_{26}\overset{\circ}{E}_1 - e_{16}\overset{\circ}{E}_2 - \eta_{12}\overset{\circ}{E}_1\overset{\circ}{E}_2)\right.
$$
$$
\left. + (c_{26} - e_{22}\overset{\circ}{E}_1 - e_{26}\overset{\circ}{E}_2 - \eta_{22}\overset{\circ}{E}_1\overset{\circ}{E}_2)\frac{\partial}{\partial X_2}\right]u_2
$$
$$
+ k\left[(-i)(e_{16} + \eta_{12}\overset{\circ}{E}_1) + (e_{26} + \eta_{22}\overset{\circ}{E}_1)\frac{\partial}{\partial X_2}\right]u_4, \qquad (6.270)
$$
$$
\Sigma_{22} = k\left[(-i)(c_{12} - e_{12}\overset{\circ}{E}_1 - e_{21}\overset{\circ}{E}_2 - \eta_{12}\overset{\circ}{E}_1\overset{\circ}{E}_2)\right.
$$
$$
\left. + (c_{26} - e_{22}\overset{\circ}{E}_1 - e_{26}\overset{\circ}{E}_2 - \eta_{22}\overset{\circ}{E}_1\overset{\circ}{E}_2)\frac{\partial}{\partial X_2}\right]u_1
$$
$$
+ k\left[(-i)(c_{26} + \overset{\circ}{S}_{12} - e_{12}\overset{\circ}{E}_2 - e_{26}\overset{\circ}{E}_2 - \eta_{12}\overset{\circ}{E}_2{}^2)\right.
$$
$$
\left. + (c_{22} + \overset{\circ}{S}_{22} - 2e_{22}\overset{\circ}{E}_2 - \eta_{22}\overset{\circ}{E}_2{}^2)\frac{\partial}{\partial X_2}\right]u_2
$$
$$
+ k\left[(-i)(e_{12} + \eta_{12}\overset{\circ}{E}_2) + (e_{22} + \eta_{22}\overset{\circ}{E}_2)\frac{\partial}{\partial X_2}\right]u_4,
$$
$$
\Sigma_{23} = k\left[(-i)(c_{45} + \overset{\circ}{S}_{12}) + (c_{44} + \overset{\circ}{S}_{22})\frac{\partial}{\partial X_2}\right]u_3.
$$

Consequently, the mechanical boundary conditions on the plane $x_2 = 0$, under the previous conditions, reduce the equalities (6.270), for given stresses Σ_{2i} with $i = \overline{1,3}$. For the $\overset{\circ}{P}_2$ wave we have relations (6.270) with $u_3 = 0$ (it yields $\Sigma_{23} = 0$ for this wave), whereas for the $\overset{\circ}{TH}$ wave we obtain the boundary conditions from (6.270) with $u_1 = u_2 = u_4 = 0$ (it results in $\Sigma_{12} = \Sigma_{22} = 0$ in this case).

We conclude that the stresses on the horizontal surface $x_2 = 0$, associated with the guided waves polarised in the sagittal plane (i.e., $\overset{\circ}{P}_2$ and $\overline{\overset{\circ}{P}_2}$), become decoupled from those associated with transverse horizontal waves (i.e., $\overline{\overset{\circ}{TH}}$ and $\overset{\circ}{TH}$), when x_3 is a dyad axis normal to the sagittal plane $x_1 x_2$. Our results generalise the classical boundary conditions for piezoelectric guided waves without initial fields, as described in [14].

6.5.2.2 Electrical Boundary Conditions

On the boundary surface of the domain we suppose the electrical boundary condition of the type:

$$\Delta_n = \Delta_k n_k = -\overline{w}, \quad \text{with} \quad k = \overline{1,3}, \tag{6.271}$$

where the normal component of the electrical displacement Δ_n is related to the surface density of electric charge $\overline{w}$.

In our case, as the boundary of the domain $x_2 > 0$ is the plane $x_2 = 0$, the previous boundary condition becomes:

$$\Delta_2 = \overline{w} \quad \text{on} \quad x_2 = 0. \tag{6.272}$$

Using the constitutive equation (6.62) and the derivation rules, we find that

$$\Delta_2 = \overset{\circ}{\Lambda}_{2nm}\, u_{m,n} - \overset{\circ}{\epsilon}_{2l}\, \phi_{,l} = k\left(-i\,\overset{\circ}{\Lambda}_{211} + \overset{\circ}{\Lambda}_{221}\,\frac{\partial}{\partial X_2} \right) u_1$$

$$+ k\left(-i\,\overset{\circ}{\Lambda}_{212} + \overset{\circ}{\Lambda}_{222}\,\frac{\partial}{\partial X_2} \right) u_2 + k\left(-i\,\overset{\circ}{\Lambda}_{213} + \overset{\circ}{\Lambda}_{223}\,\frac{\partial}{\partial X_2} \right) u_3$$

$$+ k\left(i\,\overset{\circ}{\epsilon}_{12} - \overset{\circ}{\epsilon}_{22}\,\frac{\partial}{\partial X_2} \right) u_4. \tag{6.273}$$

For a sagittal plane normal to a direct axis of order two and if $\overset{\circ}{E}_1 = \overset{\circ}{E}_2 = 0$ we obtain the following electrical boundary condition

$$k\left[(-i\, e_{25} + e_{24}) + \overset{\circ}{E}_3\left(-i\, \eta_{12} + \eta_{22} \right)\frac{\partial}{\partial X_2} \right] u_3 + k\left(i\,\overset{\circ}{\epsilon}_{12} - \overset{\circ}{\epsilon}_{22}\,\frac{\partial}{\partial X_2} \right) u_4 = \overline{w}, \tag{6.274}$$

on $x_2 = 0$. It is obvious that this type of boundary condition suits the $\overline{\overset{\circ}{TH}}$ wave only.

For a sagittal plane parallel to a mirror plane and if $\overset{\circ}{E}_3 = 0$ we derive the following electrical boundary condition.

$$k\left[(-\mathrm{i}\,e_{21} + e_{26}) + \overset{\circ}{E}_1\left(-\mathrm{i}\,\eta_{12} + \eta_{22}\frac{\partial}{\partial X_2}\right)\right]u_1 + k\left[(-\mathrm{i}\,e_{26} + e_{22})\right.$$
$$\left.+ \overset{\circ}{E}_2\left(-\mathrm{i}\,\eta_{12} + \eta_{22}\frac{\partial}{\partial X_2}\right)\right]u_2 + k\left(\mathrm{i}\,\overset{\circ}{\epsilon}_{12} - \overset{\circ}{\epsilon}_{22}\frac{\partial}{\partial X_2}\right)u_4 = \overline{w}, \tag{6.275}$$

on $x_2 = 0$. It is evident that this kind of boundary condition is specific to the wave $\overline{\overset{\circ}{P}_2}$ only.

6.6 Conclusions

The present chapter analyses the basic elements concerning the problem of wave propagation in prestrained and prepolarised solid media. In the first part we derive the field and constitutive equations, as well as the boundary conditions, related to the behaviour of incremental fields superposed on large static deformation and electric fields. We obtain the dynamic and static energy balance equations and we present the static and dynamic local stability criteria.

In the second part of the work we find the conditions of propagation for plane harmonic waves in various types of crystals subject to initial electromechanical fields. In this framework we show the electrostrictive effect, we define and analyse generalised anisotropy factors and coupling coefficients, and we demonstrate the influence of initial fields on the shape of slowness surfaces. We also present the problem of attenuated wave propagation in isotropic solids and in cubic crystals subject to initial electromechanical fields. Finally, we study the coupling conditions of waveguide propagation in monoclinic piezoelectric crystals subject to initial fields.

The wide range of crystal classes analysed here demonstrates the unity of the mathematical method used (i.e., spectral properties of acoustic tensor), and the variety of the physical significance (resp., of practical applications) of the solutions found. We expect to obtain results for the problem of guided wave (of Rayleigh and Love type) propagation in special classes of crystals, subject to initial fields.

References

[1]   Baesu, E, On electroacoustic energy flux, *ZAMP*, **54**, 1001–1009 (2003).

[2]   Baesu, E, Fortuné, D and Soós, E, Incremental behaviour of hyperelastic dielectrics and piezoelectric crystals, *ZAMP*, **54**, 160–178 (2003).

[3]   Baumhauer, JC and Tiersten, HF, Nonlinear electrostatics equations for small fields superimposed on a bias, *J. Acoust. Soc. Amer.*, **54**, 1017–1034 (1973).

[4]   Cristescu, ND, Craciun, EM and Soós, E, *Mechanics of Elastic Composites*, Chapmann & Hall/CRC, Boca Raton, FL, 2004.

[5]   Chai, JF and Wu, TT, Propagation of surface waves in a prestressed piezoelectric materials, *J. Acoust. Soc. Amer.*, **100**, 2112–2122 (1996).

[6]   Eringen, AC and Maugin, GA, *Electrodynamics of Continua*, vol. I, Springer, New York, 1990.

[7]   Fedorov, FI, *Theory of Elastic Waves in Crystals*, Plenum Press, New York, 1968.

[8]   Guz, AN, *Fundamentals of the Three-Dimensional Theory of Stability of Deformable Bodies*, Viska Schola, Kiev, 1986 (in Russian).

[9]   Holliday, D and Resnick, R, *Physics, Part II*, John Wiley, New York, 1966.

[10] Hu, Y, Yang, J and Jiang, Q, Surface waves in electrostrictive materials under biasing fields, *ZAMP*, **55**, 678–700 (2004).

[11] Landau, L and Lifchitz, E, *Électrodynamique des milieux continus*, MIR, Moscou, 1969.

[12] Malvern, LE, *Introduction to the Mechanics of a Continuum Medium*, Prentice-Hall, Englewood Cliffs, NJ, 1969.

[13] Ogden, RW, *Non-Linear Elastic Deformations*, John Wiley, New York, 1984.

[14] Royer, D and Dieulesaint, E, *Elastic Waves in Solids, Vol. I - Free and Guided Propagation*, Springer, Berlin, 2000.

[15] Simionescu-Panait, O, The influence of initial fields on wave propagation in piezoelectric crystals, *Int. J. Appl. Electromag. Mech.*, **12**, 241–252 (2000).

[16] Simionescu-Panait, O, Progressive wave propagation in the meridian plane of a 6mm-type piezoelectric crystal subject to initial fields, *Math. Mech. Solids*, **6**, 661–670 (2001).

[17] Simionescu-Panait, O, The electrostrictive effect on wave propagation in isotropic solids subject to initial fields, *Mech. Res. Comm.*, **28**, 685–691 (2001).

[18] Simionescu-Panait, O, Wave propagation in cubic crystals subject to initial mechanical and electric fields, *ZAMP*, **53**, 1038–1051 (2002).

[19] Simionescu-Panait, O, Propagation of attenuated waves in isotropic solids subject to initial electro-mechanical fields, in: *Proc. of Int. Conf. New Trends in Continuum Mechanics*, 267–275, Ed. Theta, Bucharest, 2005.

[20] Simionescu-Panait, O, Attenuated wave propagation on a face of a cubic crystal subject to initial electro-mechanical fields, *Int. J. Appl. Electromag. Mech.*, **22**, 111–120 (2005).

[21] Simionescu-Panait, O, Initial fields impact on attenuated wave propagation in isotropic solids, *Math. Reports*, **8(58)**, 239–250 (2006).

[22] Simionescu-Panait, O, The influence of initial fields on the propagation of attenuated waves along an edge of a cubic crystal, in: *Proc. of Fourth Workshop Mathematical Modelling of Environmental and Life Sciences Problems*, 231–242, Ed. Academiei Române, Bucharest, 2006.

[23] Simionescu-Panait, O, A study of initial fields influence on propagation of attenuated waves on a face of a cubic crystal, *Rev. Roumaine Math. Pures Appl.*, **51**, 379–390 (2006).

[24] Simionescu-Panait, O, Propagation of attenuated waves along an edge of a cubic crystal subject to initial electro-mechanical fields, *Math. Mech. Solids*, **12**, 107–118 (2007).

[25] Simionescu-Panait, O, Decomposition of waveguides propagating in piezoelectric crystals subject to initial fields, in: *Proc. of Fifth Workshop Mathematical Modelling of Environmental and Life Sciences Problems*, 191–199, Ed. Academiei Române, Bucharest, 2007.

[26] Simionescu-Panait, O and Ana, I, On the coupling of guided waves propagating in piezoelectric crystals subject to initial fields, *Math. Mech. Solids*, published online doi:10.1177/1081286507086518 (2008) (in press).

[27] Simionescu-Panait, O and Soós, E, Wave propagation in piezoelectric crystals subjected to initial deformations and electric fields, *Math. Mech. Solids*, **6**, 437–446 (2001).

[28] Sirotin, II and Shaskolskaya, MP, *Crystal Physics*, Nauka, Moscow, 1975 (in Russian).

[29] Soós, E, Stability, resonance, and stress concentration in prestressed piezoelectric crystals containing a crack, *Int. J. Eng. Sci.*, **34**, 1647–1673 (1996).

[30] Tiersten, HF, *A Development of the Equations of Electromagnetism in Material Continua*, Springer, New York, 1990.

[31] Tiersten, HF, On the accurate description of piezoelectric resonators subject to biasing deformations, *Int. J. Eng. Sci.*, **33**, 2239–2259 (1995).

[32] Yang, JS, Bleustein-Gulyaev waves in strained piezoelectric ceramics, *Mech. Res. Comm.*, **28**, 679–683 (2001).

[33] Yang, J and Hu, Y, Mechanics of electroelastic bodies under biasing fields, *Appl. Mech. Rev.*, **57**, 173–189 (2004).

Chapter 7
Fully Dynamic Theory

Jiashi Yang

7.1 Introduction

The theory of piezoelectricity is based on a quasistatic approximation [1]. As a result, in this theory, although the mechanical equations are dynamic, the electromagnetic equations are static, and the electric field and the magnetic field are not dynamically coupled. Therefore, the theory of piezoelectricity does not describe the wave behavior of electromagnetic fields. For many applications in piezoelectric acoustic wave devices, the quasistatic theory is sufficient; but there are situations in which full electromagnetic coupling needs to be considered. When electromagnetic waves are involved, the complete set of Maxwell equations needs to be used, coupled to the mechanical equations of motion. Such a fully dynamic theory has been called piezoelectromagnetism by some researchers.

Solutions for the propagation of plane waves in an unbounded piezoelectromagnetic medium were obtained in [2]. In addition to waves that are essentially acoustic, there are also waves that are essentially electromagnetic. These two groups of modes interact through piezoelectric coupling. Effects of viscosity and conductivity on plane waves were analyzed in [3]. Results on plane waves can also be found in [4–6]. Surface waves were studied in [7, 8] for hexagonal crystals and in [9] for lithium niobate. Wave scattering around a circular cylinder was treated in [10]. Transient antiplane or shear-horizontal (SH) surface waves in a ceramic half-space under surface load were analyzed in [11, 12]. More references on early contributions to variational formulations of the theory of piezoelectromagnetism and electromagnetic radiation from a vibrating piezoelectric body are given in relevant sections later.

Jiashi Yang
Department of Engineering Mechanics University of Nebraska, Lincoln, NE 68588–0526, USA e-mail: Jyang1@unl.edu

J. Yang (ed.), *Special Topics in the Theory of Piezoelectricity*, DOI: 10.1007/978-0-387-89498-0_7, 247
© Springer Science + Business Media, LLC 2009

7.2 Governing Equations

The three-dimensional equations of linear piezoelectromagnetism consist of
the equations of motion and Maxwell equations

$$T_{ji,j} = \rho \ddot{u}_i,$$
$$\varepsilon_{ijk} E_{k,j} = -\dot{B}_i,$$
$$\varepsilon_{ijk} H_{k,j} = \dot{D}_i + J_i,$$
$$B_{i,i} = 0, \qquad D_{i,i} = q, \tag{7.1}$$

as well as the following constitutive relations,

$$T_{ij} = c_{ijkl} S_{kl} - e_{kij} E_k,$$
$$D_i = e_{ijk} S_{jk} + \varepsilon_{ij} E_j,$$
$$B_i = \mu_{ij} H_j, \tag{7.2}$$

where B_i is the magnetic induction, H_i is the magnetic field, J_i is the electric
current, and μ_{ij} is the magnetic permeability. When the material is nonmagnetizable, we have $\mu_{ij} = \mu_0 \delta_{ij}$, where μ_0 is the magnetic permeability of free
space. With Equation (7.2), for a nonmagnetizable body in a source-free
region, Equation (7.1) can be written as the following two equations for
u and **E**,

$$c_{ijkl} u_{k,li} = \rho \ddot{u}_j + e_{kij} E_{k,i},$$
$$E_{i,kk} - E_{k,ki} = \mu_0 \varepsilon_{ik} \ddot{E}_k + \mu_0 e_{ikl} \ddot{u}_{k,l}, \tag{7.3}$$

which shows the piezoelectric coupling between acoustic and electromagnetic
waves. With the introduction of the usual vector potential **A**, and scalar
potential ϕ for electromagnetic fields by

$$E_k = -\phi_{,k} - \dot{A}_k, \quad B_k = \varepsilon_{kij} A_{j,i}, \tag{7.4}$$

Equations $(7.1)_{2,4}$ are identically satisfied. Equations $(7.1)_{3,5}$ can be written
as equations in terms of the potentials.

On the boundary surface of a finite body, the mechanical displacement or
traction can be prescribed as boundary conditions. The boundary conditions
for the electromagnetic fields are usually in terms of tangential **E**, **H** and
normal **B**, **D**. Consider a finite dielectric body occupying a region V (see
Figure 7.1). The boundary surface of V is denoted by S, with a unit exterior
normal **n**. For electromagnetic boundary conditions, we consider the following
partitions of S (see Figure 7.1),

$$S_\phi \cup S_D = S_A \cup S_H = S,$$
$$S_\phi \cap S_D = S_A \cap S_H = 0. \tag{7.5}$$

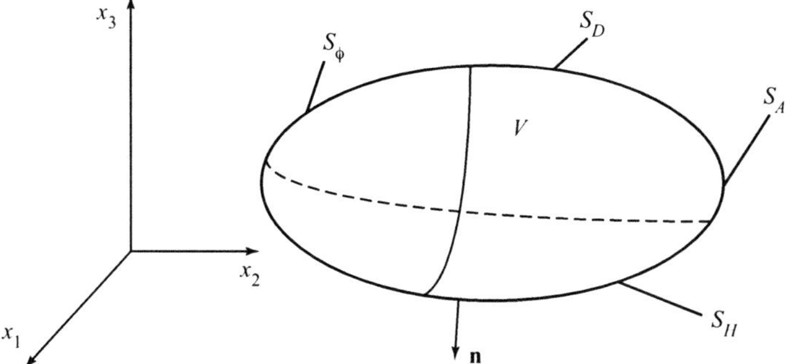

Fig. 7.1 A piezoelectric body and partitions of its boundary.

On S, we may prescribe

$$\phi = \overline{\phi} \quad \text{on} \quad S_\phi, \quad D_i n_i + \overline{d} = 0 \quad \text{on} \quad S_D,$$

$$\varepsilon_{ijk} n_j A_k = \overline{a}_i \quad \text{on} \quad S_A, \quad \varepsilon_{ijk} n_j H_k = \overline{h}_i \quad \text{on} \quad S_H, \tag{7.6}$$

where $\overline{\phi}$, $\overline{d}$, $\overline{a}_i$, and $\overline{h}$ are known boundary data.

7.3 Variational Formulation

Variational formulations for coupled electromagnetic and acoustic waves were studied in [1, 13]. A variational principle for piezoelectromagnetism in a compound continuum representing a diatomic material was given in [14]. Piezoelectromagnetic fields inside and outside a finite body with continuity conditions at the interface between the body and free space can be found in [15]. A so-called generalized variational principle with all mechanical and electromagnetic fields as independent variables was given in [16]. Variational principles and generalized variational principles for the eigenvalue problem of free vibrations of a piezoelectromagnetic body were also obtained [13, 17, 18]. Discontinuous fields were considered variationally in [19]. Variational formulations were also used in the derivation of piezoelectromagnetic plate equations [20, 21].

Consider the following variational functional [1, 15],

$$\Pi(\mathbf{A}, \phi) = \int_{t_0}^{t_1} dt \int_V \left[\frac{1}{2} (\varepsilon_{ij} E_i E_j - \mu_{ij}^{-1} B_i B_j) + J_i A_i - q\phi \right] dV$$

$$- \int_{t_0}^{t_1} dt \int_{S_D} \overline{d}\phi dS - \int_{t_0}^{t_1} dt \int_{S_H} \overline{h}_i A_i dS, \tag{7.7}$$

with the admissible functions satisfying

$$\phi = \overline{\phi} \quad \text{on} \quad S_\phi$$
$$\varepsilon_{ijk} n_j A_k = \overline{a} \quad \text{on} \quad S_A,$$
$$A_i(\mathbf{x},\, t_0) = A_i^0,$$
$$A_i(\mathbf{x},\, t_1) = A_i^1 \quad \text{in } V, \tag{7.8}$$

where A_i^0 and A_i^1 are prescribed data at t_0 and t_1. Then,

$$\delta\Pi = \int_{t_0}^{t_1} dt \int_V \left[(D_{i,i} - q)\delta\phi - \left(\varepsilon_{ijk} H_{k,j} - \dot{D}_i - J_i \right) \delta A_i \right] dV$$
$$- \int_{t_0}^{t_1} dt \int_{S_D} (D_i n_i + \overline{d})\delta\phi dS + \int_{t_0}^{t_1} dt \int_{S_H} (\varepsilon_{ijk} n_j H_k - \overline{h}_i)\delta A_i dS. \tag{7.9}$$

Therefore the stationary condition of Equation (7.7) yields Equations $(7.1)_{3,5}$ and $(7.6)_{2,4}$.

7.4 Quasistatic Approximation

The quasistatic approximation in the theory of piezoelectricity can be considered as the lowest-order approximation of a perturbation procedure based on the fact that the acoustic wave speed is much smaller than the speed of light [1]. To see this, we consider a nonmagnetizable body in a source-free region. First, we write Equations (7.1) and (7.2) into the following three equations for $\mathbf{u}$, $\mathbf{E}$, and $\mathbf{B}$:

$$c_{ijkl} u_{k,li} - e_{kij} E_{k,i} = \rho \ddot{u}_j,$$
$$\varepsilon_{ijk} E_{k,j} = -\dot{B}_i,$$
$$\frac{1}{\mu_0} \varepsilon_{ijk} B_{k,j} = e_{ikl} \dot{u}_{k,l} + \varepsilon_{ik} \dot{E}_k. \tag{7.10}$$

Consider an acoustic wave with frequency ω in a piezoelectric crystal of size L. We scale the various independent and dependent variables with respect to characteristic quantities

$$\xi_i = \frac{x_i}{L}, \qquad \tau = \omega t, \qquad U_i = \frac{u_i}{L}, \qquad b_i = c_0 B_i, \tag{7.11}$$

where

$$c_0 = \frac{1}{\sqrt{\varepsilon_0 \mu_0}} \tag{7.12}$$

is the speed of light in free space, and the scaling yields a $\mathbf{b}$ in the same units as $\mathbf{E}$. Then Equation (7.10) takes the following form

$$\frac{1}{L}c_{ijkl}\frac{\partial^2 U_k}{\partial\xi_l\partial\xi_i} - \frac{1}{L}e_{kij}\frac{\partial E_k}{\partial\xi_i} = \rho\omega^2 L\frac{\partial^2 U_j}{\partial\tau^2},$$

$$\frac{1}{L}\varepsilon_{ijk}\frac{\partial E_k}{\partial\xi_j} = -\frac{\omega}{c_0}\frac{\partial b_i}{\partial\tau},$$

$$\frac{1}{c_0 L\mu_0}\varepsilon_{ijk}\frac{\partial b_k}{\partial\xi_j} = \omega\varepsilon_0\frac{e_{ikl}}{\varepsilon_0}\frac{\partial^2 U_k}{\partial\xi_l\partial\tau} + \omega\varepsilon_0\frac{\varepsilon_{ik}}{\varepsilon_0}\frac{\partial E_k}{\partial\tau}, \qquad (7.13)$$

or

$$c_{ijkl}\frac{\partial^2 U_k}{\partial\xi_l\partial\xi_i} - e_{kij}\frac{\partial E_k}{\partial\xi_i} = \rho\omega^2 L^2\frac{\partial^2 U_j}{\partial\tau^2}, \qquad \varepsilon_{ijk}\frac{\partial E_k}{\partial\xi_j} = -\eta\frac{\partial b_i}{\partial\tau},$$

$$\varepsilon_{ijk} = \frac{\partial b_k}{\partial\xi_j} = \eta\left(\frac{e_{ikl}}{\varepsilon_0}\frac{\partial^2 U_k}{\partial\xi_l\partial_\tau} + \frac{\varepsilon_{ik}}{\varepsilon_0}\frac{\partial E_k}{\partial\tau}\right), \qquad (7.14)$$

where

$$\eta = \frac{\omega L}{c_0} \ll 1. \qquad (7.15)$$

To the lowest order

$$c_{ijkl}\frac{\partial^2 U_k}{\partial\xi_l\partial\xi_i} - e_{kij}\frac{\partial E_k}{\partial\xi_i} = \rho\omega^2 L^2\frac{\partial^2 U_j}{\partial\tau^2},$$

$$\varepsilon_{ijk}\frac{\partial E_k}{\partial\xi_j} = 0, \qquad \varepsilon_{ijk}\frac{\partial b_k}{\partial\xi_j} = 0, \qquad (7.16)$$

or

$$c_{ijkl}u_{k,li} - e_{kij}E_{k,i} = \rho\ddot{u}_j, \qquad \varepsilon_{ijk}E_{k,j} = 0, \qquad \varepsilon_{ijk}H_{k,j} = 0, \qquad (7.17)$$

which are the equations for linear piezoelectricity.

7.5 Antiplane Problems of Polarized Ceramics

The equations of piezoelectromagnetism are rather complicated. In the rest of this chapter we study antiplane problems or SH waves in polarized ceramics that are mathematically relatively simple and yet can show the basic physics of piezoelectromagnetism well. For antiplane problems in polarized ceramics, the electromagnetic fields are the so-called transverse magnetic (TM) fields in electromagnetism and can be described by one magnetic field component. Therefore, one displacement component and one magnetic field component are sufficient to describe the antiplane mechanical and electromagnetic fields in polarized ceramics. This is simpler than using the usual scalar and vector potentials for electromagnetic fields. For antiplane problems in polarized ceramics [11, 12],

$$u_1 = u_2 = 0, \qquad u_3 = u_3(x_1, x_2, t), \qquad E_1 = E_1(x_1, x_2, t),$$
$$E_2 = E_2(x_1, x_2, t), \qquad E_3 = 0, \qquad H_1 = H_2 = 0, \qquad H_3 = H_3(x_1, x_2, t). \tag{7.18}$$

The nonvanishing components of S_{ij}, T_{ij}, D_i, and B_i are

$$S_4 = u_{3,2}, \qquad S_5 = u_{3,1},$$
$$T_4 = c_{44}u_{3,2} - e_{15}E_2, \qquad T_5 = c_{44}u_{3,1} - e_{15}E_1,$$
$$D_1 = e_{15}u_{3,1} + \varepsilon_{11}E_1, \qquad D_2 = e_{15}u_{3,2} + \varepsilon_{11}E_2,$$
$$B_3 = \mu_0 H_3. \tag{7.19}$$

The nontrivial ones of the equations of motion and Maxwell equations take the following form

$$c_{44}(u_{3,11} + u_{3,22}) - e_{15}(E_{1,1} + E_{2,2}) = \rho \ddot{u}_3,$$
$$e_{15}(u_{3,11} + u_{3,22}) + \varepsilon_{11}(E_{1,1} + E_{2,2}) = 0,$$
$$E_{2,1} - E_{1,2} = -\mu_0 \dot{H}_3,$$
$$H_{3,2} = e_{15}\dot{u}_{3,1} + \varepsilon_{11}\dot{E}_1,$$
$$-H_{3,1} = e_{15}\dot{u}_{3,2} + \varepsilon_{11}\dot{E}_2. \tag{7.20}$$

In Equations (7.19) and (7.20), body source and magnetization are not considered. Eliminating the electric field components from Equations $(7.20)_{1,2}$,

$$\bar{c}_{44}(u_{3,11} + u_{3,22}) = \rho \ddot{u}_3, \tag{7.21}$$

where

$$\bar{c}_{44} = c_{44} + \frac{e_{15}^2}{\varepsilon_{11}} = c_{44}(1 + k_{15}^2),$$
$$k_{15}^2 = \frac{e_{15}^2}{\varepsilon_{11}c_{55}}. \tag{7.22}$$

Differentiating Equation $(7.20)_3$ with respect to time once and substituting from Equations $(7.20)_{4,5}$, we have

$$H_{3,11} + H_{3,22} = \varepsilon_{11}\mu_0 \ddot{H}_3. \tag{7.23}$$

The above equations can be written in coordinate independent forms as

$$v_T^2 \nabla^2 u_3 = \ddot{u}_3, \qquad c^2 \nabla^2 H_3 = \ddot{H}_3, \qquad \dot{\mathbf{D}} = -\mathbf{i}_3 \times \nabla H_3, \tag{7.24}$$

where

$$v_T^2 = \frac{\bar{c}_{44}}{\rho}, \qquad c^2 = \frac{1}{\varepsilon_{11}\mu_0}. \tag{7.25}$$

ν_T and c are the speed of plane shear waves and the speed of light propagating in the x_1 direction. ∇ and ∇^2 are the two-dimensional gradient operator and Laplacian, respectively. $\mathbf{D}$ is the electric displacement in the (x_1, x_2) plane. $\mathbf{i}_3$ is the unit vector in the x_3 direction. Equations $(7.24)_{1,2}$ govern the displacement and magnetic fields and are uncoupled. However, coupling between u_3 and H_3 exists in Equation $(7.24)_3$. In this formulation, u_3 and H_3 are the primary unknowns. Once u_3 and H_3 are determined, D_1 and D_2 can be obtained from Equation $(7.24)_3$. Then the electric field and the stress components can be obtained from constitutive relations.

7.6 A Moving Dislocation

Consider a screw dislocation moving at a constant speed V along the x_1-axis [22]. At $t = 0$, the dislocation occupies the negative x_1-axis (see Figure 7.2). Choose a moving coordinate system (x, y) as follows,

$$x = x_1 - Vt,$$
$$y = x_2. \tag{7.26}$$

In the moving frame (x, y), Equations $(7.24)_{1,2}$ become

$$\beta^2 \frac{\partial^2 u_3}{\partial x^2} + \frac{\partial^2 u_3}{\partial y^2} = 0,$$

$$\gamma^2 \frac{\partial^2 H_3}{\partial x^2} + \frac{\partial^2 H_3}{\partial y^2} = 0, \tag{7.27}$$

where

$$\beta^2 = 1 - \frac{V^2}{\nu_T^2},$$

$$\gamma^2 = 1 - \frac{V^2}{c^2}. \tag{7.28}$$

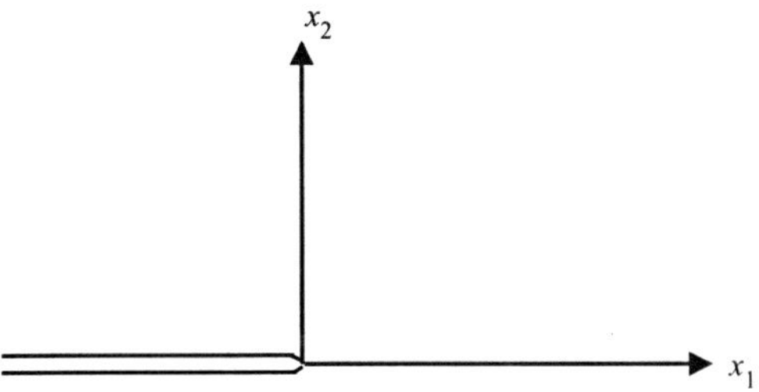

Fig. 7.2 A moving screw dislocation.

Consider Equation $(7.27)_1$ first. Introduce a new coordinate system (ξ, η) by

$$\xi = x, \quad \eta = \beta y. \tag{7.29}$$

Then Equation $(7.27)_1$ can be written as

$$\frac{\partial^2 u_3}{\partial \xi^2} + \frac{\partial^2 u_3}{\partial \eta^2} = 0. \tag{7.30}$$

In a polar coordinate system (r, θ) defined by

$$\xi = r \cos \theta,$$
$$\eta = r \sin \theta, \quad -\pi \le \theta \le \pi, \tag{7.31}$$

Equation(7.30) takes the following form,

$$\frac{\partial^2 u_3}{\partial r^2} + \frac{1}{r} \frac{\partial u_3}{\partial r} + \frac{1}{r^2} \frac{\partial^2 u_3}{\partial \theta^2} = 0. \tag{7.32}$$

Similarly, in another polar coordinate system (ρ, σ) defined by

$$x_1 - Vt = \rho \cos \sigma,$$
$$\gamma x_2 = \rho \sin \sigma, \quad -\pi \le \sigma \le \pi, \tag{7.33}$$

Equation $(7.27)_2$ becomes

$$\frac{\partial^2 H_3}{\partial \rho^2} + \frac{1}{\rho} \frac{\partial H_3}{\partial \rho} + \frac{1}{\rho^2} \frac{\partial^2 H_3}{\partial \sigma} = 0. \tag{7.34}$$

For a screw dislocation, we are interested in solutions to Equation (7.32) that are r-independent. Denoting the displacement discontinuity of the dislocation at $\theta = \pi$ by b, a constant, we have the displacement field of the dislocation as

$$u_3 = \frac{b}{2\pi} \theta = \frac{b}{2\pi} \tan^{-1} \frac{\eta}{\xi} = \frac{b}{2\pi} \tan^{-1} \frac{\beta y}{x} = \frac{b}{2\pi} \tan^{-1} \frac{\beta x_2}{x_1 - Vt}. \tag{7.35}$$

If we understand a dislocation as a pure mechanical concept as described by Equation (7.35), the electromagnetic fields associated with the dislocation are not unique, depending on how we prescribe electromagnetic boundary/continuity conditions at the dislocation. One natural consideration is that we also allow a discontinuity of H_3 at the dislocation. Then, similar to Equation (7.35), we have the following solution to (7.34),

$$H_3 = \frac{h}{2\pi} \tan^{-1} \frac{\gamma x_2}{x_1 - Vt}, \tag{7.36}$$

where h is a constant. The strain components determined by Equation (7.35) are

$$S_4 = u_{3,2} = \frac{b}{2\pi} = \frac{\beta(x_1 - Vt)}{(x_1 - Vt)^2 + (\beta x_2)^2},$$

$$S_5 = u_{3,1} = -\frac{b}{2\pi} = \frac{\beta x_2}{(x_1 - Vt)^2 + (\beta x_2)^2}. \tag{7.37}$$

From Equations $(7.24)_3$ and (7.36) we have

$$\dot{D}_1 = H_{3,2} = \frac{h}{2\pi} \frac{\gamma(x_1 - Vt)}{(x_1 - Vt)^2 + (\gamma x_2)^2},$$

$$\dot{D}_2 = -H_{3,1} = \frac{h}{2\pi} \frac{\gamma x_2}{(x_1 - Vt)^2 + (\gamma x_2)^2}, \tag{7.38}$$

which can be integrated with respect to time to yield

$$D_1 = -\frac{h\gamma}{4\pi V}\ln[(x_1 - Vt)^2 + (\gamma x_2)^2],$$

$$D_2 = \frac{h\gamma}{2\pi V}\left(\tan^{-1}\frac{\gamma x_2}{x_1 - Vt} - \tan^{-1}\frac{\gamma x_2}{x_1}\right), \tag{7.39}$$

where the integration constants have been dropped. Then from the constitutive relations in Equation (7.19), we obtain

$$E_1 = -\frac{h\gamma}{4\pi\varepsilon_{11}V}\ln[(x_1 - Vt)^2 + (\gamma x_2)^2] + \frac{e_{15}b}{2\pi\varepsilon_{11}}\frac{\beta x_2}{(x_1 - Vt)^2 + (\beta x_2)^2},$$

$$E_2 = \frac{h\gamma}{2\pi\varepsilon_{11}V}\tan^{-1}\frac{\gamma x_2}{x_1 - Vt} - \frac{e_{15}b}{2\pi\varepsilon_{11}}\frac{\beta(x_1 - Vt)}{(x_1 - Vt)^2 + (\beta x_2)^2}, \tag{7.40}$$

and

$$T_{23} = \bar{c}_{44}\frac{b}{2\pi}\frac{\beta(x_1 - Vt)}{(x_1 - Vt)^2 + (\beta x_2)^2} - \frac{e_{15}h}{2\pi\varepsilon_{11}V}\tan^{-1}\frac{\gamma x_2}{x_1 - Vt},$$

$$T_{13} = -\bar{c}_{44}\frac{b}{2\pi}\frac{\beta x_2}{(x_1 - Vt)^2 + (\beta x_2)^2} + \frac{e_{15}h\gamma}{4\pi\varepsilon_{11}V}\ln[(x_1 - Vt)^2 + (\gamma x_2)^2]. \tag{7.41}$$

For the same mechanical discontinuity at a dislocation, we may prescribe different boundary or continuity conditions for the electromagnetic fields at the dislocation, resulting in different electromagnetic fields. Consider the possibility of the following fields,

$$u_3 = \frac{b}{2\pi}\tan^{-1}\frac{\beta x_2}{x_1 - Vt},$$

$$H_3 = \frac{\bar{h}}{2\pi}\frac{x_1 - Vt}{(x_1 - Vt)^2 + (\gamma x_2)^2}, \tag{7.42}$$

where $\overline{h}$ is a constant. Equation $(7.42)_1$ is the same as (7.35). It can be verified by direct differentiation that the H_3 in Equation $(7.42)_2$ satisfies $(7.24)_2$. The strain field is still given by Equation (7.37). From Equation $(7.24)_3$, we obtain

$$D_1 = -\frac{\overline{h}}{2\pi V}\frac{\gamma^2 x_2}{(x_1 - Vt)^2 + (\gamma x_2)^2}, \qquad D_2 = \frac{\overline{h}}{2\pi V}\frac{x_1 - Vt}{(x_1 - Vt)^2 + (\gamma x_2)^2}. \tag{7.43}$$

Then from the constitutive relations the stress and electric fields can be obtained. For comparison, the solution from the quasi-static theory of piezo-electricity [23] is summarized below. The displacement and strain fields are the same as Equations (7.35) and (7.37). The electric displacement field is given by

$$D_1 = \varepsilon_{11}\frac{a}{2\pi}\frac{x_2}{(x_1 - Vt)^2 + (x_2)^2}, \qquad D_2 = -\varepsilon_{11}\frac{a}{2\pi}\frac{x_1 - Vt}{(x_1 - Vt)^2 + (x_2)^2}, \tag{7.44}$$

where a is a constant. From the dynamic solution in Equations (7.35), (7.37), and (7.43) and the quasi-static solution in Equations (7.35), (7.37), and (7.44), we see that the displacement and strain fields of the dynamic and quasi-static solutions are the same. The other fields are different. In the limit of $c \to \infty$, we have $\gamma \to 1$. If we identify the following from Equations (7.43) and (7.44),

$$\frac{\overline{h}}{V} = -\varepsilon_{11}a, \tag{7.45}$$

then the two equations become the same. Therefore, the quasi-static solution is an approximation of the dynamic solution when the speed of light approaches infinity. Piezoelectromagnetic fields associated with moving or stationary cracks were studied in [24, 25].

7.7 Surface Waves

Piezoelectromagnetic surface waves were studied in [26] using the scalar and vector potentials of electromagnetic fields and in [27] using the fields directly. The following is based on [27]. Consider a ceramic half-space poled in the x_3-direction (see Figure 7.3).

For surface waves propagating in the x_1–direction, we have

$$u_3 = U\exp(-\xi_2 x_2)\cos(\xi_1 x_1 - \omega t), \qquad H_3 = H\exp(-\eta_2 x_2)\cos(\xi_1 x_1 - \omega t), \tag{7.46}$$

where $U, H, \xi_1, \xi_2, \eta_2,$ and ω are undetermined constants. Substitution of Equation (7.46) into $(7.24)_{1,2}$ results in

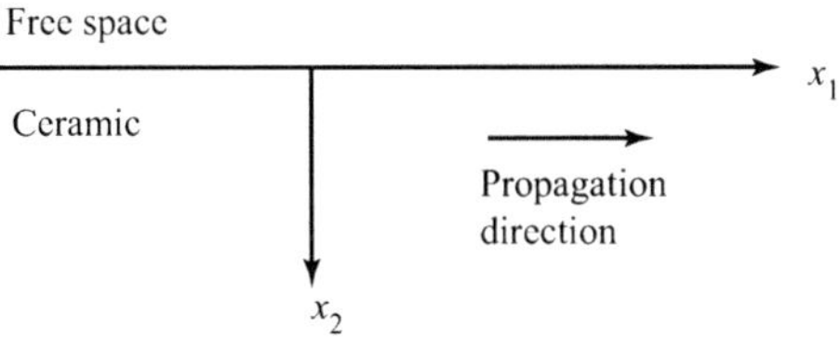

Fig. 7.3 A ceramic half-space.

$$\xi_2^2 = \xi_1^2 - \frac{\rho\omega^2}{\bar{c}_{44}} = \xi_1^2\left(1 - \frac{v^2}{v_T^2}\right) > 0,$$

$$\eta_2^2 = \xi_1^2 - \varepsilon_{11}\mu_0\omega^2 = \xi_1^2\left(1 - \frac{v^2}{c^2}\right) > 0, \tag{7.47}$$

where

$$v^2 = \frac{\omega^2}{\xi_1^2}, \qquad v_T^2 = \frac{\bar{c}_{44}}{\rho}, \qquad c^2 = \frac{1}{\varepsilon_{11}\mu_0}. \tag{7.48}$$

v is the surface wave speed. The inequalities are for decaying behavior from the surface. The stress and electric field components produced by Equation (7.46) relevant to boundary conditions are

$$T_4 = -\frac{1}{\varepsilon_{11}\omega}(\varepsilon_{11}\bar{c}_{44}\xi_2 U e^{-\xi_2 x_2} + e_{15}\xi_1 H e^{-\eta_2 x_2})\cos(\xi_1 x_1 - \omega t),$$

$$E_1 = \frac{1}{\varepsilon_{11}\omega}(e_{15}\omega\xi_1 U e^{-\xi_2 x_2} + \eta_2 H e^{-\eta_2 x_2})\sin(\xi_1 x_1 - \omega t). \tag{7.49}$$

7.7.1 An Electroded Half-Space

First consider the case when the surface at $x_2 = 0$ is electroded with a perfect conductor for which we have $E_1 = 0$. The electrode is assumed to be very thin, with negligible mass. Hence, we have the traction-free condition $T_4 = 0$ on the surface. Then from Equation (7.49) we can write

$$e_{15}\omega\xi_1 U + \eta_2 H = 0,$$

$$\varepsilon_{11}\bar{c}_{44}\omega\xi_2 U + e_{15}\xi_1 H = 0. \tag{7.50}$$

For nontrivial solutions of U and/or H, the determinant of the coefficient matrix has to vanish, which leads to

$$\sqrt{1 - \frac{v^2}{v_T^2}}\sqrt{1 - \alpha n^2 \frac{v^2}{v_T^2}} = \bar{k}_{15}^2, \tag{7.51}$$

where

$$\alpha = \frac{v_T^2}{c_0^2}, \qquad c_0^2 = \frac{1}{\varepsilon_0 \mu_0}, \qquad n^2 = \frac{\varepsilon_{11}}{\varepsilon_0}, \qquad \overline{k}_{15}^2 = \frac{e_{15}^2}{\varepsilon_{11}\overline{c}_{44}} = \frac{k_{15}^2}{1 + k_{15}^2}. \quad (7.52)$$

In Equation (7.52), α is the square of the ratio between the acoustic and light wave speeds which is normally a very small number. n is the refractive index in the x_1 direction. Equation (7.51) is an equation for the surface wave speed v. Waves with their speed determined by (7.51) are clearly nondispersive. Because it is very small, it is simpler and more revealing to examine the following perturbation solution of Equation (7.51) for small α:

$$v^2 \cong v_T^2(1 - \overline{k}_{15}^4)(1 - \alpha n^2 \overline{k}_{15}^4). \quad (7.53)$$

Equation (7.53) shows that the effect of electromagnetic coupling on the acoustic wave speed is of the order of $\alpha n^2 \overline{k}_{15}^4$. As a numerical example, we consider polarized ceramic PZT-7A. Calculation shows that

$$\overline{k}_{15} = 0.671, \qquad n^2 = 460, \qquad \alpha = 6.85 \times 10^{-9}, \qquad \alpha n^2 \overline{k}_{15}^4 = 6.38 \times 10^{-7}. \quad (7.54)$$

Hence, the modification of the acoustic wave speed due to electromagnetic effects is very small and is negligible in most applications. When α is set to zero, or when the speed of light approaches infinity, Equation (7.53) reduces to the speed of the well-known Bleustein–Gulyaev waves [28, 29] in quasi-static piezoelectricity.

7.7.2 An Unelectroded Half-Space

When the traction-free surface of the half-space at $x_2 = 0$ is unelectroded, electromagnetic waves also exist in the free space of $x_2 < 0$. The solution for the free space $x_2 < 0$ can be written as

$$H_3 = \overline{H} \exp(\overline{\eta}_2 x_2) \cos(\xi_1 x_1 - \omega t), \quad (7.55)$$

where $\overline{H}$ and $\overline{\eta}_2$ are undetermined constants. Substitution of Equation (7.55) into $(7.24)_2$, with ε_{11} replaced by ε_0 for free space, we obtain

$$\overline{\eta}_2^2 = \xi_1^2 - \varepsilon_0 \mu_0 \omega^2 > 0. \quad (7.56)$$

The electric field generated by H_3 is given by

$$E_1 = -\frac{1}{\varepsilon_0 \omega} \overline{\eta}_2 \overline{H} \exp(\overline{\eta}_2 x_2)\sin(\xi_1 x_1 - \omega t),$$

$$E_2 = \frac{1}{\varepsilon_0 \omega} \xi_1 \overline{H} \exp(\overline{\eta}_2 x_2)\cos(\xi_1 x_1 - \omega t). \quad (7.57)$$

We require the continuity of E_1 and H_3 at $x_2 = 0$ as well as the vanishing of the shear stress T_4. This implies that

$$\frac{1}{\varepsilon_{11}\omega}(e_{15}\omega\xi_1 U + \eta_2 H) + \frac{1}{\varepsilon_0\omega}\overline{\eta}_2\overline{H} = 0,$$

$$H - \overline{H} = 0, \quad \varepsilon_{11}\overline{c}_{44}\omega\xi_2 U + e_{15}\xi_1 H = 0. \tag{7.58}$$

For nontrivial solutions the determinant of the coefficient matrix of Equation (7.58) has to vanish, which results in

$$\sqrt{1 - \frac{v^2}{v_T^2}}\left(\sqrt{1 - \alpha n^2 \frac{v^2}{v_T^2}} + n^2\sqrt{1 - \alpha\frac{v^2}{v_T^2}}\right) = \overline{k}_{15}^2. \tag{7.59}$$

Equation (7.59) is an equation for v. Again, the waves are nondispersive. When α is set to zero, the corresponding result of [28, 29] is recovered. A perturbation solution of Equation (7.59) to the first order of α is

$$v^2 \cong v_T^2\left[1 - \frac{\overline{k}_{15}^4}{(1+n^2)^2}\right]\left[1 - \alpha 2n^2\frac{\overline{k}_{15}^4}{(1+n^2)^3}\right], \tag{7.60}$$

and calculation shows that, for PZT-7A,

$$\alpha 2n^2\frac{k_{15}^4}{(1+n^2)^3} = 1.30 \times 10^{-16}. \tag{7.61}$$

7.8 Waves in a Plate Between Two Half-Spaces

A piezoelectric interface wave solution between two ceramic halfspaces was given in [30]. Waves in plates were studied in [31] and [32] using potential and field formulations, respectively. Love waves in a plate over a half-space were obtained in [33]. Because electromagnetic fields can exist in free space, air gaps are often present in piezoelectric devices. Gap waves in two half-spaces at a finite distance apart were analyzed in [34]. Waves guided by the surface of a circular ceramic cylinder were obtained in [35]. Waves in polarized ceramics were treated in a general manner in [36]. Some of these results are special cases of waves propagating in a plate between two half-spaces. Therefore, in this section we examine waves in the structure shown in Figure 7.4 [37].

7.8.1 Dispersion Relations

Waves propagating in the x_1-direction in the structure shown can be classified as guided or radiating modes depending on their behavior in the x_2-direction. Guided modes have exponentially decaying behavior when $|x_2|$ is large.

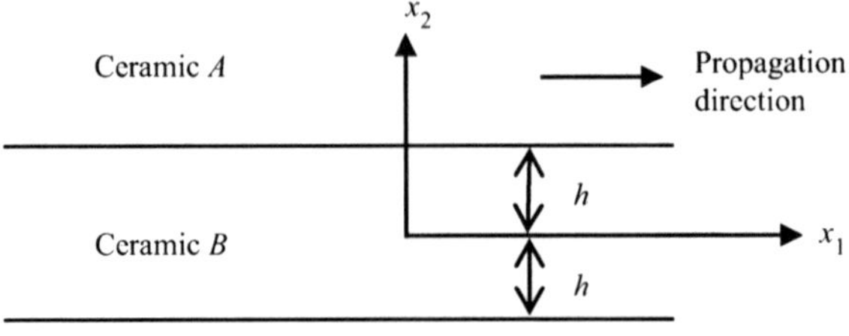

Fig. 7.4 A ceramic plate between two ceramic half-spaces.

7.8.1.1 Fields in the Upper Half-Space

Consider the following waves propagating in the x_1 direction with

$$u_3 = U_A \exp[-\eta_A(x_2 - h)]\cos(\xi x_1 - \omega t),$$
$$H_3 = H_A \exp[-\varsigma_A(x_2 - h)]\cos(\xi x_1 - \omega t), \tag{7.62}$$

where $U_A, H_A, \xi, \eta_A, \varsigma_A$, and ω are undetermined constants. The subscript in these constants indicates that they are for ceramic A. Substitution of Equation (7.62) into $(7.24)_{1,2}$ results in

$$\eta_A^2 = \xi^2 - \frac{\rho_A \omega^2}{\bar{c}_A} = \xi^2\left(1 - \frac{v^2}{v_A^2}\right) > 0,$$

$$\varsigma_A^2 = \xi^2 - \varepsilon_A \mu_0 \omega^2 = \xi^2\left(1 - \frac{v^2}{c_A^2}\right) > 0, \tag{7.63}$$

where

$$v^2 = \frac{\omega^2}{\xi^2}, \qquad v_A^2 = \frac{\bar{c}_A}{\rho_A}, \qquad c_A^2 = \frac{1}{\varepsilon_A \mu_0}. \tag{7.64}$$

In Equation (7.64), v is the wave speed that is to be determined, v_A is the speed of plane shear waves in the x_1-direction, and c_A is the speed of light in the x_1-direction. The inequalities in Equation (7.63) are for guided waves with decaying behavior from $x_2 = h$. If one or both of the inequalities is violated, the modes become radiating. From Equation $(7.24)_3$ and piezoelectromagnetic constitutive relations, we obtain

$$T_4 = -\frac{1}{\varepsilon_A \omega}\{\varepsilon_A \bar{c}_A \omega \eta_A U_A \exp[-\eta_A(x_2 - h)]$$
$$+ e_A \xi H_A \exp[-\varsigma_A(x_2 - h)]\}\cos(\xi x_1 - \omega t),$$
$$E_1 = \frac{1}{\varepsilon_A \omega} e_A \omega \xi U_A \exp[-\eta_A(x_2 - h)]$$
$$+ \varsigma_A H_A \exp[-\varsigma_A(x_2 - h)]\}\sin(\xi x_1 - \omega t), \tag{7.65}$$

which will be needed for interface continuity conditions.

7.8.1.2 Fields in the Lower Half-Space

Similarly, for the lower half-space (C) occupying $x_2 < -h$, we have

$$u_3 = U_C\exp[\eta_C(x_2 + h)]\cos(\xi x_1 - \omega t),$$
$$H_3 = H_C\exp[\varsigma_C(x_2 + h)]\cos(\xi x_1 - \omega t), \qquad (7.66)$$

where

$$\eta_C^2 = \xi^2 - \frac{\rho_C\omega^2}{\bar{c}_C} = \xi^2\left(1 - \frac{v^2}{v_C^2}\right) > 0,$$

$$\varsigma_C^2 = \xi^2 - \varepsilon_C\mu_0\omega^2 = \xi^2\left(1 - \frac{v^2}{c_C^2}\right) > 0, \qquad (7.67)$$

$$v_C^2 = \frac{\bar{c}_C}{\rho_C}, \qquad c_C^2 = \frac{1}{\varepsilon_C\mu_0}. \qquad (7.68)$$

We also obtain

$$T_4 = -\frac{1}{\varepsilon_C\omega}\{-\varepsilon_C\bar{c}_C\omega\eta_C U_C \exp[\eta_C(x_2 + h)]$$
$$+ e_C\xi H_C \exp[\varsigma_C(x_2 + h)]\}\cos(\xi x_1 - \omega t),$$

$$E_1 = \frac{1}{\varepsilon_C\omega}\{e_C\omega\xi U_C \exp[\eta_C(x_2 + h)]$$
$$- \varsigma_C H_C \exp[\varsigma_C(x_2 + h)]\}\sin(\xi x_1 - \omega t). \qquad (7.69)$$

7.8.1.3 Fields in the Plate

For the ceramic plate (B), the fields can be represented by

$$u_3 = (U_B\cos\eta_B x_2 + V_B\sin\eta_B x_2)\cos(\xi x_1 - \omega t),$$
$$H_3 = (G_B\cosh\varsigma_B x_2 + H_B\sinh\varsigma_B x_2)\cos(\xi x_1 - \omega t), \qquad (7.70)$$

where

$$\eta_B^2 = \frac{\rho_B\omega^2}{\bar{c}_B} - \xi^2 = \xi^2\left(\frac{v^2}{v_B^2} - 1\right),$$

$$\varsigma_B^2 = \xi^2 - \varepsilon_B\mu_0\omega^2 = \xi^2\left(1 - \frac{v^2}{c_B^2}\right), \qquad (7.71)$$

$$v_B^2 = \frac{\bar{c}_B}{\rho_B}, \qquad c_B^2 = \frac{1}{\varepsilon_B\mu_0}. \qquad (7.72)$$

If the η_B^2 and/or ς_B^2 in Equation (7.71) become negative, the fields in the plate change from sinusoidal to exponential in the x_2-direction or vice versa.

The field components relevant to boundary conditions are

$$T_4 = (-\bar{c}_B \eta_B U_B \sin \eta_B x_2 + \bar{c}_B \eta_B V_B \cos \eta_B x_2$$
$$- \frac{e_B}{\varepsilon_B \omega} G_B \xi \cosh \varsigma_B x_2 - \frac{e_B}{\varepsilon_B \omega} H_B \xi \sinh \varsigma_B x_2)\cos(\xi x_1 - \omega t),$$

$$E_1 = \frac{1}{\varepsilon_B \omega}(e_B \omega \xi U_B \cos \eta_B x_2 + e_B \omega \xi V_B \sin \eta_B x_2$$
$$- G_B \varsigma_B \sinh \varsigma_B x_2 - H_B \varsigma_B \cosh \varsigma_B x_2)\sin(\xi x_1 - \omega t). \qquad (7.73)$$

7.8.1.4 Continuity Conditions and Dispersion Relation

At the interfaces $x_2 = \pm h$, the continuity of u_3, T_4, H_3, and E_1 need to be imposed:

$$\begin{aligned}
u_3(h^+) &= u_3(h^-), & u_3(-h^-) &= u_3(-h^+), \\
H_3(h^+) &= H_3(h^-), & H_3(-h^-) &= H_3(-h^+), \\
T_4(h^+) &= T_4(h^-), & T_4(-h^-) &= T_4(-h^+), \\
E_1(h^+) &= E_1(h^-), & E_1(-h^-) &= E_1(-h^+),
\end{aligned} \qquad (7.74)$$

which represent eight linear, homogeneous equations for $U_A, U_B, U_C, H_A, H_B,$ H_C, V_B, and G_B. For nontrivial solutions of these constants to exist, the determinant of the coefficient matrix has to vanish. This yields the frequency equation. The expansion of the determinant of the coefficient matrix yields a rather long expression and, therefore, is not presented here. When viewed as an equation for ω, a root of the frequency equation determines a relation between ω and ξ (the dispersion relation). The presence of trigonometric functions suggests that the dispersion relations determined may be multivalued. The existence of roots can be seen in various special cases and the numerical solutions discussed below.

7.8.2 Special Cases

In this section, we examine a few special cases of the dispersion relations of antiplane waves in the structure shown in Figure 7.4.

7.8.2.1 Symmetric Waves

When the two half-spaces are of the same material, that is,

$$\rho_A = \rho_C, \quad \bar{c}_A = \bar{c}_C, \quad e_A = e_C, \quad \varepsilon_A = \varepsilon_C, \qquad (7.75)$$

the waves can be separated into symmetric and antisymmetric ones. For symmetric waves, consider

$$U_A = U_C, \qquad H_A = -H_C,$$
$$V_B = 0, \qquad G_B = 0. \tag{7.76}$$

In this case, Equations $(7.74)_{2,4,6,8}$ become the same as $(7.74)_{1,3,5,7}$. The dispersion relations assume the following simple form,

$$\left(\frac{\varsigma_A}{\varepsilon_A} \tanh \varsigma_B h + \frac{\varsigma_B}{\varepsilon_B} \right) (\bar{c}_A \eta_A - \bar{c}_B \eta_B \tan \eta_B h) = \left(\frac{e_A}{\varepsilon_A} - \frac{e_B}{\varepsilon_B} \right)^2 \xi^2 \tanh \varsigma_B h. \tag{7.77}$$

From Equation (7.77), we can see that the waves are dispersive in general. The right-hand side of (7.77), which is responsible for the coupling between acoustic and electromagnetic waves, depends on the difference of the ratio of the piezoelectric and dielectric constants. If the plate and the half-spaces are ceramics poled in the same direction, the coupling is not as strong as when they are poled in opposite directions when their piezoelectric constants have opposite signs.

When the two half-spaces are free space, that is, $c_A = 0$, $e_A = 0$, and $\epsilon_A = \epsilon_0$, Equation (7.77) becomes

$$\left(\frac{\varepsilon_B}{\varepsilon_0} \varsigma_A \tanh \varsigma_B h + \varsigma_B \right) \eta_B \tan \eta_B h = -\frac{e_B^2}{\varepsilon_B \bar{c}_B} \xi^2 \tanh \varsigma_B h, \tag{7.78}$$

which is the same as the result in [32] for piezoelectromagnetic waves in an unelectroded ceramic plate.

When the materials are nonpiezoelectric, that is, $e_A = e_B = 0$, Equation (7.77) factors into

$$\tan \varsigma_B h = -\frac{\varepsilon_A \varsigma_B}{\varepsilon_B \varsigma_A} \quad \text{and} \quad \tan \eta_B h = \frac{\bar{c}_A \eta_A}{\bar{c}_B \eta_B} \tag{7.79}$$

We recognize Equation $(7.79)_1$ as the frequency equation that determines the dispersion relations for guided electromagnetic waves in an unelectroded dielectric plate. In terms of the wave speed v, Equation $(7.79)_1$ can be written as

$$\tanh \xi h \sqrt{1 - \frac{v^2}{c_B^2}} = -\frac{\varepsilon_A}{\varepsilon_B} \frac{\sqrt{1 - \frac{v^2}{c_B^2}}}{\sqrt{1 - \frac{v^2}{c_A^2}}}, \tag{7.80}$$

which shows the dependence of the wave speed v on the wave number ξ (dispersion). The dispersion relations determined by Equation (7.80) are with an infinite number of branches. It can be expected that the behavior of (7.77) will be more complicated. In terms of the velocity, v, $(7.79)_2$ can be written as

$$\tan \xi h \sqrt{\frac{v^2}{v_B^2} - 1} = \frac{\bar{c}_A}{\bar{c}_B} \frac{\sqrt{\frac{v^2}{v_A^2} - 1}}{\sqrt{\frac{v^2}{v_B^2} - 1}}, \tag{7.81}$$

which is the frequency equation that determines the dispersion relations for SH waves in an elastic plate between two elastic half-spaces.

For the special case of an elastic plate alone without the half-spaces ($\bar{c}_A = 0$), Equation (7.81) further reduces to

$$\tan \xi h \sqrt{\frac{v^2}{v_B^2} - 1} = 0, \tag{7.82}$$

which determines the velocities of elastic waves in a plate.

When the speed of light goes to infinity (i.e., $c_A \to \infty$ and $c_B \to \infty$), from Equations (7.63) and (7.71), we have $\varsigma_A \to \xi$ and $\varsigma_B \to \xi$. Then Equation (7.77) reduces to

$$\left(\frac{1}{\varepsilon_A} \tanh \xi h + \frac{1}{\varepsilon_B} \right) \left(\bar{c}_A \frac{\eta_A}{\xi} - \bar{c}_B \frac{\eta_B}{\xi} \tan \eta_B h \right) = \left(\frac{e_A}{\varepsilon_A} - \frac{e_B}{\varepsilon_B} \right)^2 \tanh \xi h, \tag{7.83}$$

which determines the dispersion relations for symmetric, guided quasi-static piezoelectric waves in a ceramic plate between two half-spaces.

7.8.2.2 Antisymmetric Waves

For antisymmetric waves, consider

$$U_A = -U_C, \qquad H_A = H_C, \qquad U_B = 0, \qquad H_B = 0. \tag{7.84}$$

In this case the dispersion relation is determined by

$$\left(\frac{\varsigma_A}{\varepsilon_A} \coth \varsigma_B h + \frac{\varsigma_B}{\varepsilon_B} \right) \left(\bar{c}_A \eta_A + \bar{c}_B \eta_B \cot \eta_B h \right) = \left(\frac{e_A}{\varepsilon_A} - \frac{e_B}{\varepsilon_B} \right)^2 \xi^2 \coth \varsigma_B h. \tag{7.85}$$

Observations similar to Equations (7.78)–(7.83) can also be made.

7.8.2.3 Love Waves

When the upper half-space is a vacuum, Equation (7.74)$_1$ should be dropped and (7.74)$_5$ should be replaced by $0 = T_4(h^-)$. We also have

$$\bar{c}_A = 0, \qquad e_A = 0, \qquad \varepsilon_A = \varepsilon_0. \tag{7.86}$$

The dispersion relation has the following form

$$
\begin{aligned}
&\frac{\varsigma_B\cosh 2\varsigma_B h}{\varepsilon_B}\left\{-\bar{c}_B\eta_B\cos 2\eta_B h\left[\bar{c}_C\eta_C\left(\frac{\varsigma_A}{\varepsilon_0}+\frac{\varsigma_C}{\varepsilon_C}\right)\right.\right.\\
&\left.\left.-\xi^2\left(\frac{2e_B^2}{\varepsilon_B^2}-\frac{2e_Be_C}{\varepsilon_B\varepsilon_C}+\frac{e_C^2}{\varepsilon_C^2}\right)\right]\right.\\
&\left.+\sin 2\eta_B h\left[\bar{c}_B^2\eta_B^2\left(\frac{\varsigma_A}{\varepsilon_0}+\frac{\varsigma_C}{\varepsilon_C}\right)+\frac{\bar{c}_C\eta_C e_B^2\xi^2}{\varepsilon_B^2}\right]\right\}\\
&+\sinh 2\varsigma_B h\left\langle-\bar{c}_B\eta_B\cos 2\eta_B h\left\{\bar{c}_C\eta_C\left(\frac{\varsigma_B^2}{\varepsilon_B^2}+\frac{\varsigma_A\varsigma_C}{\varepsilon_0\varepsilon_C}\right)\right.\right.\\
&\left.-\xi^2\left[-\frac{2e_Be_C\varsigma_A}{\varepsilon_0\varepsilon_B\varepsilon_C}+\frac{e_C^2\varsigma_A}{\varepsilon_0\varepsilon_C^2}+\frac{e_B^2}{\varepsilon_B^2}\left(\frac{\varsigma_A}{\varepsilon_0}+\frac{\varsigma_C}{\varepsilon_C}\right)\right]\right\}\\
&+\sin 2\eta_B h\left\{\bar{c}_B^2\eta_B^2\left(\frac{\varsigma_B^2}{\varepsilon_B^2}+\frac{\varsigma_A\varsigma_C}{\varepsilon_0\varepsilon_C}\right)\right.\\
&\left.\left.+\frac{e_B^2\xi^2}{\varepsilon_B^2}\left[\frac{\bar{c}_C\eta_C\varsigma_C}{\varepsilon_C}-\xi^2\left(\frac{e_B}{\varepsilon_B}-\frac{e_C}{\varepsilon_C}\right)^2\right]\right\}\right\rangle\\
&=\frac{2\bar{c}_Be_B\varsigma_B\eta_B\xi^2}{\varepsilon_B^2}\left(\frac{e_B}{\varepsilon_B}-\frac{e_C}{\varepsilon_C}\right).
\end{aligned}
\tag{7.87}
$$

Equation (7.87) is not comparable, but is an addition to the results for the Love waves analyzed in [33] where the plate is either a perfect conductor itself or it carries a perfect conductor electrode at $x_2 = h$.

In the special case when all materials are nonpiezoelectric, Equation (7.87) can be factored into

$$
\left[\frac{\varsigma_B}{\varepsilon_B}\left(\frac{\varsigma_A}{\varepsilon_0}+\frac{\varsigma_C}{\varepsilon_C}\right)+\left(\frac{\varsigma_B^2}{\varepsilon_B^2}+\frac{\varsigma_A\varsigma_C}{\varepsilon_0\varepsilon_C}\right)\tanh 2\varsigma_B h\right]\times\left(\bar{c}_B\eta_B\tan 2\eta_B h-\bar{c}_C\eta_C\right)=0.
\tag{7.88}
$$

The two factors are for uncoupled electromagnetic and elastic waves, respectively. The second factor is the well-known frequency equation that determines the speed of elastic Love waves in a layer over a half-space.

7.8.2.4 Surface Waves

When the upper ceramic half-space (A) and the layer (B) do not exist, we have surface waves. Setting the relevant material constants of A and B to zero and letting $h = 0$ in Equation (7.87), we obtain

$$\frac{\eta_C}{\xi}\left(\frac{\varsigma_C}{\xi} + \frac{\varepsilon_C}{\varepsilon_0}\frac{\varsigma_A}{\xi}\right) = \frac{e_C^2}{\varepsilon_C \bar{c}_C},\tag{7.89}$$

which can be shown to be the same as Equation (7.59).

7.8.3 Numerical Results for Acoustic Modes

For some numerical examples, we examine the dispersion relations of the symmetric waves determined by Equation (7.77). The material constants used in the calculations are given in Table 7.1.

A few branches of the dispersion relation are plotted in Figure 7.5 when all ceramics are poled in the same direction. The waves are clearly dispersive. From a plate point of view, the lowest branch of the dispersion relation of symmetric waves is called a face-shear wave with a finite phase speed when $\xi h = 0$. Other higher branches are all called thickness-twist waves with unbounded phase speed when $\xi h = 0$. If the poling direction of the plate is reversed, the dispersion curves are given in Figure 7.6. Compared to the curves in Figure 7.5, the curves in Figure 7.6 are more separated. The relevant material constants of PZT-4 and PZT-7A are not very different. The reversal of the poling direction in the plate changes the signs of the piezoelectric constants of the plate. This in some sense makes the material of the plate more different from that of the half-spaces, which has some effect on the dispersion curves.

Dispersion relations of the first symmetric mode (face-shear) for different combinations of plate and half-space ceramics are shown in Figure 7.7. The figure shows that the wave speeds at zero wave number (the intercepts with the vertical axis in the figure) depend strongly on the material.

Next, we examine Equation (7.78) for waves in plates. For device applications, long waves (wavelength $\gg$ plate thickness) are usually used.

Table 7.1 Material constants of polarized ceramics

	$\rho(\mathrm{kg/m^3})$	$c_{44}(10^{10}\ \mathrm{N/m^3})$	$e_{15}(\mathrm{C/m^2})$	$\varepsilon_{11}(10^{-8}\ \mathrm{C/Vm})$
PZT $-$ 4	7500	2.56	12.7	0.646
PZT $-$ 5H	7500	2.30	17.0	1.506
PZT $-$ 6B	7550	3.55	4.6	0.360
PZT $-$ 7A	7600	2.53	9.2	0.407
BaTiO$_3$	5700	4.39	11.4	0.982
PZT $-$ 5	7750	2.11	12.3	0.811
PZT $-$ 7	7800	2.50	13.5	1.71

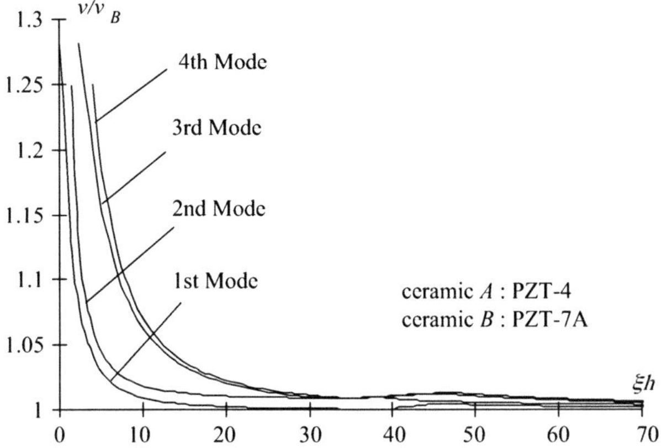

Fig. 7.5 Dispersion curves for symmetric waves (all ceramics poled in the same direction).

Therefore, we examine long waves with $\xi h \ll 1$. We focus on the lowest acoustic branch (the face-shear wave). In this case, in terms of the wave speed v, Equation (7.78) assumes the following form,

$$(1 + n^2 \xi h) \left(\frac{v^2}{v_B^2} - 1 \right) = -k^2, \tag{7.90}$$

where

$$n^2 = \frac{\varepsilon_B}{\varepsilon_0}, \qquad k^2 = \frac{e_B^2}{\varepsilon_B \bar{c}_B}. \tag{7.91}$$

Equation (7.90) shows that the originally nondispersive long face-shear wave becomes dispersive due to electromagnetic coupling. We plot Equation (7.90) for a few ceramics in Figure 7.8. The figure shows that the dispersion is noticeable only for long waves with $\xi h < 0.1$. This translates

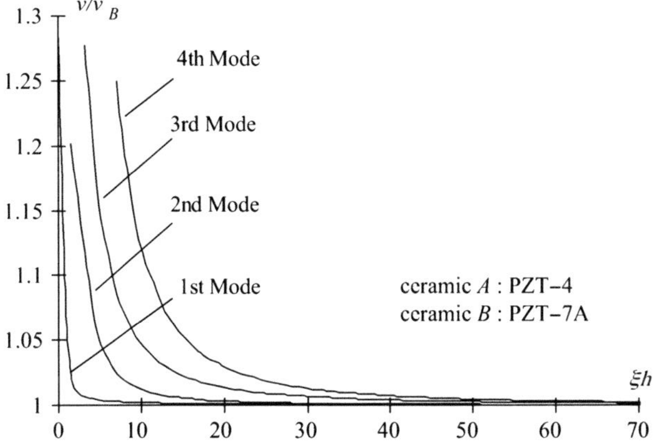

Fig. 7.6 Dispersion curves for symmetric waves (plate poling reversed).

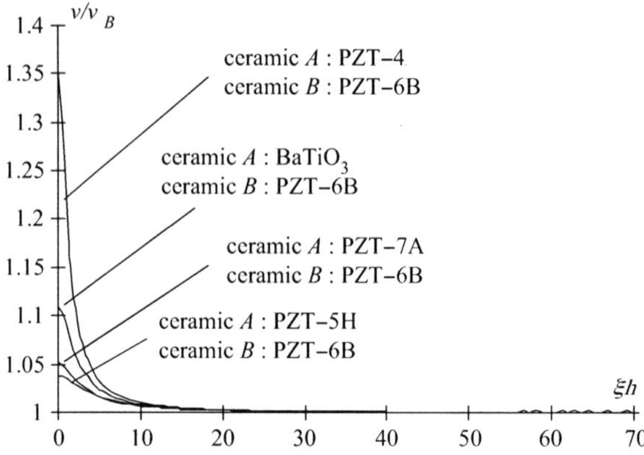

Fig. 7.7 Dispersion curves for face-shear waves (all ceramics poled in the same direction).

into $2\pi h/\lambda < 0.1$, or $\lambda/2h > 10\pi$; that is, the wavelength is about 30 times the plate thickness.

7.8.4 Electromagnetic Modes and Acoustic Leaking

Design of electromagnetic (EM) wave guides and resonators has been routinely performed using Maxwell equations. In fact, some of the materials used for dielectric devices have piezoelectric coupling. These devices are often mounted on substrates of other materials with acoustic interaction. If the device material has piezoelectric coupling, the EM waves in the device

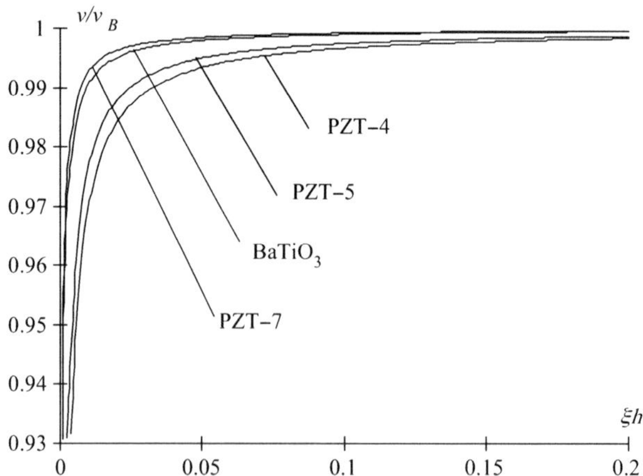

Fig. 7.8 Dispersion of long face-shear waves in a plate in free space.

are accompanied by acoustic waves interacting with the substrate. Therefore, there is a possibility of acoustic leakage or radiation of energy in an EM wave device. In this section, we examine this phenomenon [38]. Consider a plate of polarized ceramics between two half-spaces of another polarized ceramic (see Figure 7.4). For symmetric waves propagating in the x_1 direction with a factor of $\cos(\xi x_1 - \omega t)$, the corresponding dispersion relations are determined by Equation (7.77):

$$\left(\frac{\beta_A}{\varepsilon_A}\tanh\beta_B h + \frac{\beta_B}{\varepsilon_B}\right)(\bar{c}_A\alpha_A - \bar{c}_B\alpha_B\tan\alpha_B h) = \left(\frac{e_A}{\varepsilon_A} - \frac{e_B}{\varepsilon_B}\right)^2 \xi^2\tanh\beta_B h, \tag{7.92}$$

where

$$\alpha_A^2 = \xi^2 - \frac{\rho_A\omega^2}{\bar{c}_A} = \xi^2\left(1 - \frac{v^2}{v_A^2}\right), \qquad \beta_A^2 = \xi^2 - \varepsilon_A\mu_0\omega^2 = \xi^2\left(1 - \frac{v^2}{c_A^2}\right) > 0, \tag{7.93}$$

$$\alpha_B^2 = \frac{\rho_B\omega^2}{\bar{c}_B} - \xi^2 = \xi^2\left(\frac{v^2}{v_B^2} - 1\right), \qquad \beta_B^2 = \xi^2 - \varepsilon_B\mu_0\omega^2 = \xi^2\left(1 - \frac{v^2}{c_B^2}\right). \tag{7.94}$$

In Equations (7.92) through (7.94), we have changed the η and ς in (7.77) into α and β. The inequality in $(7.93)_2$ is for the real value of β_A (only the positive root is taken) so that the waves are electromagnetically guided. Whether the waves are acoustically guided depends on the sign of α_A^2.

Because the speed of light is much higher than the acoustic wave speed, it can be seen from Equation (7.93) that, in the range of $v_A < v < c_A$, α_A^2 becomes negative or α_A becomes pure imaginary and β_A is still real and positive. In this case, the EM fields of the coupled waves are still guided; but the acoustic fields are not. Consider the case when $v_A < v < c_A$. To be specific, we also limit ourselves to the case when $v > c_B$. Together with $v_A < v < c_A$, we are considering v in the range of

$$v_A < c_B < v < c_A. \tag{7.95}$$

For convenience we denote

$$\alpha_A^2 = \xi^2\left(1 - \frac{v^2}{v_A^2}\right) = -\hat{\alpha}_A^2, \qquad \alpha_A = i\hat{\alpha}_A,$$

$$\beta_B^2 = \xi^2\left(1 - \frac{v^2}{c_B^2}\right) = -\hat{\beta}_B^2, \qquad \beta_B = i\hat{\beta}_B. \tag{7.96}$$

With Equation (7.96), we can write (7.92) as

$$\frac{\beta_A}{\varepsilon_A}\tan\hat{\beta}_B h + \frac{\hat{\beta}_B}{\varepsilon_B} = \left(\frac{e_A}{\varepsilon_A} - \frac{e_B}{\varepsilon_B}\right)^2 \frac{\xi^2\tan\hat{\beta}_B h}{\bar{c}_A i\hat{\alpha}_A - \bar{c}_B\alpha_B\tan\alpha_B h}, \tag{7.97}$$

where we have also used $\tanh iZ = i \tan Z$ for a complex variable Z. The right-hand side of Equation (7.97) is due to piezoelectric coupling which may be a small effect. We use an iteration procedure to solve Equation (7.97). As the lowest order of approximation, we neglect the right-hand side of (7.97) and denote the EM frequencies of the left-hand side of (7.97) by ω_0:

$$\frac{\beta_{A0}}{\varepsilon_A} \tan \hat{\beta}_{B0} h + \frac{\hat{\beta}_{B0}}{\varepsilon_B} \cong 0, \tag{7.98}$$

where

$$\beta_{A0} = \sqrt{\xi^2 - \frac{\omega_0^2}{c_A^2}},$$

$$\hat{\beta}_{B0} = \sqrt{\frac{\omega_0^2}{c_B^2} - \xi^2}. \tag{7.99}$$

Given a wave number ξ, Equation (7.98) determines a series of frequencies $\omega_0(\xi)$ for guided EM waves in the structure. When piezoelectric coupling is considered, these EM frequencies are perturbed and are determined by Equation (7.97). Let the corresponding frequencies from Equation (7.97) be denoted by

$$\omega = \omega_0 + \Delta\omega. \tag{7.100}$$

Substituting Equation (7.100) into (7.97), we obtain the following first-order modification of the EM frequencies due to piezoelectric coupling.

$$\frac{\Delta\omega}{\omega_0} \cong \frac{1}{\omega_0^2} \left(\frac{e_A}{\varepsilon_A} - \frac{e_B}{\varepsilon_B}\right)^2 \times \frac{\xi^2 \tan(\hat{\beta}_{B0} h)[-\bar{c}_B \alpha_{B0} \tan(\alpha_{B0} h) - \bar{c}_A i \hat{\alpha}_{A0}]}{(\bar{c}_A \hat{\alpha}_{A0})^2 + [\bar{c}_B \alpha_{B0} \tan(\alpha_{B0} h)]^2}$$

$$\times \left(\frac{\beta_{A0} h}{\varepsilon_A \hat{\beta}_{B0} c_B^2 \cos^2(\hat{\beta}_{B0} h)} - \frac{\tan(\hat{\beta}_{B0} h)}{\varepsilon_A \beta_{A0} c_A^2} + \frac{1}{\varepsilon_B \hat{\beta}_{B0} c_B^2}\right)^{-1}, \tag{7.101}$$

where

$$\hat{\alpha}_{A0} = \sqrt{\frac{\omega_0^2}{v_A^2} - \xi^2},$$

$$\alpha_{B0} = \sqrt{\frac{\omega_0^2}{v_B^2} - \xi^2}. \tag{7.102}$$

Equation (7.101) gives the frequency perturbation of EM waves due to piezoelectric coupling. It is a complex number. In addition to its real part representing a frequency shift or additional dispersion, its imaginary part describes damped waves due to acoustic radiation. As a numerical

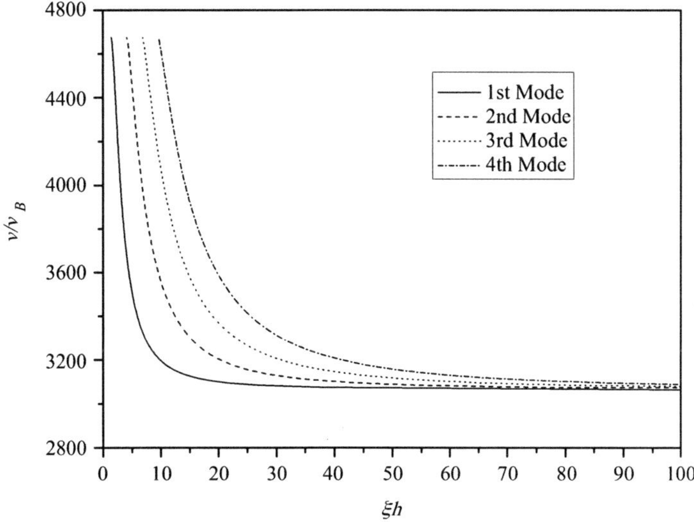

Fig. 7.9 Dispersion relations of guided EM waves from Equation (7.98).

example, consider PZT-4 and PZT-5H for materials A and B that satisfy Equation (7.95).

The first few branches of the dispersion relation for guided EM waves, that is, the solution to Equation (7.98), are shown in Figure 7.9. These waves are clearly dispersive. For short waves with a large wave number, the dispersion is small, and the wave speed of various branches seems to approach the same constant. The range of the wave speed is bounded by Equation (7.95) from below and above.

When the waves represented by the dispersion relations in Figure 7.9 are substituted into the right-hand side of Equation (7.101), the imaginary part of the left-hand side of (7.101), which represents dissipation, is shown in Figure 7.10. The most basic point to note is that the curves in Figure 7.10 are always positive, indicating damped waves rather than growing waves. The decay is of the order of 10^{-5}. The way we order the modes is such that higher-order modes have higher frequencies.

Figure 7.10 is for a very small range of wave speed and the curves already have quite a few oscillations. Curves corresponding to higher-order modes with higher frequencies have more oscillations. We are interested in the behavior of these curves over the entire range as bounded by Equation (7.95). In Figure 7.11 we only plot the peak values of the curves in Figure 7.10 versus wave speed over the entire range, otherwise the figure would be too crowded. The figure shows that the dissipation is sensitive to the wave speed. For the modes considered, there exists a maximum dissipation at a particular wave speed. At the lower bound of the wave speed, all waves approach the

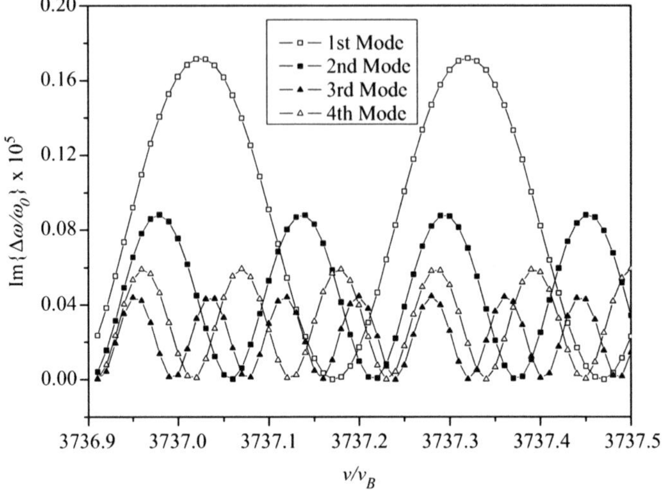

Fig. 7.10 Dissipation of guided EM waves due to acoustic leakage.

same speed, as shown in Figure 7.9, and the corresponding dissipation also approaches a common value which is zero.

Figure 7.12 shows the real part of the left-hand side of Equation (7.101) which represents additional dispersion induced by coupling to acoustic waves. As in Figure 7.11, only the maximum (and minimum) values of sinusoidal curves are shown in Figure 7.12. The most basic qualitative difference between Figures 7.11 and 7.12 is that the maximum and minimum values

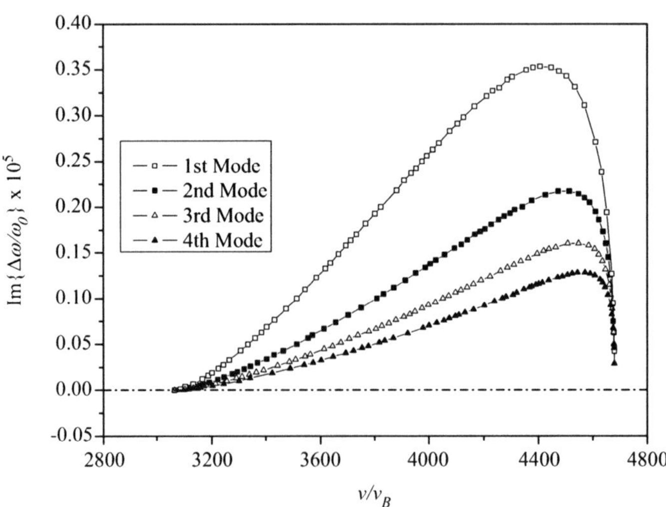

Fig. 7.11 Dissipation of guided EM waves due to acoustic leakage (peak values).

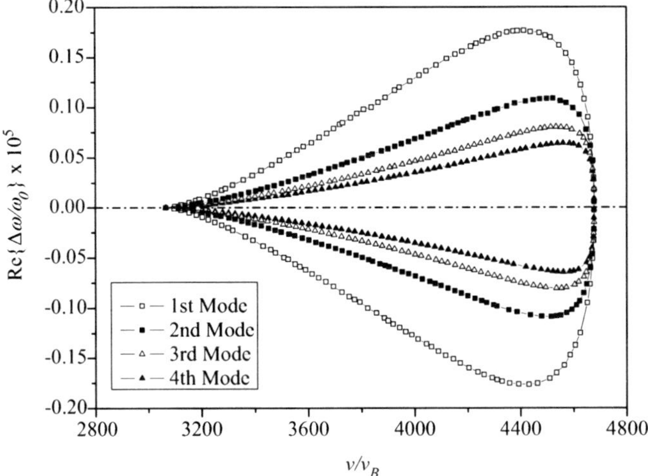

Fig. 7.12 Frequency shift (additional dispersion) of guided EM waves due to acoustic leakage (peak values).

of the additional dispersion in Figure 7.12 are with alternating signs. The additional dispersion also has a maximum which seems to be at the same location as the maximum dissipation.

7.9 Electromagnetic Radiation from a Vibrating Cylinder

A vibrating piezoelectric body radiates electromagnetic power. This phenomenon is out of the classical theory of quasi-static piezoelectricity and can only be described using the theory of piezoelectromagnetism. Radiated electromagnetic power from an AT-cut quartz plate mechanically forced into thickness-shear vibrations was calculated in [39]. Power radiation was also analyzed in [40] for an AT-cut quartz plate in thickness-shear vibration under lateral excitation and in [41], in general, for a piezoelectric plate of an arbitrary crystal in mechanically forced thickness vibrations. Next, we consider electromagnetic radiation from a vibrating circular cylinder of ceramics poled in the x_3-direction, as shown in Figure 7.13. The cylinder is mechanically driven at $r = b$ by a time-harmonic shear stress τ. The surface at $r = b$ is unelectroded. Electromagnetic waves propagate away from the cylinder (radiation).

For the special case of a solid cylinder ($a = 0$), the governing equations and boundary conditions are:

$$v_T^2 \nabla^2 u_3 = \ddot{u}_3, \qquad c^2 \nabla^2 H_3 = \ddot{H}_3, \qquad r < a, \qquad c_0^2 \nabla^2 H_3 = \ddot{H}_3, \qquad r > a,$$

$$(7.103)$$

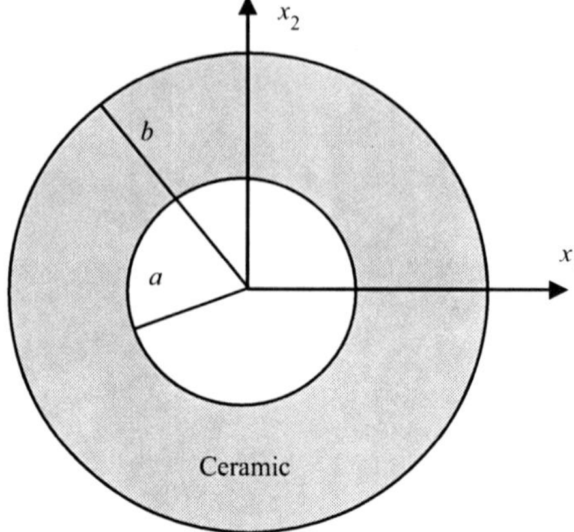

Fig. 7.13 A ceramic circular cylinder with axial poling.

and

$$u_3, \ H_3 \text{ are finite}, \qquad r = 0,$$
$$H_3 \text{ is outgoing}, \qquad r \to \infty,$$
$$T_{r3} = \tau \sin v\theta \exp(-i\omega t), \qquad r = b,$$
$$H_3, \ E_\theta \text{ are continuous}, \qquad r = b. \tag{7.104}$$

For the fields inside the cylinder, in polar coordinates, we have

$$v_T^2 \left(\frac{\partial^2 u_3}{\partial r^2} + \frac{1}{r}\frac{\partial u_3}{\partial r} + \frac{1}{r^2}\frac{\partial^2 u_3}{\partial \theta^2} \right) = \ddot{u}_3,$$
$$c^2 \left(\frac{\partial^2 H_3}{\partial r^2} + \frac{1}{r}\frac{\partial H_3}{\partial r} + \frac{1}{r^2}\frac{\partial^2 H_3}{\partial \theta^2} \right) = \ddot{H}_3, \tag{7.105}$$

and

$$\varepsilon_{11}\dot{E}_r = \frac{1}{r}H_{3,\theta} - e_{15}\dot{u}_{3,r}, \qquad \varepsilon_{11}\dot{E}_\theta = -H_{3,r} - e_{15}\frac{1}{r}\dot{u}_{3,\theta}. \tag{7.106}$$

Consider

$$u_3(r, \ \theta, \ t) = u(r)\sin v\theta \exp(-i\omega t), \qquad H_3(r, \ \theta, \ t) = H(r)\cos v\theta \exp(-i\omega t), \tag{7.107}$$

where v is a real number (for solutions periodic in θ, v must be an even integer). Substitution of Equation (7.107) into (7.105) results in

$$\frac{\partial^2 u}{\partial r^2} + \frac{1}{r}\frac{\partial u}{\partial r} + \left(\alpha^2 - \frac{v^2}{r^2} \right) u = 0, \qquad \frac{\partial^2 H}{\partial r^2} + \frac{1}{r}\frac{\partial H}{\partial r} + \left(\beta^2 - \frac{v^2}{r^2} \right) H = 0, \tag{7.108}$$

where we have denoted

$$\alpha = \frac{\omega}{v_T}, \qquad \beta = \frac{\omega}{c}. \tag{7.109}$$

Equation (7.108) can be written as Bessel's equations of order v. Then general solutions for u_3 and H_3 can be written as

$$u_3 = [C_1 J_v(\alpha r) + C_2 Y_v(\alpha r)]\sin v\theta \exp(-i\omega t),$$
$$H_3 = [C_3 J_v(\beta r) + C_4 Y_v(\beta r)]\cos v\theta \exp(-i\omega t), \tag{7.110}$$

where J_v and Y_v are the vth order Bessel functions of the first and second kind. C_1 through C_4 are undetermined constants. From Equation (7.108), we obtain the following expressions that are useful for boundary and/or continuity conditions.

$$D_r = \frac{v}{i\omega r}[C_3 J_v(\beta r) + C_4 Y_v(\beta r)]\sin v\theta \exp(-i\omega t),$$
$$D_\theta = \frac{\beta}{i\omega}[C_3 J_v'(\beta r) + C_4 Y_v'(\beta r)]\cos v\theta \exp(-i\omega t), \tag{7.111}$$

$$E_r = \left\{ \frac{v}{\varepsilon_{11}\omega r}[C_3 J_v(\beta r) + C_4 Y_v(\beta r)] \right.$$
$$\left. - \frac{e_{15}\alpha}{\varepsilon_{11}}[C_1 J_v'(\alpha r) + C_2 Y_v'(\alpha r)] \right\} \sin v\theta \exp(-i\omega t), \tag{7.112}$$

$$E_\theta = \left\{ \frac{\beta}{\varepsilon_{11}i\omega}[C_3 J_v'(\beta r) + C_4 Y_v'(\beta r)] \right.$$
$$\left. - \frac{e_{15}v}{\varepsilon_{11}r}[C_1 J_v(\alpha r) + C_2 Y_v(\alpha r)] \right\} \cos v\theta \exp(-i\omega t),$$

$$T_{rz} = \left\{ \bar{c}_{44}\alpha[C_1 J_v'(\alpha r) + C_2 Y_v'(\alpha r)] \right.$$
$$\left. - \frac{e_{15}v}{\varepsilon_{11}i\omega r}[C_3 J_v(\beta r) + C_4 Y_v(\beta r)] \right\} \sin v\theta \exp(-i\omega t),$$

$$T_{\theta z} = \left\{ \frac{\bar{c}_{44}v}{r}[C_1 J_v(\alpha r) + C_2 Y_v(\alpha r)] \right.$$
$$\left. - \frac{e_{15}}{i\omega\varepsilon_{11}}[C_3 J_v'(\beta r) + C_4 Y_v'(\beta r)] \right\} \cos v\theta \exp(-i\omega t), \tag{7.113}$$

where a superimposed prime indicates differentiation with respect to the whole argument of a function.

In the free space of $r > b$, the electromagnetic fields are given by

$$H_3 = [C_5 H_v^{(1)}(\gamma r) + C_6 H_v^{(2)}(\gamma r)]\cos v\theta \exp(-i\omega t),$$

$$D_r = \frac{v}{i\omega r}[C_5 H_v^{(1)}(\gamma r) + C_6 H_v^{(2)}(\gamma r)]\sin v\theta \exp(-i\omega t),$$

$$D_\theta = \frac{\gamma}{i\omega}[C_5 H_v^{(1)'}(\gamma r) + C_6 H_v^{(2)'}(\gamma r)]\cos v\theta \exp(-i\omega t),$$

$$E_r = \frac{v}{i\omega r \varepsilon_0}[C_5 H_v^{(1)}(\gamma r) + C_6 H_v^{(2)}(\gamma r)]\sin v\theta \exp(-i\omega t),$$

$$E_\theta = \frac{\gamma}{i\omega\varepsilon_0}[C_5 H_v^{(1)'}(\gamma r) + C_6 H_v^{(2)'}(\gamma r)]\cos v\theta \exp(-i\omega t), \tag{7.114}$$

where $H_v^{(1)}$ and $H_v^{(2)}$ are the vth order Hankel's functions of the first and second kind, and

$$\gamma = \frac{\omega}{c_0}. \tag{7.115}$$

Because Y_v is singular at the origin, terms associated with C_2 and C_4 have to be dropped. To satisfy the radiation condition at $r \to \infty$, we must have $C_6 = 0$. What need to be satisfied at $r = b$ are that

$$T_{rz}(b) = \bar{c}_{44}\alpha C_1 J_v'(\alpha b) - \frac{e_{15}v}{\varepsilon_{11}i\omega b}C_3 J_v(\beta b) = \tau,$$

$$H_3(b^-) = C_3 J_v(\beta b) = C_5 H_v^{(1)}(\gamma b) = H_3(b^+),$$

$$E_\theta(b^-) = \frac{\beta}{\varepsilon_{11}i\omega}C_3 J_v'(\beta b) - \frac{e_{15}v}{\varepsilon_{11}b}C_1 J_v(\alpha b)$$

$$= \frac{\gamma}{i\omega\varepsilon_0}C_5 H_v^{(1)'}(\gamma b) = E_\theta(b^+). \tag{7.116}$$

Note that when $v = 0$ (axisymmetric), Equation $(7.116)_1$ becomes uncoupled from $(7.116)_{2,3}$. In this case, H_3 cannot be excited by τ. Hence, there is no radiation. In the following, we consider the case of $v \neq 0$. From Equation (7.116), we obtain

$$C_1 = b\frac{\beta b J_v'(\beta b)H_v^{(1)}(\gamma b) - \frac{\varepsilon_{11}}{\varepsilon_0}\gamma b J_v(\beta b)H_v^{(1)'}(\gamma b)}{\Delta}\frac{\tau}{\bar{c}_{44}},$$

$$C_3 = i\omega e_{15}b\frac{v J_v(\alpha b)H_v^{(1)}(\gamma b)}{\Delta}\frac{\tau}{\bar{c}_{44}}, \quad C_5 = i\omega e_{15}b\frac{v J_v(\alpha b)J_v(\beta b)}{\Delta}\frac{\tau}{\bar{c}_{44}}, \tag{7.117}$$

where

$$\Delta = \alpha b\beta b J_v'(\alpha b)J_v'(\beta b)H_v^{(1)}(\gamma b) - \bar{k}_{15}^2 v^2 J_v(\alpha b)J_v(\beta b)H_v^{(1)}(\gamma b)$$

$$- \frac{\varepsilon_{11}}{\varepsilon_0}\alpha b\gamma b J_v'(\alpha b)J_v(\beta b)H_v^{(1)'}(\gamma b). \tag{7.118}$$

$\Delta = 0$ yields the frequency equation that determines resonances.

We calculate the radiation at far fields with large r using the following asymptotic expressions of Bessel functions with large arguments,

$$H_v^{(1)}(x) \cong \sqrt{\frac{2}{\pi x}}\, \exp i\left(x - \frac{v\pi}{2} - \frac{\pi}{4}\right),$$

$$H_v^{(1)'}(x) \cong i\sqrt{\frac{2}{\pi x}}\, \exp i\left(x - \frac{v\pi}{2} - \frac{\pi}{4}\right). \tag{7.119}$$

Then

$$H_3 \cong \sqrt{\frac{2}{\pi \gamma r}}\, \exp i\left(\gamma r - \frac{v\pi}{2} - \frac{\pi}{4}\right) \cos v\theta \exp(-i\omega t),$$

$$E_\theta \cong \frac{\gamma}{\omega \varepsilon_0} C_5 \sqrt{\frac{2}{\pi \gamma r}}\, \exp i\left(\gamma r - \frac{v\pi}{2} - \frac{\pi}{4}\right) \cos v\theta \exp(-i\omega t), \tag{7.120}$$

which are clearly outgoing. To calculate the radiated power, we need the radial component of the Poynting vector which, when averaged over a period of time, with the complex notation, is given by

$$S_r = \frac{1}{2}(\mathbf{E}^* \times \mathbf{H})_r = \frac{1}{2}\mathrm{Re}\{E_\theta^* H_3\} = \frac{C_5 C_5^*}{\pi \omega \varepsilon_0 r} \cos^2 v\theta, \tag{7.121}$$

where an asterisk represents the complex conjugate. Equation (7.121) shows that the energy flux is inversely proportional to r. It also shows the angular distribution of the power radiation. The radiated power per unit length of the cylinder is

$$S = \int_0^{2\pi} S_r r d\theta = \frac{C_5 C_5^*}{2\pi \omega \varepsilon_0}\left(2\pi + \frac{1}{2v}\sin 4v\pi\right). \tag{7.122}$$

We are interested in the frequency range of acoustic waves. Therefore, αb is finite, $\beta b \ll 1$, and $\gamma b \ll 1$. For small arguments, we have

$$J_v(x) \cong \frac{x^v}{2^v \Gamma(1+v)}, \qquad H_v^{(1)}(x) \cong -i\frac{2^v \Gamma(v)}{\pi x^v},$$

$$\frac{x J_v'(x)}{J_v(x)} \cong v, \qquad \frac{x H_v^{(1)'}(x)}{H_v^{(1)}(x)} \cong -v. \tag{7.123}$$

Then, approximately,

$$C_5 = \frac{i\omega e_{15} b J_v(\alpha b)}{(1 + \frac{\varepsilon_{11}}{\varepsilon_0})\alpha b J_v'(\alpha b) - \bar{k}_{15}^2 v J_v(\alpha b)} \frac{\tau}{\bar{c}_{44}} \frac{1}{H_v^{(1)}(\gamma b)}. \tag{7.124}$$

In this approximate form, the denominator of the first factor of Equation (7.124) represents the frequency equation for quasi-static electromechanical resonances in piezoelectricity. With Equation (7.124), the radiated power can be written as

$$S = \frac{\omega e_{15}^2 b^2}{2\pi\varepsilon_0} \left| \frac{J_v(\alpha b)}{(1 + \frac{\varepsilon_{11}}{\varepsilon_0})\alpha b J_v'(\alpha b) - \overline{k}_{15}^2 v J_v(\alpha b)} \frac{\tau}{\overline{c}_{44}} \right|^2 \times \frac{2\pi + \frac{\sin 4v\pi}{2v}}{H_v^{(1)}(\gamma b)[H_v^{(1)}(\gamma b)]^*}.$$

$$(7.125)$$

Equation (7.125) shows that S is large near resonant frequencies. S is singular at these frequencies unless some damping is present. In the limit of $\omega \to 0$, we have α, β, and γ all $\to 0$. In this case $S \to 0$ as expected. S is proportional to the square of the piezoelectric constant. For materials with strong piezoelectric coupling, the radiated power is stronger.

References

[1]   Nelson DF (1979) *Electric, Optic and Acoustic Interactions in Crystals*. Wiley, New York

[2]   Kyame JJ (1949) Wave propagation in piezoelectric crystals. *J Acoust Soc Am* 21: 159–167

[3]   Kyame JJ (1953) Conductivity and viscosity effects on wave propagation in piezoelectric crystals. *J Acoust Soc Am* 26: 990–993

[4]   Pailloux PMH (1958) Piezoelectricite calcul des vitesses de propagation. Le Journal de *Physique et le Radium* 19: 523–526

[5]   Hruska H (1966) The rate of propagation of ultrasonic waves in ADP and in Voigt's theory. *Czech J Phys B* 16: 446–453

[6]   Hruska H (1966) Relation between the general and the simplified condition for the velocity of propagation of ultrasonic waves in a piezoelectric medium. *Czech J Phys B* 18: 214–221

[7]   Tseng CC, White PM (1967) Propagation of piezoelectric and elastic surface waves on the basal plane of hexagonal piezoelectric crystals. *J Appl Phys* 38: 4274–4280

[8]   Tseng CC (1967) Elastic surface waves on free surface and metallized surface of CdS, ZnO, and PZT-4. *J Appl Phys* 38: 4281–4284

[9]   Spaight RN, Koerber GG (1971) Piezoelectric surface waves on LiNbO$_3$. *IEEE Trans Son Ultrason* 18: 237–238

[10]   Moon FC (1970) Scattering of waves by a cylindrical piezoelectric inclusion. *J Acoust Soc Am* 48: 253–262

[11]   Sedov A, Schmerr LW Jr (1986) Some exact solutions for the propagation of transient electroacoustic waves I: piezoelectric half-space. *Int J Engng Sci* 24: 557–568

[12]   Schmerr LW Jr, Sedov A (1986) Some exact solutions for the propagation of transient electroacoustic waves II: plane interface between two piezoelectric media. *Int J Engng Sci* 24: 921–932

[13]   Morishita K, Kumagai N (1978) Systematic derivation of variational expressions for electromagnetic and/or acoustic waves. *IEEE Trans Microwave Theor Technol* 26: 684–689

[14]   Mindlin RD (1978) A variational principle for the equations of piezoelectromagnetism in a compound medium. In: *Complex Variable Analysis and Its Applications* (I. N. Vekua 70th Birthday Volume). Academy of Sciences USSR, Nauka, Moscow

[15]   Lee PCY (1991) A variational principle for the equations of piezoelectromagnetism in elastic dielectric crystals. *J Appl Phys* 69: 7470–7473

[16] Yang JS (1991) A generalized variational principle for piezoelectromagnetism in an elastic medium. *Arch Mech* 43: 795–798

[17] Yang JS (1993) Variational principles for the vibration of an elastic dielectric. *Arch Mech* 45: 279–284

[18] Yang JS, Wu XY (1995) The vibration of an elastic dielectric with piezoelectromagnetism. *Quart Appl Math* 53: 753–760

[19] Altay GA, Dökmeci MC (2004) Fundamental equations of certain electromagnetic-acoustic discontinuous fields in variational form. *Continuum Mech Thermodyn* 16: 53–71

[20] Altay GA, Dökmeci MC (2005) Variational principles and vibrations of a functionally graded plate. *Computer Struct* 83: 1340–1354

[21] Yang JS, Zhou HG (2005) Two-dimensional equations for electromagnetic waves in multi-layered thin films. *Int J Solids Struct* 42: 6662–6679

[22] Yang JS (2004) A moving dislocation in piezoelectromagnetic ceramics. *Acta Meach* 172: 123–129

[23] Wang X, Zhong Z (2002), A moving piezoelectric screw dislocation. *Mech Res Commun* 29: 425–429

[24] Yang JS (2004) Effects of electromagnetic coupling on a moving crack in polarized ceramics. *Int J. of Fract* 126: L83–L88

[25] Li XF, Yang JS (2005) Electromagnetoelastic behavior induced by a crack under antiplane mechanical and inplane electric impacts. *Int J Fract* 132: 49–65

[26] Li S (1996) The electromagneto-acoustic surface wave in a piezoelectric medium: the Bleustein-Gulyaev mode. *J Appl Phys* 80: 5264–5269

[27] Yang JS (2000) Bleustein-Gulyaev waves in piezoelectromagnetic materials. *Int J Appl Electromag and Mech* 12: 235–240

[28] Bleustein JL (1968) A new surface wave in piezoelectric materials. *Appl Phys Lett* 13: 412–413

[29] Gulyaev YuV (1969) Electroacoustic surface waves in solids. Sov Phys JETP Letters 9: 37–38

[30] Yang JS, Zhou HG (2005) An interface wave in piezoelectromagnetic materials. *Int J Appl Electromag and Mech* 21: 63–68

[31] To AC, Glaser SD (2005) On the quasi-static assumption in modeling shear horizontal (SH) waves in a transversely isotropic (6mm) medium. http://www.ce.berkeley.edu/ ~ albertto/piezo.pdf. Accessed 2005

[32] Yang JS (2004) Piezoelectromagnetic waves in a ceramic plate. *IEEE Trans Ultrason Ferroelect Freq Contr* 51: 1035–1039

[33] Yang JS (2004) Love waves in piezoelectromagnetic materials. *Acta Mech* 168: 111–117

[34] Yang JS (2006) Acoustic gap waves in piezoelectromagnetic materials. *Math Mech Solids* 11: 451–458

[35] Yang JS, Guo SH (2006): Piezoelectromagnetic waves guided by the surface of a ceramic cylinder. *Acta Mech* 181: 199–205

[36] Iadonisi G, Perroni CA, Cantele G et al (2008) General solutions to the equations of piezoelectromagnetism: infinite medium and slab. *J Appl Phys* 63: 064109

[37] Jiang SN, Jiang Q, Li XF et al (2006) Piezoelectromagnetic waves in a ceramic plate between two ceramic half-spaces. *Int J Solids Struct* 43: 5799–5810

[38] Yang JS, Chen XH, Soh AK (2007) Acoustic leakage in electromagnetic waveguides made from piezoelectric materials. *J Appl Phys* 101: 066105

[39] Mindlin RD (1972) Electromagnetic radiation from a vibrating quartz plate. *Int J Solids Struct* 9: 697–702

[40] Lee PCY (1989) Electromagnetic radiation from an AT-cut quartz plate under lateral-field excitation. *J Appl Phys* 65: 1395–1399

[41] Lee PCY, Kim YG, Prevost JH (1990) Electromagnetic radiation from doubly rotated piezoelectric crystal plates vibrating at thickness frequencies. *J Appl Phys* 67: 6633–6642

Chapter 8
Nonlocal and Gradient Effects

Jiashi Yang

8.1 Introduction

Classical continuum theories such as elasticity, electrostatics, and piezoelectricity are the long-wave and low-frequency limits of the equations of lattice dynamics. The partial differential equations of these continuum theories can be obtained from the finite difference equations of lattice dynamics by Taylor expansions and truncations. The equations of classical continuum theories are accurate for phenomena with a characteristic length much larger than microscopic characteristic lengths, for example, the distance between neighboring atoms in a lattice. When the characteristic length of a problem is not much larger than the microscopic characteristic length, classical continuum theories do not predict results consistent with lattice dynamics, and hence are no longer valid. For example, lattice waves are dispersive but the theory of elasticity only predicts nondispersive plane waves which are the long wave limit of lattice waves.

There are different ways to modify the classical continuum theories so that their range of applicability can be extended to problems with smaller characteristic lengths, with results closer to lattice dynamics in a wider range of wave lengths.

One is to generalize the constitutive relations in classical continuum theories from local to nonlocal [1–5]. In classical continuum theories, the constitutive relations are for stress and strain (or electric field and polarization) at the same material point, with a zero interaction distance (local theories). These constitutive relations neglect the microstructure of real materials and lead to macroscopic theories for large characteristic lengths. In nonlocal constitutive relations, a finite microscopic interaction distance is

Jiashi Yang
Department of Engineering Mechanics,University of Nebraska, Lincoln, NE 68588-0526, USA, e-mail: Jyang1@unl.edu

J. Yang (ed.), *Special Topics in the Theory of Piezoelectricity*, DOI: 10.1007/978-0-387-89498-0_8,   281
© Springer Science + Business Media, LLC 2009

introduced which can represent, for example, the distance between neighboring atoms in a lattice.

Another approach is to keep higher-order terms in the Taylor expansions of the finite difference equations of lattice dynamics when deriving continuum equations. This results in the so-called gradient theories of various orders [1, 6], with the classical continuum theories as the lowest-order gradient theory. The higher-order gradients can be from mechanical or electrical origins, resulting in strain gradient or polarization gradient theories. Gradient continuum theories lead to higher-order differential equations and require more boundary conditions than classical continuum theories. This chapter only considers gradient effects of electric variables.

Nonlocal and gradient theories can describe size effects that are important in small-scale problems. They also have important consequences in problems with singularities such as concentrated sources and defects, and can describe surface and boundary layer phenomena. Nonlocal and gradient theories are closer to microscopic theories such as lattice dynamics than classical continuum theories. They are still applicable when the characteristic length of a problem is so small that classical continuum theories begin to fail. The development of new technologies results in very thin electromechanical films or wires and very small electronic devices. The study of these small devices presents new problems that classical theories may not be able to describe. Theories with nonlocal or gradient effects may allow us to go a little further in these problems than classical theories.

8.2 Nonlocal Theory

A theory for nonlocal piezoelectricity was given in [7]. Consider a piezoelectric body V. The nonlocal constitutive relations are:

$$T_{ij}(\mathbf{x}) = \int_V [c_{ijkl}(\mathbf{x}, \mathbf{x}')S_{kl}(\mathbf{x}') - e_{kij}(\mathbf{x}, \mathbf{x}')E_k(\mathbf{x}')]dV(\mathbf{x}'),$$

$$D_i(\mathbf{x}) = \int_V [e_{ikl}(\mathbf{x}, \mathbf{x}')S_{kl}(\mathbf{x}') + \varepsilon_{ik}(\mathbf{x}, \mathbf{x}')E_k(\mathbf{x}, \mathbf{x}')]dV(\mathbf{x}'). \qquad (8.1)$$

As a special case, when the nonlocal material moduli are Dirac delta functions, Equation (8.1) reduces to the classical constitutive relations in linear piezoelectricity. Substitution of Equation (8.1) into the equation of motion and the charge equation of electrostatics results in integral-differential equations that are usually mathematically challenging. The potential field due to a point charge was obtained in [7], which differs from the classical Coulomb field. It was also shown in [7] that, different from the classical theory of piezoelectricity, short plane waves are dispersive.

8.3 Thin-Film Capacitance by Nonlocal Theory

In this section we give an example of a simple nonlocal problem [8]. Consider an unbounded dielectric plate as shown in Figure 8.1. The plate is electroded and a voltage is applied. We want to obtain its capacitance from the nonlocal theory of electrostatics.

The problem is one-dimensional. The boundary value problem consists of the following equations:

$$\frac{dD}{dx} = 0, \qquad 0 < x < h, \qquad D = \varepsilon_0 E + P, \qquad 0 < x < h,$$

$$P = \varepsilon_0 \chi \int_0^h K(x', x) E(x') dx', \qquad 0 < x < h,$$

$$E = -\frac{d\phi}{dx}, \qquad 0 < x < h, \tag{8.2}$$

and boundary conditions

$$\phi(0) = 0 \qquad \phi(h) = V. \tag{8.3}$$

When the kernel function has the following special form,

$$K(x', x) = \delta(x' - x), \tag{8.4}$$

Equation (8.2) reduces to the classical form. χ is the dimensionless relative electric susceptibility. The dielectric material of the capacitor is assumed to be homogeneous and isotropic. Hence $K(x', x)$ must be invariant under translation and inversion. We have

$$K(x', x) = K(x' - x) = K(x - x'). \tag{8.5}$$

$K(x', x)$ should have a localized behavior, large near $x' = x$ and decaying away from there. We chose the following kernel function,

$$K(x' - x) = \frac{1}{2\alpha} e^{-(|x' - x|/\alpha)}, \qquad \alpha > 0, \tag{8.6}$$

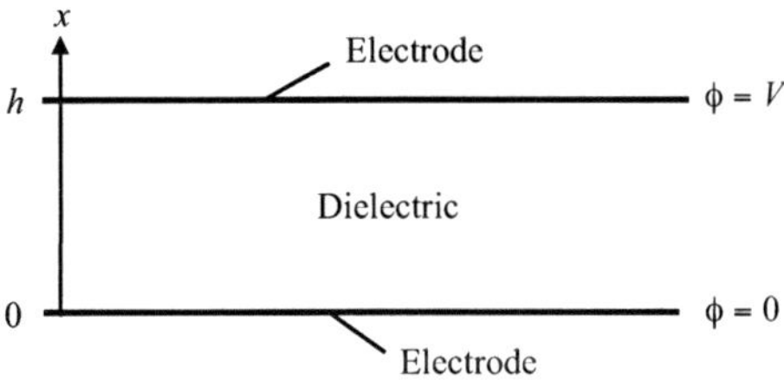

Fig. 8.1 A thin dielectric plate.

where α is a material parameter with the dimension of length. It is a characteristic length of microscopic interactions. It is easy to verify that $K(x', x)$ has the following properties,

$$\lim_{\substack{\alpha \to 0^+ \\ x \neq x'}} K = 0, \qquad \int_{-\infty}^{+\infty} K\, dx = 1. \tag{8.7}$$

Hence

$$\lim_{\alpha \to 0^+} K = \delta(x' - x), \tag{8.8}$$

which shows that $K(x', x)$ does include the local form as a limit case. The above $K(x', x)$ is the fundamental solution of the following differential operator,

$$-\alpha^2 \frac{\partial^2 K}{\partial x^2} + K = \delta(x' - x). \tag{8.9}$$

Integrating Equation $(8.2)_1$ once, with Equations $(8.2)_{2,3}$, we obtain

$$D = \varepsilon_0 E(x) + \varepsilon_0 \chi \int_0^h K(x', x) E(x') dx' = -\sigma_e, \tag{8.10}$$

where σ_e is an integration constant that physically represents the surface free charge density on the electrode at $x = h$. Equation (8.10) can be written as

$$E(x) = -\chi \int_0^h K(x', x) E(x') dx' - \frac{\sigma}{\varepsilon_0}, \tag{8.11}$$

which is a Fredholm integral equation of the second kind for the electric field E. Instead of solving Equation (8.11) directly, we proceed as follows. With Equation (8.9), we differentiate (8.11) with respect to x twice and obtain

$$\begin{aligned}
\frac{d^2 E(x)}{dx^2} &= -\chi \int_0^h \frac{\partial^2 K(x', x)}{\partial x^2} E(x') dx' \\
&= -\chi \int_0^h \frac{1}{\alpha^2} \left[K(x', x) - \delta(x' - x) \right] E(x') dx' \\
&= \frac{1}{\alpha^2} \left[-\chi \int_0^h K(x', x) E(x') dx' - \frac{\sigma_e}{\varepsilon_0} \right] \\
&\quad + \frac{1}{\alpha^2} \chi \int_0^h \delta(x' - x) E(x') dx' + \frac{1}{\alpha^2} \frac{\sigma_e}{\varepsilon_0} \\
&= \frac{1}{\alpha^2} \left[E(x) + \chi E(x) + \frac{\sigma_e}{\varepsilon_0} \right].
\end{aligned} \tag{8.12}$$

Hence a solution E of the integral Equation (8.11) also satisfies the following differential equation,

$$\alpha^2 \frac{d^2 E}{dx^2} - (1 + \chi)E = \frac{\sigma_e}{\varepsilon_0}. \tag{8.13}$$

The general solution to Equation (8.13) can be obtained easily. It has two exponential terms from the corresponding homogeneous equation, and a constant term which is the particular solution. The general solution contains two new integration constants. These two constants result from the differentiation in obtaining Equation (8.13) from (8.11). Hence the solution to Equation (8.13) may not satisfy (8.11). Therefore we substitute the general solution to Equation (8.13) back into (8.11), which determines the two new integration constants. Then, with the boundary conditions in Equation (8.3), we can determine σ_e and another integration constant resulting from integrating E for ϕ, and thus obtain the nonlocal electric potential distribution ϕ as

$$\phi = \left[\frac{x}{h} + \frac{\chi}{kh} \frac{\sinh k(x - \frac{h}{2}) + \sinh \frac{kh}{2}}{\cosh \frac{kh}{2} + k\alpha \sinh \frac{kh}{2}}\right] \left(1 + \frac{2\chi}{kh} \frac{\tanh \frac{kh}{2}}{1 + k\alpha \tanh \frac{kh}{2}}\right)^{-1} V,$$

$$\phi_0 = \frac{x}{h} V, \tag{8.14}$$

where ϕ_0 is the classical local solution, and

$$k = \sqrt{\frac{1 + \chi}{\alpha}}. \tag{8.15}$$

The nonlocal electric field distribution E and the local solution E_0 are:

$$E = \left[1 + \chi \frac{\cosh k(x - \frac{h}{2})}{\cosh \frac{kh}{2} + k\alpha \sinh \frac{kh}{2}}\right] \left(1 + \frac{2\chi}{kh} \frac{\tanh \frac{kh}{2}}{1 + k\alpha \tanh \frac{kh}{2}}\right)^{-1} E_0,$$

$$E_0 = -\frac{V}{h}. \tag{8.16}$$

Denoting the capacitance per unit electrode area from the local theory by C_0 and the one from the nonlocal theory by C, we have

$$C = \left(1 + \frac{2\chi}{kh} \frac{\tanh \frac{kh}{2}}{1 + k\alpha \tanh \frac{kh}{2}}\right)^{-1} C_0, \qquad C_0 = \frac{\varepsilon_0(1 + \chi)}{h}. \tag{8.17}$$

With the expression of k in Equation (8.15), we write Equation (8.17)$_1$ in the following form,

$$\frac{C}{C_0} = \left(1 + \frac{\chi}{\sqrt{1 + \chi}\frac{h}{2\alpha}} \frac{\tanh(\sqrt{1 + \chi}\frac{h}{2\alpha})}{1 + \sqrt{1 + \chi}\tanh(\sqrt{1 + \chi}\frac{h}{2\alpha})}\right)^{-1}. \tag{8.18}$$

The thin-film capacitance from the nonlocal theory differs from the result of the local theory. The nonlocal solution depends on the ratio $h/2\alpha$ of the film thickness to the microscopic characteristic length. From Equation (8.18) we immediately have (when $\chi > 0$):

$$C/C_0 < 1, \tag{8.19}$$

which shows that the nonlocal result is smaller than the local result. From (8.18) we also have the following limit behavior,

$$\lim_{\frac{h}{\alpha} \to \infty} \frac{C}{C_0} = 1, \tag{8.20}$$

which shows that when the film thickness is large compared to the microscopic characteristic length, the nonlocal solution approaches the local solution. We also have the limit

$$\lim_{\frac{h}{\alpha} \to 0} \frac{C}{C_0} = \frac{1}{1 + \chi} < 1, \tag{8.21}$$

which shows that the nonlocal and local solutions differ more for materials with a larger χ. We plot C/C_0 from Equation (8.18) as a function of $h/2\alpha$ for values of $\chi = 1, 10$, and 100 in Figure 8.2. The figure shows that for a film with a moderate value of $\chi = 100$, when the thickness $h/2\alpha \approx 10$, there is a deviation of about 10% from the local theory which has a fixed value of 1. The figure also shows that $C/C_0 < 1$ and the deviation from 1 becomes larger as h becomes smaller.

The spatial distribution of the electric field for $\chi = 10$ and for two values of $h/2\alpha = 1$ and 5, respectively, is shown in Figure 8.3. It is interesting to see that the field is large near the electrodes compared to the local solution with the fixed value of 1. The curve with $h/2\alpha = 5$ has a larger electric field near the electrodes than the curve with $h/2\alpha = 1$. This is a boundary effect exhibited by the nonlocal theory. Even for a thick capacitor, Equation (8.16) still yields

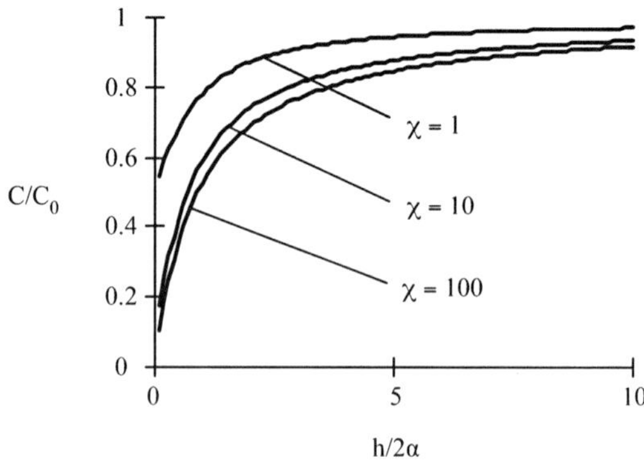

Fig. 8.2 Capacitance for $\chi = 1, 10$, and 100.

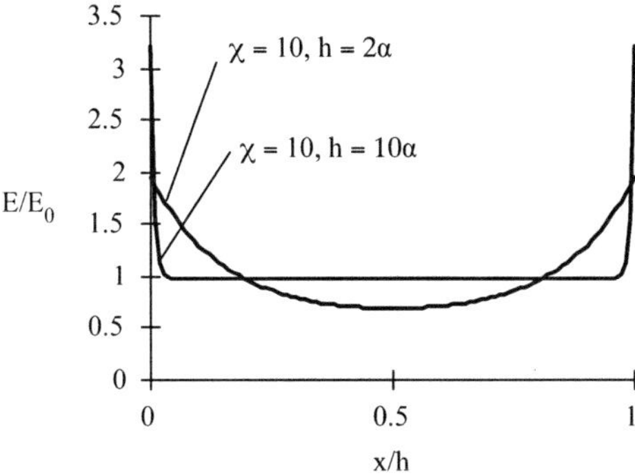

Fig. 8.3 Electric field distribution for $\chi = 10$, $h/2\alpha = 1$, and 5.

$$\lim_{\frac{h}{\alpha} \to \infty} \frac{E(h)}{E_0} = 1 + \frac{\chi}{1 + \sqrt{1 + \chi}} > 1. \tag{8.22}$$

When $\chi = 10$, Equation (8.22) yields a limit value of 3.32. For materials with a large χ the value of Equation (8.22) can be large. Because E is larger near the electrodes and D is a constant, P must be smaller near the electrodes than near the center of the plate.

The spatial distribution of the normalized deviation of the electric potential from the local solution for $\chi = 10$ and for two values of $h/2\alpha = 1$ and 5, respectively, are shown in Figure 8.4. The curve with $h/2\alpha = 5$ shows a smaller deviation.

Finally, we note that in Equation (8.13) the small parameter α appears as the coefficient of the term with the highest derivative. Hence, when α tends to zero, we have a singular perturbation problem of boundary layer type for a differential equation. For this type of problem, when

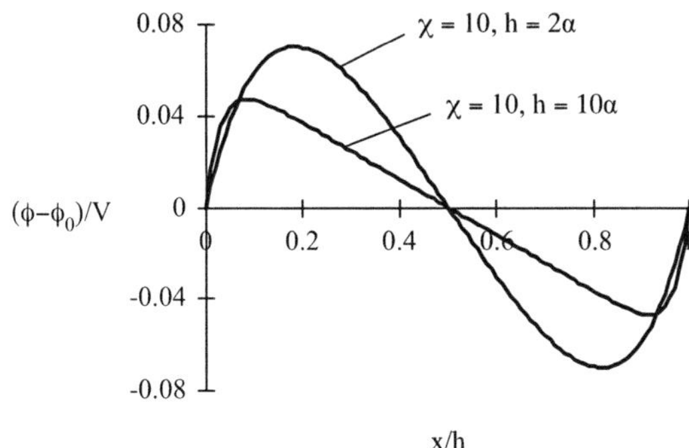

Fig. 8.4 Electric potential deviation for $\chi = 10$; $h/2\alpha = 1$, and 5.

the small parameter is set to zero, certain boundary conditions have to be dropped because the order of the differential equation is lowered. Equation (8.13) is a consequence of an integral-differential equation of ϕ defined by Equation (8.2) which only needs two boundary conditions. In the solution procedure, two of the integration constants in the general solution to Equation (8.13) were determined by the integral equation in (8.11). However, if we take Equation (8.13) as our starting point, we need two more boundary conditions. This is because (8.13) is a fourth-order differential equation for ϕ (considering it has already been integrated once with an integration constant σ_e). Then when α is set to zero, two boundary conditions have to be dropped.

8.4 Electromechanical Coupling by Nonlocal Theory

Electromechanical coupling factors of piezoelectric materials or devices depend on materials, structural shapes, and deformation modes. In this section we examine the effect of nonlocality on electromechanical coupling factors of thin films (see Figure 8.5) [9]. The surfaces of the film are electroded. The electrodes can be shorted or open.

We consider a one-dimensional piezoelectric model [9]:

$$\frac{dT}{dx} = 0, \qquad \frac{dD}{dx} = 0, \qquad T = cS - eE, \qquad D = \varepsilon_0 E + P,$$

$$S = \frac{du}{dx}, \qquad E = -\frac{d\phi}{dx}, \qquad P = \varepsilon_0 \chi \int K(x, x') E(x') dx', \qquad (8.23)$$

where T is the relevant stress component, D is the electric displacement, S is the strain, E is the electric field, and P is the polarization. $c, e,$ and $\varepsilon_0(1 + \chi) = \varepsilon$ are the usual elastic, piezoelectric, and dielectric constants. We consider the case when electrical nonlocal effects are present but mechanically the material is still local. We still use the following kernel function from Equation (8.6),

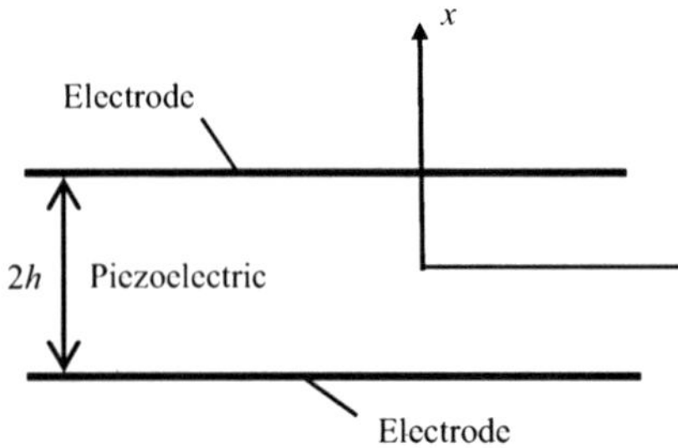

Fig. 8.5 A thin piezoelectric film and coordinate system.

$$K(x' - x) = \frac{1}{2\alpha}e^{-(|x'-x|)/\alpha}, \qquad \alpha > 0,$$

$$P = \varepsilon_0 \chi \int_{-h}^{h} K(x, x')E(x')dx'. \tag{8.24}$$

Our one-dimensional model above can be used to model thickness-stretch or thickness-shear deformations of a film, which are common operating modes for devices.

To calculate electromechanical coupling factors we slowly apply a pair of equal and opposite tractions $\overline{T}$ at $x = \pm h$, and calculate the work done by these tractions in two separate processes with shorted electrodes or open electrodes.

8.4.1 Shorted Electrodes

The boundary conditions are

$$T(\pm h) = \overline{T}, \qquad \phi(\pm h) = 0. \tag{8.25}$$

The solution to Equations (8.23) and (8.25) is simply

$$\phi = 0 \qquad u = \frac{\overline{T}}{c}x, \tag{8.26}$$

which is in fact the classical piezoelectric solution. The work done by the surface tractions per unit area in this process is

$$W_1 = 2\left[\frac{1}{2}\overline{T}u(h)\right] = 2\left(\frac{1}{2}\overline{T}\frac{\overline{T}}{c}h\right) = \frac{\overline{T}^2}{c}h. \tag{8.27}$$

In this case, because there is no electric field, the related piezoelectric stiffening effect does not exist.

8.4.2 Open Electrodes

For open electrodes, the boundary conditions are:

$$T(\pm h) = \overline{T}, \qquad D(\pm h) = 0. \tag{8.28}$$

From Equations (8.23) and (8.28),

$$T = cS - eE = \overline{T}, \qquad D = eS + \varepsilon_0 E + P = -\sigma_e = 0, \tag{8.29}$$

where σ_e is an integration constant that physically represents the surface free charge density of the electrode at $x = h$. In the case we are considering in fact $\sigma_e = 0$. Eliminating S from the two equations in (8.29), we obtain

$$\left(\varepsilon_0 + \frac{e^2}{c}\right) E(x) + \varepsilon_0 \chi \int_{-h}^{h} K(x, x') E(x') dx' = -\left(\sigma_e + e\frac{\overline{T}}{c}\right), \qquad (8.30)$$

which is a Fredholm integral equation of the second kind for the electric field E. With Equation (8.9), we differentiate (8.30) with respect to x twice and obtain

$$\alpha^2 \frac{d^2 E}{dx^2} - \beta^2 E = \gamma, \qquad (8.31)$$

where we have denoted

$$\beta^2 = \frac{\overline{\varepsilon}}{\overline{\varepsilon}_0}, \qquad \overline{\varepsilon} = \varepsilon + \frac{e^2}{c} = \varepsilon(1 + k^2), \qquad k^2 = \frac{e^2}{\varepsilon c}.$$

$$\overline{\varepsilon}_0 = \varepsilon_0 + \frac{e^2}{c}, \qquad \gamma = \frac{\sigma_e + e\overline{T}/c}{\varepsilon_0 + e^2/c} = \frac{e}{\overline{\varepsilon}_0} \frac{\overline{T}}{c}. \qquad (8.32)$$

The general solution to Equation (8.31) can be obtained in a straightforward manner. It has two integration constants. We substitute the general solution to Equation (8.31) back into the integral equation in Equation (8.30) and the two integration constants can be determined. E must be an even function of x, thus one of the integration constants is zero. We have

$$E = -\left(1 + A \cosh\frac{\beta}{\alpha}x\right) \frac{e}{\varepsilon} \frac{\overline{T}}{\overline{c}} = E_0 + \Delta E, \qquad (8.33)$$

where

$$A = \frac{1}{\cosh\frac{\beta h}{\alpha} + \beta \sinh\frac{\beta h}{\alpha}} \frac{\varepsilon_0 \chi}{\overline{\varepsilon}_0}, \qquad \overline{c} = c(1 + k^2),$$

$$E_0 = -\frac{e}{\varepsilon} \frac{\overline{T}}{\overline{c}}, \qquad \Delta E = -A \cosh\left(\frac{\beta}{\alpha}x\right) \frac{e}{\varepsilon} \frac{\overline{T}}{\overline{c}}. \qquad (8.34)$$

$\overline{c}$ is a piezoelectrically stiffened elastic constant. E_0 is the classical solution. When $\alpha \to 0$, Equation (8.34) implies that $A \to 0$ and Equation (8.33) reduces to the classical solution. Integrating (8.33), we obtain the electric potential as

$$\phi = \left(x + A\frac{\alpha}{\beta} \sinh\frac{\beta}{\alpha}x\right) \frac{e}{\varepsilon} \frac{\overline{T}}{\overline{c}} = \phi_0 + \Delta\phi,$$

$$\phi_0 = x\frac{e}{\varepsilon} \frac{\overline{T}}{\overline{c}}, \qquad \Delta\phi = A\frac{\alpha}{\beta} \sinh\left(\frac{\beta}{\alpha}x\right) \frac{e}{\varepsilon} \frac{\overline{T}}{\overline{c}}. \qquad (8.35)$$

ϕ is an odd function of x. From Equation $(8.29)_1$ we can find the displacement field as

$$u = \frac{\overline{T}}{\overline{c}} \left(x - k^2 \frac{\alpha}{\beta} A \sinh \frac{\beta}{\alpha} x \right) = u_0 + \Delta u,$$

$$u_0 = \frac{\overline{T}}{\overline{c}} x,$$

$$\Delta u = -\frac{\overline{T}}{\overline{c}} k^2 \frac{\alpha}{\beta} A \sinh \frac{\beta}{\alpha} x. \tag{8.36}$$

The work done by the surface traction per unit area is

$$W_2 = \frac{\overline{T}^2}{\overline{c}} (h - k^2 \frac{\alpha}{\beta} A \sinh \frac{\beta}{\alpha} h). \tag{8.37}$$

When the electrodes are open, there exists an electric field which causes the well-known piezoelectric stiffening effect. This is exhibited by the presence of $\overline{c}$ in Equation (8.36) as compared to c in Equation (8.26), and that $\overline{c} > c$. In addition to piezoelectric stiffening, the nonlocal term in (8.36) also makes the displacement field smaller than the classical solution. Therefore we expect $W_2 < W_1$.

8.4.3 Electromechanical Coupling Factor

We define the electromechanical coupling factor as

$$k^2_{\text{Nonlocal}} = \frac{W_1 - W_2}{W_1}. \tag{8.38}$$

Substituting from Equations (8.27) and (8.37), we find

$$k^2_{\text{Nonlocal}} = k^2_{\text{Local}} \left(1 + \frac{\alpha}{\beta h} A \sinh \frac{\beta}{\alpha} h \right), \tag{8.39}$$

where

$$k^2_{\text{Local}} = \frac{k^2}{1 + k^2} \tag{8.40}$$

is the coupling factor predicted by the classical theory of piezoelectricity. Compared to Equation (8.40), the nonlocal solution in Equation (8.39) has a dependence on h (size effect) among other parameters. Equation (8.39) also shows that the nonlocal coupling factor is larger than the classical local coupling factor.

8.4.4 Numerical Results

For numerical results we consider polarized ceramics PZT-5H. For thickness-stretch we identify:

$$c = c_{33}, \qquad e = e_{33}, \qquad \varepsilon = \varepsilon_{33}. \tag{8.41}$$

For thickness-shear:

$$c = c_{44}, \qquad e = e_{15}, \qquad \varepsilon = \varepsilon_{11}. \tag{8.42}$$

In Figures 8.6 through 8.8 we plot the normalized field differences $\Delta\phi/\phi_0$, $\Delta E/E_0$, and $\Delta u/u_0$. Clearly, nonlocal polarization creates a boundary layer of electromechanical fields. In the central region of the plate the fields are essentially classical. When the film is thinner, the boundary layer has a more important role when calculating the electromechanical coupling factor which is a global quantity. For very thin plates ($h/\alpha \cong 2$) the boundary layers occupy the entire plate thickness.

The electromechanical coupling factor is shown in Figure 8.9. The factor appears larger for thin films as predicted by the nonlocal polarization law. It is in fact as expected because a nonlocal polarization law causes an effective drop of the dielectric constant as shown in the analysis of thin-film capacitance in Section 8.3 of this chapter, and the dielectric constant is in the denominator of the coupling factor. The nonlocal solution is supposed to be useful when h/α is not very large and the classical solutions begin to

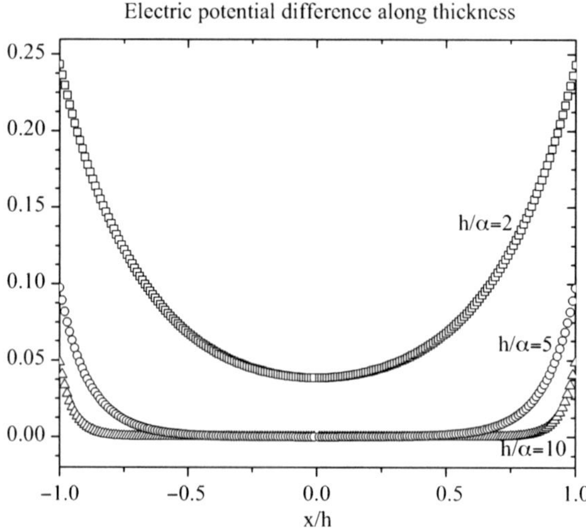

Fig. 8.6 Normalized electric potential difference along film thickness.

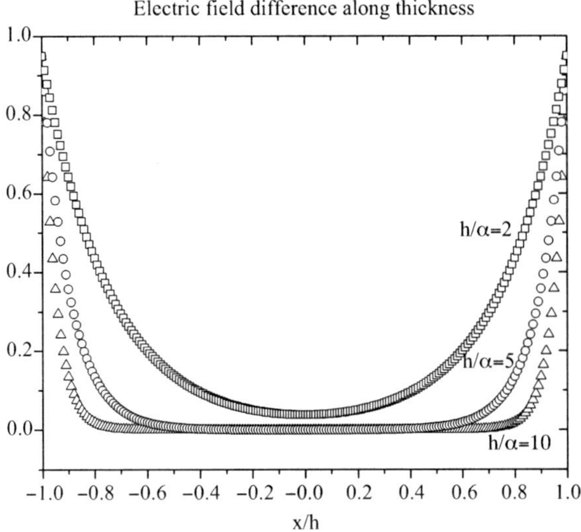

Fig. 8.7 Normalized electric field difference along film thickness.

become invalid, for example, when h/α is close to 5 or a little smaller in Figure 8.9. If h/α becomes still smaller, even the nonlocal solutions may become invalid as indicated by that for one of the curves in Figure 8.9 the nonlocal solution predicts a coupling factor greater than 1.

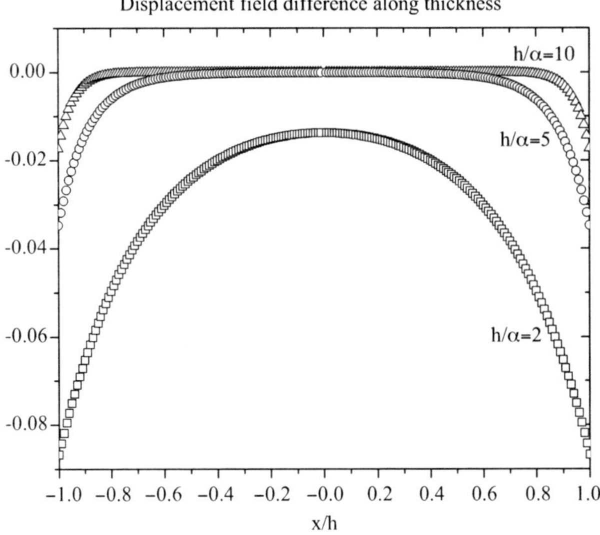

Fig. 8.8 Normalized displacement difference along film thickness.

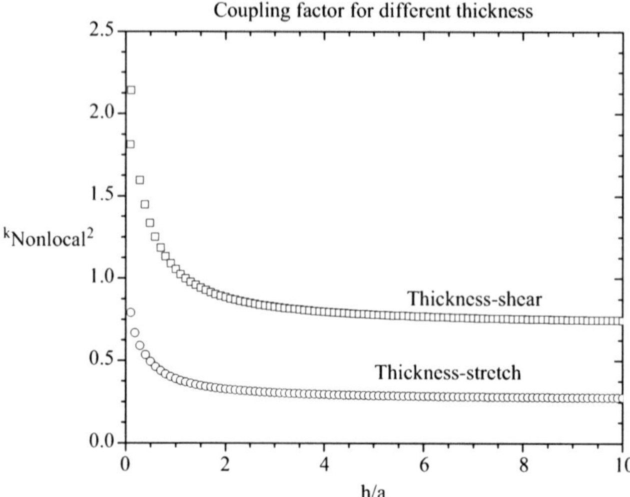

Fig. 8.9 Electromechanical coupling factor versus film thickness.

8.4.5 Comparison with Experiment

There are no experimental data available for direct comparison on electromechanical coupling factors of very thin films. Because what we are studying is in fact the effects of nonlocal electrical behavior on electromechanical
coupling factors, we compare the dielectric constant for which there are experimental data. Experiments show that the dielectric constant of a very thin
film appears to be thickness-dependent and is smaller than that of the bulk
material. From the results of Section 8.3 we can obtain an effective or apparent dielectric constant of the thin film as

$$\frac{\varepsilon^{eff}}{\varepsilon_0} = \left(1 + \frac{\chi}{\sqrt{1+\chi}\frac{h}{2\alpha}} \frac{\tanh(\sqrt{1+\chi}\frac{h}{2\alpha})}{1 + \sqrt{1+\chi}\tanh(\sqrt{1+\chi}\frac{h}{2\alpha})}\right)^{-1} \frac{\varepsilon}{\varepsilon_0}, \qquad (8.43)$$

which is thickness-dependent. Consider a thin film of zinc sulfide (ZnS). The
bulk dielectric constant is $\varepsilon/\varepsilon_0 = 8.5$. Equation (8.43) is plotted in Figure 8.10
versus experimental data [10]. A good agreement results when the parameter
α in the nonlocal theory which represents a microscopic characteristic length
is chosen to be 5 Å.

8.5 Gradient Effects as Weak Nonlocal Effects

Gradient effects in constitutive relations are related to weak nonlocal effects.
For example, consider a one-dimensional nonlocal constitutive relation

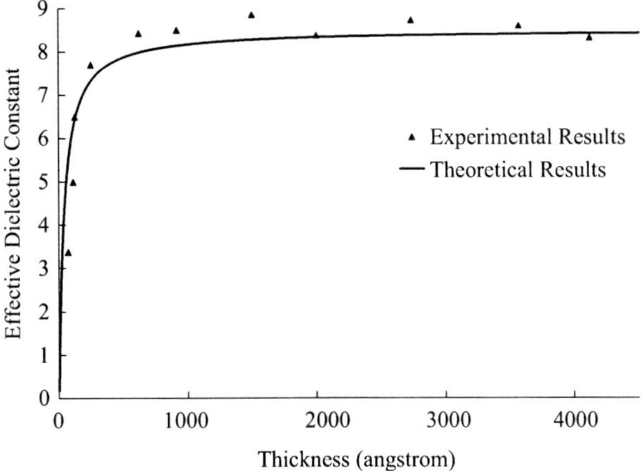

Fig. 8.10 Comparison of dielectric constant.

between Y and X in a homogeneous unbounded medium. We have

$$
\begin{aligned}
Y(x) &= \int_{-\infty}^{+\infty} K(x' - x)X(x')dx' \\
&= \int_{-\infty}^{+\infty} K(x' - x)X[x + (x' - x)]dx' \\
&= \int_{-\infty}^{+\infty} K(x' - x)[X(x) + X'(x)(x' - x) + \cdots]dx' \\
&\cong \int_{-\infty}^{+\infty} K(x' - x)[X(x) + X'(x)(x' - x)]d(x' - x) \\
&= \int_{-\infty}^{+\infty} K(x' - x)X(x)d(x' - x) \\
&\quad + \int_{-\infty}^{+\infty} K(x' - x)X'(x)(x' - x)d(x' - x) \\
&= X(x)\int_{-\infty}^{+\infty} K(x' - x)d(x' - x) \\
&\quad + X'(x)\int_{-\infty}^{+\infty} K(x' - x)(x' - x)d(x' - x) \\
&= aX(x) + bX'(x),
\end{aligned}
\tag{8.44}
$$

where

$$
a = \int_{-\infty}^{+\infty} K(x' - x)\,d(x' - x), \qquad b = \int_{-\infty}^{+\infty} K(x' - x)(x' - x)\,d(x' - x).
\tag{8.45}
$$

Therefore, to the lowest order of approximation, the nonlocal relation reduces to a local one, and to the next order a gradient term appears.

8.6 Gradient Effects and Lattice Dynamics

Gradient terms can also be introduced in the following procedure. Consider the extensional motion of a one-dimensional spring-mass system (see Figure 8.11).

The motion of the ith particle is governed by the finite difference equation

$$m\ddot{u}(i) = k[u(i+1) - u(i)] - k[u(i) - u(i-1)], \tag{8.46}$$

or, with the introduction of x,

$$
\begin{aligned}
m\ddot{u}(x) &= k[u(x+a) + u(x-a) - 2u(x)] \\
&= k\left[u(x) + u'(x)a + \frac{1}{2}u''(x)a^2 + \frac{1}{6}u'''(x)a^3 + \frac{1}{24}u''''(x)a^4 + \cdots \right. \\
&\quad + u(x) - u'(x)a + \frac{1}{2}u''(x)a^2 - \frac{1}{6}u'''(x)a^3 \\
&\quad \left. + \frac{1}{24}u''''(x)a^4 + \cdots - 2u(x) \right] \cong k\left[u''(x)a^2 + \frac{1}{12}u''''(x)a^4 \right],
\end{aligned}
\tag{8.47}
$$

which is a fourth-order equation and can be viewed as the result of a gradient theory. It should be noted that, according to [11], an elasticity theory with the first strain gradient in the constitutive relation is fundamentally flawed in that it is qualitatively inconsistent with lattice dynamics and the second strain gradient needs to be included to correct the inconsistency.

8.7 Polarization Gradient Theory

Mindlin [12] generalized the theory of piezoelectricity by allowing the energy density to depend on the polarization gradient $P_{j,i}$ in addition to

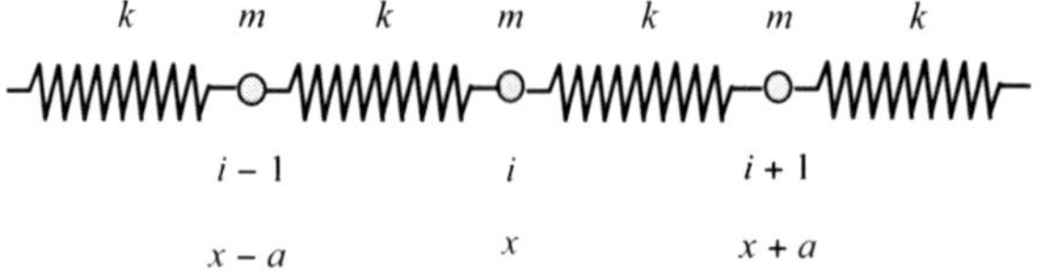

Fig. 8.11 A spring-mass system.

the polarization itself and strain:

$$\Pi(u_i, P_i, \phi) = \int_V \left[W(S_{ij}, P_i, P_{j,i}) - \frac{1}{2}\varepsilon_0 \phi_{,i}\phi_{,i} + \phi_{,i}P_i \right] dV, \qquad (8.48)$$

where boundary terms are dropped for simplicity. The stationary conditions of the above functional for independent variations of u_i, ϕ and P_i are

$$\left(\frac{\partial W}{\partial S_{ij}} \right)_{,i} = 0, \quad -\varepsilon_0 \phi_{,ii} + P_{i,i} = 0, \quad -\frac{\partial W}{\partial P_i} + \left(\frac{\partial W}{\partial P_{j,i}} \right)_{,j} - \phi_{,i} = 0. \quad (8.49)$$

Equation (8.49) represents seven equations for u_i, P_i and ϕ. If the dependence of W on the polarization gradient is dropped, Equation (8.49) reduces to the theory of linear piezoelectricity. The polarization gradient theory represents the long-wave, low-frequency limit of the finite difference equations for a lattice of shell-model atoms [11–14]. What motivated Mindlin to study the effects of the polarization gradient was the capacitance of a very thin dielectric film. Experiments show that the capacitance of a very thin film is systematically smaller than the classical prediction. Using his polarization gradient theory, Mindlin [12] showed that when the film thickness becomes comparable to a material parameter in the polarization gradient theory, the gradient solution can capture the trend of the deviation from the classical prediction. He also showed that the polarization gradient solution of thin-film capacitance agrees with the prediction from lattice dynamics. Mindlin [15] also studied the electric potential of a point charge. The classical solution diverges at the point charge. In fact, at a point very close to the charge, the charge can no longer be considered as a point charge and its distribution has to be taken into consideration. The gradient theory yields a solution that differs from the classical solution only at the close range of the source point, and is valid at a closer distance to the source point than the classical solution. Mindlin [16] showed that in a material with centrosymmetry without piezoelectric coupling, linear electromechanical coupling can still exist due to the polarization gradient. He also studied the polarization gradient effect and electromagnetic fields in diatomic dielectrics [17], and electromagnetic radiation from a vibrating sphere [18].

The polarization gradient theory has also been used to study surface effects and crack problems [19], stress functions and fields due to a concentrated force [20], a point charge in a half-space [21] and a concentrated force on a half-space (the Boussinesq problem) [22], and acceleration waves [23, 24]. Shock waves were investigated in [25]. Conservation laws for the polarization gradient theory were derived [26] from the invariance of the variational integral.

Mindlin's polarization gradient theory was extended in several different directions. The nonlinear version of the theory was first given in [27]. Thermal coupling was included in the theories in [28]–[30]. Fully electromagnetic coupling with complete Maxwell equations was considered in [31] which is a very general theory also including magnetization. In studying certain phenomena in ferroelectric crystals, polarization inertia [32, 33] needs to be considered. A theory including both the polarization gradient and inertia was given in [34] which has support from lattice dynamics [35]–[37], and has been used to study various modes and waves in ferroelectrics [38]–[41]. A general theory including the polarization gradient and inertia as well as the strain gradient was given in [42]. A theory the including polarization gradient and inertia effects in diatomic elastic dielectrics was derived in [43]. A lattice dynamics approach of diatomic dielectrics can be found in [44]. A systematic presentation of the polarization gradient theory was given in [45]. The polarization gradient has also been included in the Landau–Ginsburg functional for studying phase transition [46] and composites [47].

8.8 Electric Field Gradient Theory

For dielectrics, instead of the polarization gradient, the electric field gradient can also be used as constitutive variables [1]. The resulting theory is called dielectrics with spatial dispersion and can be used to describe optical activity in dielectrics. A more physical view of such a theory is the electric quadrupole, because the electric field gradient is the thermodynamic conjugate of the electric quadrupole. The electric quadruple is an old subject in electrodynamics of rigid dielectrics [48]. Theories for elastic dielectrics with electric quadrupoles were developed in [49]–[52]. There are also various extensions to include thermal effects [53], polarization gradient [54], and strain gradient [55]. The relation between the theory of electric field gradient (or quadrupole) and the theory of polarization gradient is discussed in [56].

Consider the following variational functional with electric field gradient [57],

$$\Pi(u_i, \phi) = \int_V \left[W(S_{ij}, E_i, E_{i,j}) - \frac{1}{2}\varepsilon_0 E_i E_i - f_i u_i + q\phi \right] dV$$

$$- \int_S \left[\bar{t}_i u_i + \bar{d}\phi + \bar{\pi}\frac{\partial \phi}{\partial \mathbf{n}} \right] dS, \tag{8.50}$$

where $\bar{d}$ is related to the surface free charge. The presence of the $\bar{\pi}$ term is variationally consistent. We choose

$$W(S_{ij}, E_i, E_{i,j}) - \frac{1}{2}\varepsilon_0 E_i E_i = H(S_{ij}, E_i) - \frac{1}{2}\varepsilon_0 \alpha_{ijkl} E_{i,j} E_{k,l}, \tag{8.51}$$

where H is the usual electric enthalpy function of piezoelectric materials:

$$H(S_{ij}, E_i) = \frac{1}{2}c_{ijkl}S_{ij}S_{kl} - e_{ikl}E_i S_{kl} - \frac{1}{2}\varepsilon_{ij}E_i E_j. \tag{8.52}$$

γ_{ijk} and α_{ijkl} are new material constants due to the introduction of the electric field gradient into the energy density function. γ_{ijk} has the dimension of length. α_{ijkl} has the dimension of $(\text{length})^2$. Physically they may be related to characteristic lengths of microstructural interactions of the material. Because $E_{i,j} = E_{j,i}$, α_{ijkl} has the same structure as c_{ijkl} as required by crystal symmetry, and γ_{ijk} has the same structure as e_{ijk}. For W to be negative definite in the case of pure electric phenomena without mechanical fields, we require α_{ijkl} to be positive definite.

With the following variational constraints:

$$S_{ij} = \frac{(u_{i,j} + u_{j,i})}{2}, \qquad E_i = -\phi_{,i}, \tag{8.53}$$

from the variational functional in Equation (8.50), for independent variations of u_i and ϕ in V, we have

$$T_{ji,j} + f_i = 0, \qquad D_{i,i} = q, \tag{8.54}$$

where we have denoted

$$T_{ij} = \frac{\partial W}{\partial S_{ij}} = c_{ijkl} S_{kl} - e_{kij} E_k,$$

$$D_i = \varepsilon_0 E_i + P_i = \varepsilon_{ij} E_j + e_{ikl} S_{kl} - \varepsilon_0 \alpha_{ijkl} E_{k,lj},$$

$$P_i = \Pi_i - \Pi_{ij,j} = \varepsilon_0 \chi_{ij} E_j + e_{ikl} S_{kl} - \varepsilon_0 \alpha_{ijkl} E_{k,lj},$$

$$\Pi_i = -\frac{\partial W}{\partial E_i} = e_{ikl} S_{kl} + \varepsilon_0 \chi_{ij} E_j, \qquad \Pi_{ij} = -\frac{\partial W}{\partial E_{i,j}} = \varepsilon_0 \alpha_{ijkl} E_{k,l}, \tag{8.55}$$

and $\varepsilon_{ij} = \varepsilon_0(\delta_{ij} + \chi_{ij})$. χ_{ij} is the relative electric susceptibility. When the energy density does not depend on the electric field gradient, the equations reduce to the theory of piezoelectricity. The first variation of the functional in Equation (8.50) also implies the following as possible forms of boundary conditions on S.

$$T_{ji} n_j = \bar{t}_i \quad \text{or} \quad \delta u_i = 0, \qquad \int_S \left[(D_i n_i - \bar{d})\delta\phi + \Pi_{ij} n_j (\nabla_s \delta\phi)_i \right] dS = 0,$$

$$\Pi_{ij} n_j n_i = \bar{\pi} \quad \text{or} \quad \delta\left(\frac{\partial \phi}{\partial \mathbf{n}}\right) = 0, \tag{8.56}$$

where ∇_s is the surface gradient operator. One obvious possibility of Equation (8.56)$_2$ is $\delta\phi = 0$ on S. With substitutions from Equations (8.53) and (8.55), Equation (8.54) can be written as four equations for u_i and ϕ:

$$c_{ijkl} u_{k,lj} + e_{kij}\phi_{,kj} + f_i = \rho\ddot{u}_i, \quad e_{ikl} u_{k,li} - \varepsilon_{ij}\phi_{,ij} + \varepsilon_0 \alpha_{ijkl}\phi_{,ijkl} = q, \tag{8.57}$$

where we have added the acceleration term.

8.9 Antiplane Problems of Polarized Ceramics

Antiplane problems of polarized ceramics are relatively simple mathematically and can still show the basic physical picture [57]. For antiplane motions of polarized ceramics, Equation (8.57) reduces to a much simpler form. Consider

$$u_1 = u_2 = 0, \qquad u_3 = u_3(x_1, x_2, t), \qquad \phi = \phi(x_1, x_2, t). \qquad (8.58)$$

The nonzero strain and electric field components are

$$\begin{Bmatrix} S_5 \\ S_4 \end{Bmatrix} = \nabla u, \qquad \begin{Bmatrix} E_1 \\ E_2 \end{Bmatrix} = -\nabla \phi. \qquad (8.59)$$

For ceramics poled in the x_3-direction, the nontrivial components of T_{ij} and D_i are

$$\begin{Bmatrix} T_5 \\ T_4 \end{Bmatrix} = c\nabla u + e\nabla\phi, \qquad \begin{Bmatrix} D_1 \\ D_2 \end{Bmatrix} = e\nabla u - \varepsilon\nabla\phi + \varepsilon_0\alpha\nabla(\nabla^2\phi), \qquad (8.60)$$

where ∇^2 is the two-dimensional Laplacian, $c = c_{44}$, $e = e_{15}$, $\varepsilon = \varepsilon_{11}$, and $\alpha = \alpha_{11}$. The nontrivial ones of Equation (8.57) take the form

$$c\nabla^2 u + e\nabla^2\phi + f = \rho\ddot{u}, \qquad e\nabla^2 u - \varepsilon\nabla^2\phi + \varepsilon_0\alpha\nabla^2\nabla^2\phi = q, \qquad (8.61)$$

where $f = f_3$. For static problems, Equation (8.61) can be decoupled into

$$\bar{c}\nabla^2 u + f + \frac{\varepsilon_0}{\varepsilon}\alpha\nabla^2(-c\nabla^2 u - f) = \frac{e}{\varepsilon}q, \qquad \nabla^2\phi = \frac{1}{\varepsilon}(-c\nabla^2 u - f), \quad (8.62)$$

or

$$-\bar{\varepsilon}\nabla^2\phi + \varepsilon_0\alpha\nabla^2\nabla^2\phi = q + \frac{e}{c}f, \qquad \nabla^2 u = -\frac{1}{c}(e\nabla^2\phi + f), \qquad (8.63)$$

where

$$\bar{c} = c(1 + k^2), \qquad \bar{\varepsilon} = \varepsilon(1 + k^2), \qquad k^2 = \frac{e^2}{(\varepsilon c)}. \qquad (8.64)$$

Equation (8.63) was used to study thin-film capacitance in [58]. Results similar to those from the nonlocal theory or the polarization gradient theory were obtained.

8.10 A General Solution in Polar Coordinates

We consider static problems with $q = 0$ and $f = 0$. Let

$$F = \nabla^2\phi, \qquad \beta^2 = \frac{\bar{\varepsilon}}{\varepsilon_0\alpha}. \qquad (8.65)$$

Equation $(8.63)_1$ becomes

$$\nabla^2 F - \beta^2 F = 0. \tag{8.66}$$

In a polar coordinate system defined by $x_1 = r\cos\theta$ and $x_2 = r\sin\theta$, by separation of variables, the general solution for F periodic in θ can be found as

$$F = \sum_{n=0}^{\infty}(a_n\cos n\theta + b_n\sin n\theta)[c_n I_n(\beta r) + d_n K_n(\beta r)], \tag{8.67}$$

where a_n, b_n, c_n, and d_n are undetermined constants. I_n and K_n are modified Bessel functions of order n of the first and second kind. Then from Equation $(8.65)_1$ the general solution for ϕ is

$$\phi = a_0\left[g_0 + h_0\ln r + \frac{c_0}{\beta^2}I_0(\beta r) + \frac{d_0}{\beta^2}K_0(\beta r)\right] + \sum_{n=1}^{\infty}(a_n\cos n\theta + b_n\sin n\theta)$$

$$\times\left[g_n r^n + h_n r^{-n} + \frac{c_n}{\beta^2}I_n(\beta r) + \frac{d_n}{\beta^2}K_n(\beta r)\right], \tag{8.68}$$

where g_n and h_n are undetermined constants. Once ϕ is known, from Equation $(8.63)_2$, u is given by

$$u = a_0\left[l_0 + p_0\ln r - \frac{e}{c}\frac{c_0}{\beta^2}I_0(\beta r) - \frac{e}{c}\frac{d_0}{\beta^2}K_0(\beta r)\right]$$

$$+ \sum_{n=1}^{\infty}(a_n\cos n\theta + b_n\sin n\theta)$$

$$\times\left[l_n r^n + p_n r^{-n} - \frac{e}{c}\frac{c_n}{\beta^2}I_n(\beta r) - \frac{e}{c}\frac{d_n}{\beta^2}K_n(\beta r)\right], \tag{8.69}$$

where l_n and p_n are undetermined constants. The above general solution was used to study the concentration of electromechanical fields near a small hole [59].

8.11 A Line Source

Consider the potential field of a line charge with a density Q_e per unit length along the x_3-axis [57]. We need to solve Equation (8.63) with a concentrated electric source:

$$-\bar{\varepsilon}\nabla^2\phi + \varepsilon_0\alpha\nabla^2\nabla^2\phi = Q_e\delta(x_1, x_2). \tag{8.70}$$

Equation (8.70) can be rewritten as

$$(-\bar{\varepsilon} + \varepsilon_0\alpha\nabla^2)\nabla^2\phi = Q_e\delta(x_1, x_2). \tag{8.71}$$

Therefore $\nabla^2\phi$ is the fundamental solution of the differential operator in (8.71), which is known. Hence

$$\nabla^2\phi = \frac{1}{r}\frac{d}{dr}\left(r\frac{d\phi}{dr}\right) = -\frac{Q_e}{2\pi\varepsilon_0\alpha}K_0(\beta r), \tag{8.72}$$

where K_0 is the zero-order modified Bessel function of the second kind. Because

$$xK_0(x) = -\frac{d}{dx}[xK_1(x)], \qquad K_1(x) = -\frac{d}{dx}[K_0(x)], \tag{8.73}$$

integrating Equation (8.72) twice we obtain

$$\phi = -\frac{Q_e}{2\pi\bar\varepsilon}[\ln r + K_0(\beta r)], \qquad \beta^2 = \bar\varepsilon/(\varepsilon_0\alpha), \tag{8.74}$$

where the $\ln r$ term is the classical solution. Because

$$K_0(x) \to -\ln x, \quad x \to 0, \qquad K_0(x) \to \left(\frac{\pi}{2x}\right)^{1/2}e^{-x}, \qquad x \to \infty, \tag{8.75}$$

we have

$$\phi \to \frac{Q_e}{2\pi\bar\varepsilon}\ln\beta = \frac{Q_e}{4\pi\bar\varepsilon}\ln\frac{\bar\varepsilon}{\varepsilon_0\alpha}, \quad r \to 0, \qquad \phi \to -\frac{Q_e}{2\pi\bar\varepsilon}\ln r, \quad r \to \infty. \tag{8.76}$$

When α approaches zero, Equation (8.74) reduces to the classical result. The potential field is plotted in Figure 8.12.

The figure shows that in the far field ϕ approaches the classical solution. At the source point ϕ is not singular. This is fundamentally different from the classical solution. The curve with the larger value of β is closer to the classical solution. These qualitative behaviors are as expected.

8.12 Dispersion of Short Plane Waves

In the source-free case, from Equation (8.62) we obtain

$$\bar c\nabla^2 u + \frac{\varepsilon_0}{\varepsilon}\alpha\nabla^2(\rho\ddot u - c\nabla^2 u) = \rho\ddot u. \tag{8.77}$$

Consider the propagation of the following plane wave [57],

$$u = \exp[i(\xi x_1 - \omega t)]. \tag{8.78}$$

Substitution of Equation (8.78) into (8.77) yields the following dispersion relation,

$$\omega^2 = \frac{c}{\rho}\xi^2\frac{1 + k^2 + \frac{\varepsilon_0}{\varepsilon}\alpha\xi^2}{1 + \frac{\varepsilon_0}{\varepsilon}\alpha\xi^2}. \tag{8.79}$$

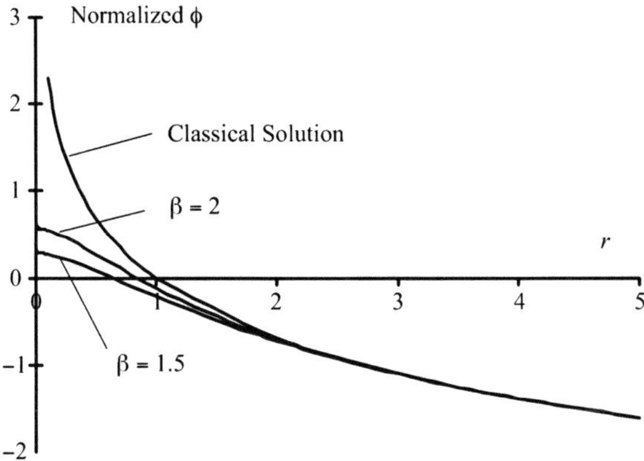

Fig. 8.12 Normalized potential field $(-2\pi\bar{\varepsilon}\phi/Q_e)$ of a line source.

Different from the plane waves in linear piezoelectricity, Equation (8.79) shows that the waves are dispersive, and the dispersion is caused by the electric field gradient through electromechanical coupling. The dispersion disappears when $k = 0$, or when there is no electromechanical coupling. We note that the dispersion is more pronounced when $\xi\sqrt{\alpha}$ is not small, or when the wavelength $2\pi/\xi$ is not large when compared to the microscopic characteristic length $\sqrt{\alpha}$. When $\xi\sqrt{\alpha}$ just begins to show its effect, Equation (8.79) can be approximated by

$$\omega^2 \cong \frac{\bar{c}}{\rho}\xi^2\left[1 - \frac{k^2}{1+k^2}\frac{\varepsilon_0}{\varepsilon}\alpha\xi^2\right]. \tag{8.80}$$

As a numerical example we consider polarized ceramics PZT-7A. We plot Equation (8.80) in Figure 8.13 for different values of $\sqrt{\alpha}$. The figure shows that larger values of $\sqrt{\alpha}$ yield more dispersion, as expected. What matters is the relative magnitude of ξ with respect to $\sqrt{\alpha}$.

8.13 Dispersion of Short Surface Waves

Consider a ceramic half-space as shown in Figure 8.14. Its surface carries an electrode of a perfect conductor. The electrode is thin and its mechanical effects can be neglected.

We are interested in antiplane surface waves propagating in the positive x_1-direction [60]. In the case of classical linear piezoelectricity these waves are called Bleustein–Gulyaev waves [61, 62]. They are nondispersive waves whose existence relies on piezoelectric coupling and do not have an elastic

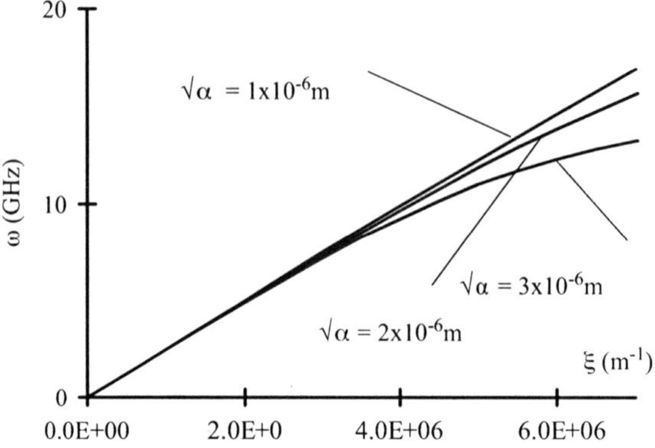

Fig. 8.13 Dispersion curves of plane waves.

counterpart. We begin with the source-free form of Equation (8.61). We write $(8.61)_2$ as

$$\nabla^2(eu - \varepsilon\phi + \varepsilon_0\alpha\nabla^2\phi) = 0. \tag{8.81}$$

Similar to [61], we introduce

$$\psi = \phi - \frac{e}{\varepsilon}u - \frac{\varepsilon_0\alpha}{\varepsilon}\nabla^2\phi. \tag{8.82}$$

Then Equation (8.61) is equivalent to the following system of second-order equations,

$$\begin{aligned}
c\nabla^2 u + e\nabla^2\phi &= \rho\ddot{u}, \\
\varepsilon_0\alpha\nabla^2\phi - \varepsilon\phi + eu + \varepsilon\psi &= 0, \\
\nabla^2\psi &= 0.
\end{aligned} \tag{8.83}$$

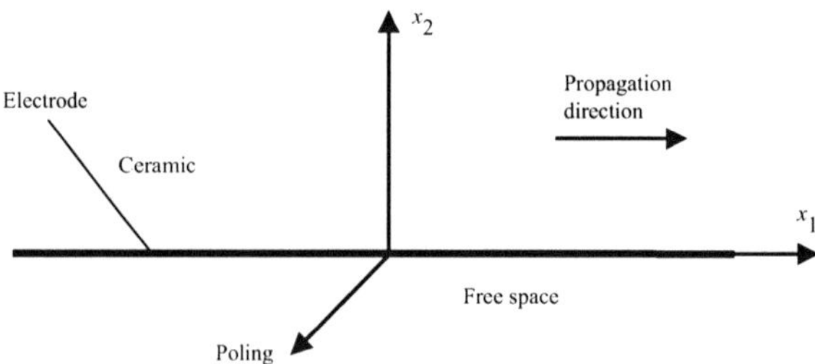

Fig. 8.14 An electroded half-space of ceramics poled in the x_3-direction.

We look for surface wave solutions in the following form,

$$\begin{Bmatrix} u \\ \phi \\ \psi \end{Bmatrix} = \begin{Bmatrix} A \\ B \\ C \end{Bmatrix} \exp(-\eta x_2)\cos(\xi x_1 - \omega t), \qquad (8.84)$$

where A, B, C, η, ξ, and ω are undetermined constants. Substitution of Equation (8.84) into (8.83) yields the following homogeneous, linear equations for A, B, and C,

$$\begin{bmatrix} c(\eta^2 - \xi^2) + \rho\omega^2 & e(\eta^2 - \xi^2) & 0 \\ e & \varepsilon_0\alpha(\eta^2 - \xi^2) - \varepsilon & \varepsilon \\ 0 & 0 & \eta^2 - \xi^2 \end{bmatrix} \begin{Bmatrix} A \\ B \\ C \end{Bmatrix} = 0. \qquad (8.85)$$

For nontrivial solutions the determinant of the coefficient matrix has to vanish, which yields a cubic equation for η^2. The three relevant roots of the cubic equation are denoted by

$$\eta_{(1)} = \xi(1 + f_{(1)})^{1/2},$$
$$\eta_{(2)} = \xi(1 + f_{(2)})^{1/2},$$
$$\eta_{(3)} = \xi, \qquad (8.86)$$

where

$$f_{(1)} = \frac{1}{2\xi^2\alpha}\left\{ n^2(1+k^2) - \xi^2\alpha\frac{V^2}{V_0^2} + \sqrt{[n^2(1+k^2) - \xi^2\alpha\frac{V^2}{V_0^2}]^2 + 4\xi^2\alpha n^2\frac{V^2}{V_0^2}} \right\},$$

$$f_{(2)} = \frac{1}{2\xi^2\alpha}\left\{ n^2(1+k^2) - \xi^2\alpha\frac{V^2}{V_0^2} - \sqrt{[n^2(1+k^2) - \xi^2\alpha\frac{V^2}{V_0^2}]^2 + 4\xi^2\alpha n^2\frac{V^2}{V_0^2}} \right\},$$

$$(8.87)$$

and

$$V = \frac{\omega}{\xi}, \qquad V_0^2 = \frac{c}{\rho}, \qquad n^2 = \frac{\varepsilon}{\varepsilon_0}. \qquad (8.88)$$

Then the general surface wave solution can be constructed as

$$u = \frac{\varepsilon_0}{e}[(n^2 - \xi^2\alpha f_{(1)})C_1\exp(-\eta_{(1)}x_2)$$
$$+ (n^2 - \xi^2\alpha f_{(2)})C_2\exp(-\eta_{(2)}x_2)]\cos(\xi x_1 - \omega t),$$
$$\phi = [C_1\exp(-\eta_{(1)}x_2) + C_2\exp(-\eta_{(2)}x_2)$$
$$+ C_3\exp(-\xi x_2)]\cos(\xi x_1 - \omega t),$$
$$\psi = C_3\exp(-\xi x_2)\cos(\xi x_1 - \omega t), \qquad (8.89)$$

where C_1, C_2, and C_3 are undetermined constants. The classical piezoelectric boundary conditions for an electroded surface are $T_4 = 0$ and $\phi = 0$. For the electric field gradient theory, an additional boundary condition is needed. Guided by the variational result in Equation (8.56)$_3$, we apply a simple boundary condition of $\partial\phi/\partial x_2 = 0$. This additional boundary condition imposes a further restriction on the electric field that already satisfies the classical boundary conditions of piezoelectricity. $\partial\phi/\partial x_2 = 0$ implies, through the constitutive relation for T_4, that $\partial u/\partial x_2 = 0$ at $x_2 = 0$. Hence,

$$\frac{\partial u}{\partial x_2} = \frac{\varepsilon_0}{e}[(n^2 - \xi^2\alpha f_{(1)})C_1(-\eta_{(1)})$$
$$+ (n^2 - \xi^2\alpha f_{(2)})C_2(-\eta_{(2)})]\cos(\xi x_1 - \omega t) = 0,$$
$$\phi = (C_1 + C_2 + C_3)\cos(\xi x_1 - \omega t) = 0,$$
$$\frac{\partial\phi}{\partial x_2} = [C_1(-\eta_{(1)}) + C_2(-\eta_{(2)}) + C_3(-\xi)]\cos(\xi x_1 - \omega t) = 0, \qquad (8.90)$$

which are three homogeneous equations for C_1, C_2, and C_3. For nontrivial solutions the determinant of the coefficient matrix has to vanish, which gives

$$\frac{\sqrt{1 + f_{(1)}}(\sqrt{1 + f_{(2)}} - 1)}{\sqrt{1 + f_{(2)}}(\sqrt{1 + f_{(1)}} - 1)} = \frac{n^2 - \xi^2\alpha f_{(2)}}{n^2 - \xi^2\alpha f_{(1)}}. \qquad (8.91)$$

Equation (8.91) is an equation for the surface wave speed V. It shows that V depends on the wave number ξ. Hence the wave is dispersive. This is fundamentally different from the corresponding Bleustein–Gulyaev waves in linear piezoelectricity.

The dispersion depends on the combination of $\xi^2\alpha$. It can be verified that when $\xi^2\alpha = 0$, Equation (8.91) gives the speed of Bleustein–Gulyaev waves as expected:

$$V^2 = V_{B-G}^2 = V_T^2(1 - \overline{k}^4), \qquad (8.92)$$

where

$$V_T^2 = \frac{\bar{c}}{\rho} = \frac{c}{\rho}(1 + k^2), \qquad \overline{k}^2 = \frac{e^2}{\varepsilon\bar{c}} = \frac{k^2}{1 + k^2}. \qquad (8.93)$$

When $\xi^2\alpha \ll 1$, we have long waves whose wave length $\lambda = 2\pi/\xi$ is large compared to the microscopic characteristic length $\sqrt{\alpha}$:

$$\xi^2\alpha \ll 1 \quad \Leftrightarrow \quad (2\pi)^2\alpha \ll \lambda^2. \qquad (8.94)$$

In this case the dispersion is small. For shorter waves, that is, λ is not much larger than α, the dispersion is stronger.

For some quantitative results we solve Equation (8.91) numerically and plot the dispersion curves for different ceramics in Figure 8.15. The curves show clearly that short waves are dispersive but long waves are less so.

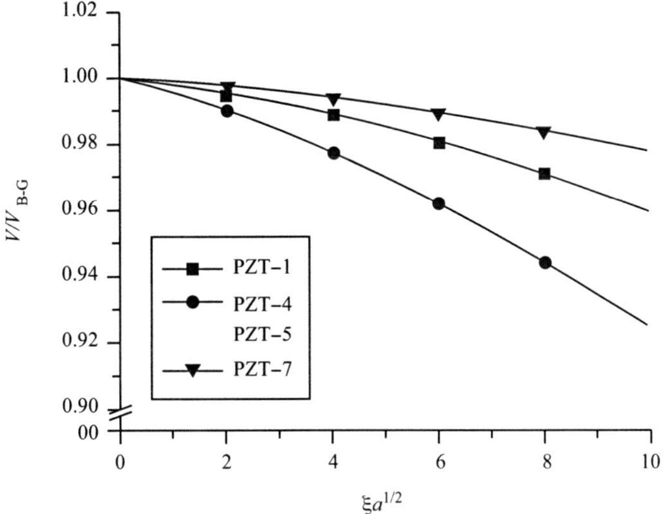

Fig. 8.15 Dispersion curves of antiplane surface waves.

More precisely, when $\xi\sqrt{\alpha} \cong 10$ or $\lambda\sqrt{\alpha} \cong 2\pi/10$, that is, the wavelength λ is of the same order of the microscopic characteristic length $\sqrt{\alpha}$, the wave speed is reduced by 7–8% due to dispersion.

8.14 A Circular Inclusion

We now consider a circular inclusion under an electric field E^0 and a shear strain S^0 at infinity (see Figure 8.16, where $\alpha = \infty$ except in a numerical example later) [63].

8.14.1 Exterior Fields

Far away from the inclusion we have

$$\phi = -E^0 x_1 = -E^0 r \cos\theta, \qquad u = S^0 x_1 = S^0 r \cos\theta. \qquad (8.95)$$

We take the following terms from the general solution in Equations (8.68) and (8.69),

$$\phi = \left[-E^0 r + \frac{h_1}{r} + \frac{d_1}{\beta^2} K_1(\beta r)\right]\cos\theta, \qquad u = \left[S^0 r + \frac{p_1}{r} - \frac{e}{c}\frac{d_1}{\beta^2} K_1(\beta r)\right]\cos\theta,$$

$$(8.96)$$

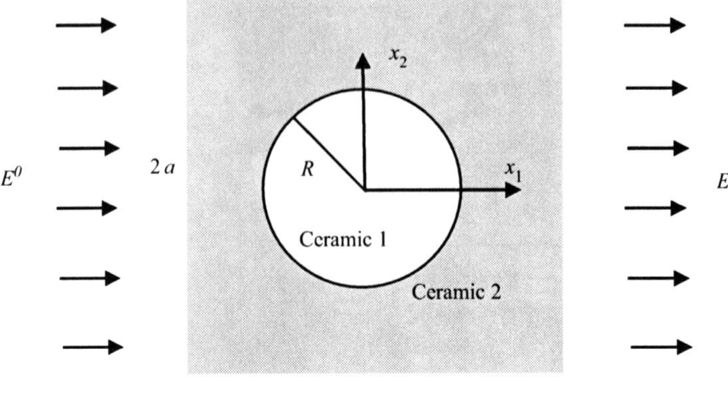

Fig. 8.16 A circular inclusion in an unbounded matrix.

which agrees with Equation (8.95) at infinity. Corresponding to Equation (8.96), we have

$$S_{rz} = \left\{ S^0 - \frac{p_1}{r^2} + \frac{ed_1}{c\beta} \left[K_0(\beta r) + \frac{K_1(\beta r)}{\beta r} \right] \right\} \cos\theta,$$

$$S_{\theta z} = -\left[S^0 + \frac{p_1}{r^2} - \frac{ed_1}{c\beta} \frac{K_1(\beta r)}{\beta r} \right] \sin\theta, \tag{8.97}$$

$$T_{rz} = \left[(cS^0 - eE^0) - (cp_1 + eh_1)\frac{1}{r^2} \right] \cos\theta,$$

$$T_{\theta z} = -\left[(cS^0 - eE^0) + (cp_1 + eh_1)\frac{1}{r^2} \right] \sin\theta, \tag{8.98}$$

$$E_r = \left\{ E^0 + \frac{h_1}{r^2} + \frac{d_1}{\beta} \left[K_0(\beta r) + \frac{K_1(\beta r)}{\beta r} \right] \right\} \cos\theta,$$

$$E_\theta = -\left[E^0 - \frac{h_1}{r^2} - \frac{d_1}{\beta} \frac{K_1(\beta r)}{\beta r} \right] \sin\theta, \tag{8.99}$$

$$D_r = \left[(eS^0 + \varepsilon E^0) - (ep_1 - \varepsilon h_1)\frac{1}{r^2} \right] \cos\theta,$$

$$D_\theta = -\left[(eS^0 + \varepsilon E^0) + (ep_1 - \varepsilon h_1)\frac{1}{r^2} \right] \sin\theta, \tag{8.100}$$

$$\Pi_{rr} = -\varepsilon_0 \left\{ \alpha_{11} \left[\frac{2h_1}{r^3} + \frac{d_1}{\beta r} K_0(\beta r) + \frac{2d_1}{(\beta r)^2} K_1(\beta r) + d_1 K_1(\beta r) \right] \right.$$

$$\left. - \alpha_{12} \left[\frac{2h_1}{r^3} + \frac{d_1}{\beta r} K_0(\beta r) + \frac{2d_1}{(\beta r)^2} K_1(\beta r) \right] \right\} \cos\theta, \tag{8.101}$$

$$\Pi_{\theta\theta} = -\varepsilon_0 \left\{ \alpha_{12} \left[\frac{2h_1}{r^3} + \frac{d_1}{\beta r} K_0(\beta r) + \frac{2d_1}{(\beta r)^2} K_1(\beta r) + d_1 K_1(\beta r) \right] \right.$$

$$\left. - \alpha_{11} \left[\frac{2h_1}{r^3} + \frac{d_1}{\beta r} K_0(\beta r) + \frac{2d_1}{(\beta r)^2} K_1(\beta r) \right] \right\} \cos\theta, \qquad (8.102)$$

$$\Pi_{r\theta} = -2\varepsilon_0 \alpha_{66} \left[\frac{2h_1}{r^3} + \frac{d_1}{\beta r} K_0(\beta r) + \frac{2d_1}{(\beta r)^2} K_1(\beta r) \right] \sin\theta. \qquad (8.103)$$

8.14.2 Interior Fields

We take the following terms from the general solution in Equations (8.68) and (8.69),

$$\phi = \left[g_1 r + \frac{c_1}{\hat{\beta}^2} I_1(\hat{\beta} r) \right] \cos\theta,$$

$$u = \left[l_1 r - \frac{\hat{e}}{\hat{c}} \frac{c_1}{\hat{\beta}^2} I_1(\hat{\beta} r) \right] \cos\theta, \qquad (8.104)$$

which are finite at the origin. Corresponding to Equation (8.104), we have

$$S_{rz} = \left\{ l_1 + \frac{\hat{e}c_1}{\hat{c}\hat{\beta}} \left[I_0(\hat{\beta} r) + \frac{I_1(\hat{\beta} r)}{\hat{\beta} r} \right] \right\} \cos\theta,$$

$$S_{\theta z} = - \left[l_1 - \frac{\hat{e}c_1}{\hat{c}\hat{\beta}} \frac{I_1(\hat{\beta} r)}{\hat{\beta} r} \right] \sin\theta, \qquad (8.105)$$

$$T_{rz} = (\hat{c} l_1 + \hat{e} g_1)\cos\theta,$$
$$T_{\theta z} = -(\hat{c} l_1 + \hat{e} g_1)\sin\theta, \qquad (8.106)$$

$$E_r = \left\{ -g_1 + \frac{c_1}{\hat{\beta}} \left[I_0(\hat{\beta} r) + \frac{I_1(\hat{\beta} r)}{\hat{\beta} r} \right] \right\} \cos\theta,$$

$$E_\theta = \left[g_1 + \frac{c_1}{\hat{\beta}} \frac{I_1(\hat{\beta} r)}{\hat{\beta} r} \right] \sin\theta, \qquad (8.107)$$

$$D_r = (\hat{e} l_1 - \hat{\varepsilon} g_1)\cos\theta,$$
$$D_\theta = -(\hat{e} l_1 - \hat{\varepsilon} g_1)\sin\theta, \qquad (8.108)$$

$$\Pi_{rr} = -\varepsilon_0 c_1 \left\{ \hat{\alpha}_{11} \left[I_1(\hat{\beta}r) - \frac{1}{\hat{\beta}r} I_0(\hat{\beta}r) + \frac{2}{(\hat{\beta}r)^2} I_1(\hat{\beta}r) \right] \right.$$
$$\left. + \hat{\alpha}_{12} \left[\frac{1}{\hat{\beta}r} I_0(\hat{\beta}r) - \frac{2}{(\hat{\beta}r)^2} I_1(\hat{\beta}r) \right] \right\} \cos\theta, \tag{8.109}$$

$$\Pi_{\theta\theta} = -\varepsilon_0 c_1 \left\{ \hat{\alpha}_{12} \left[I_1(\hat{\beta}r) - \frac{1}{\hat{\beta}r} I_0(\hat{\beta}r) + \frac{2}{(\hat{\beta}r)^2} I_1(\hat{\beta}r) \right] \right.$$
$$\left. + \hat{\alpha}_{11} \left[\frac{1}{\hat{\beta}r} I_0(\hat{\beta}r) - \frac{2}{(\hat{\beta}r)^2} I_1(\hat{\beta}r) \right] \right\} \cos\theta, \tag{8.110}$$

$$\Pi_{r\theta} = -2\varepsilon_0 \hat{\alpha}_{66} c_1 \left[\frac{1}{\hat{\beta}r} I_0(\hat{\beta}r) - \frac{2}{(\hat{\beta}r)^2} I_1(\hat{\beta}r) \right] \sin\theta. \tag{8.111}$$

8.14.3 Continuity Conditions

From Equation (8.56), we require the following continuity conditions at the interface $r = R$, which are variationally consistent.

$$u(R^-) = u(R^+), \qquad T_{rz}(R^-) = T_{rz}(R^+), \qquad \phi(R^-) = \phi(R^+),$$
$$\left[D_r - \frac{\partial \Pi_{r\theta}}{r\partial\theta} \right]\Big|_{R^-} = \left[D_r - \frac{\partial \Pi_{r\theta}}{r\partial\theta} \right]\Big|_{R^+},$$
$$\frac{\partial\phi}{\partial r}\Big|_{R^-} = \frac{\partial\phi}{\partial r}\Big|_{R^+}, \qquad \Pi_{rr}(R^-) = \Pi_{rr}(R^+). \tag{8.112}$$

Substitution of Equations (8.96) through (8.111) into Equation (8.112) gives six equations for h_1, d_1, p_1, g_1, c_1, and l_1, which are solved numerically.

8.14.4 Numerical Results

For numerical results, PZT-5A and BaTiO$_3$ are used for the matrix and the inclusion, respectively. What matters is the relative magnitude of α_{jikl} with respect to R. We are interested in some qualitative results only and artificially choose:

$$\alpha_{11} = 1.2 \times 10^{-6}, \qquad \hat{\alpha}_{11} = 1.4 \times 10^{-6},$$
$$\alpha_{12} = 0.4 \times 10^{-6}, \qquad \hat{\alpha}_{12} = 0.5 \times 10^{-6}\,\mathrm{m}^2. \tag{8.113}$$

A few different values of R are used, ranging from $2\sqrt{\alpha}$ to $10\sqrt{\alpha}$, where α stands for $\hat{\alpha}_{11}$. Only E^0 is applied. $S^0 = 0$.

Equations (8.112) are solved on a computer. Electric field distribution is shown in Figure 8.17. A fundamental difference from the classical inclusion solution is that the electric field in the inclusion is no longer uniform. Near the interface the electric field is larger than the nearly uniform electric field in the central region of the inclusion. Therefore some field concentration exists near the interface, which is as expected because gradient theories are usually associated with boundary layer effects. Field concentration is important to strength and failure considerations.

To see the effect of $\sqrt{\alpha}/R$ on the field distribution, we plot the interior fields for two different values of $\sqrt{\alpha}/R$ in Figures 8.18 and 8.19. When $\sqrt{\alpha}/R$ is relatively small, the interior field is more uniform. In Figures 8.18 and 8.19, the electric field in the almost uniform central regions are in fact about the same.

If the piezoelectric constants are set to zero, we have a pure electric inclusion problem of dielectrics. We calculated the ratio of D_1 and E_1 averaged over the square region in Figure 8.16, which represents the effective dielectric constant. The result is shown in Figure 8.20. The figure shows that when $R/\sqrt{\alpha}$ is large the effective dielectric constant is not sensitive to $R/\sqrt{\alpha}$. In this case the effective dielectric constant is basically the one predicted by the classical inclusion theory. When $R/\sqrt{\alpha}$ is not large, the effective dielectric constant is smaller (size effect) as expected because the electric field is larger near the interface due to the gradient effects.

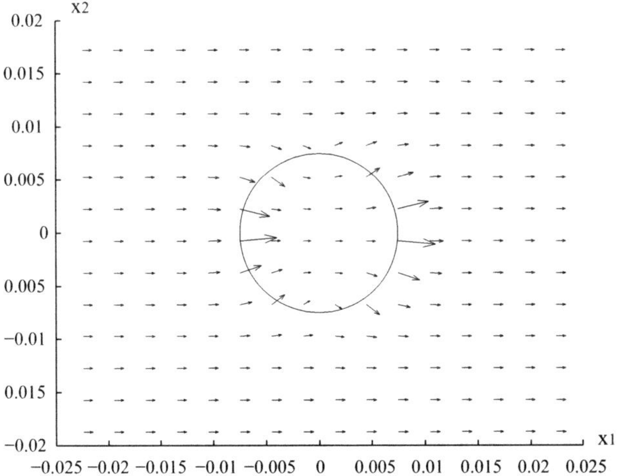

Fig. 8.17 Electric field distribution ($\sqrt{\alpha} = R/5$).

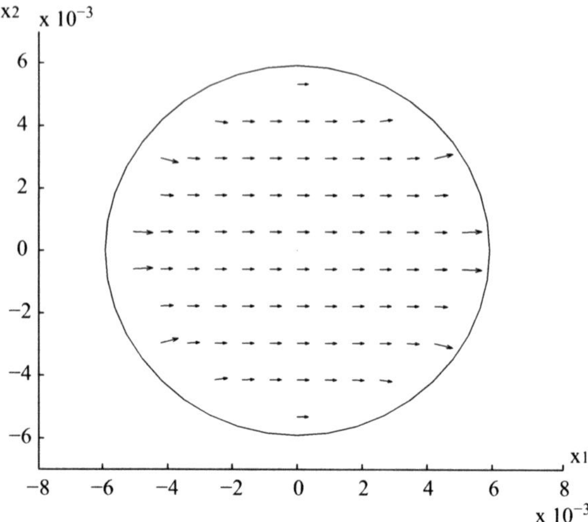

Fig. 8.18 Interior electric field ($\sqrt{\alpha} = R/5$).

In a polar coordinate system, consider a semi-infinite crack at $\theta = \pi$ in an unbounded and source-free ($q = 0$ and $f = 0$) region (see Figure 8.21) [64]. The crack faces are traction-free.

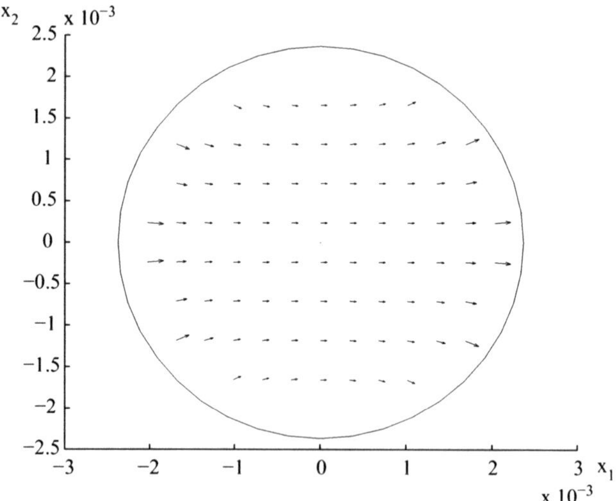

Fig. 8.19 Interior electric field ($\sqrt{\alpha} = R/3$).

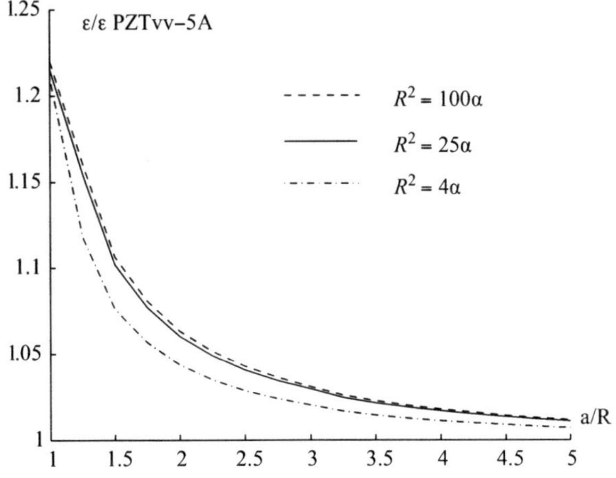

Fig. 8.20 Effective dielectric constant $(\varepsilon = \varepsilon_{11})$.

8.15 A Semi-Infinite Crack

The displacement and electric potential are governed by Equation (8.62). Let

$$F = \nabla^2 u, \qquad \beta^2 = \frac{\bar{c}\varepsilon}{c\varepsilon_0\alpha}, \tag{8.114}$$

then Equation $(8.62)_1$ becomes

$$\nabla^2 F - \beta^2 F = 0. \tag{8.115}$$

In polar coordinates, Equation (8.115) takes the following form,

$$\left(\frac{\partial^2}{\partial r^2} + \frac{\partial}{r\partial r} + \frac{1}{r^2}\frac{\partial^2}{\partial\theta^2}\right)F - \beta^2 F = 0. \tag{8.116}$$

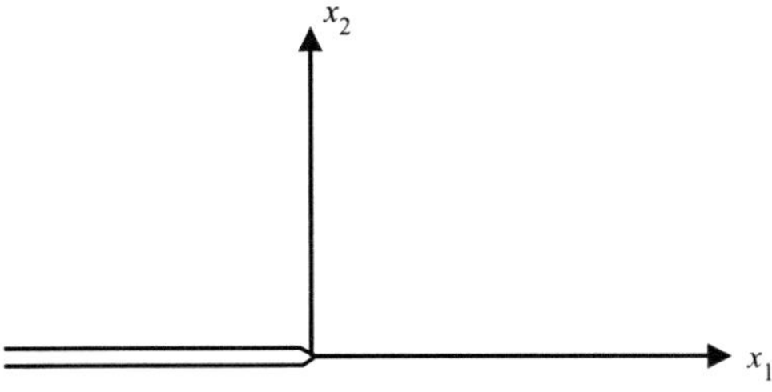

Fig. 8.21 A semi-infinite crack.

Motivated by the classical solution, we look for

$$F(r, \theta) = F(r) \sin \frac{\theta}{2}. \tag{8.117}$$

Substituting Equation (8.117) into (8.116), we obtain

$$\frac{d^2 F}{d(\beta r)^2} + \frac{1}{\beta r} \frac{dF}{d(\beta r)} - \left[1 + \frac{(1/2)^2}{(\beta r)^2}\right] F = 0, \tag{8.118}$$

which is the modified Bessel equation of order $1/2$. Its general solution is

$$F(r) = C_1 I_{1/2}(\beta r) + C_2 K_{1/2}(\beta r), \tag{8.119}$$

where $I_{1/2}$ and $K_{1/2}$ are the first and second kind modified Bessel functions of order $1/2$:

$$I_{1/2}(x) = \sqrt{\frac{2}{\pi x}} \sinh x,$$

$$K_{1/2}(x) = \sqrt{\frac{2}{\pi x}} e^{-x}. \tag{8.120}$$

C_1 and C_2 are undetermined constants. Because $I_{1/2}$ is divergent for large arguments, we choose $C_1 = 0$. To find u we now need to solve

$$\left(\frac{\partial^2}{\partial r^2} + \frac{\partial}{r \partial r} + \frac{1}{r^2} \frac{\partial^2}{\partial \theta^2}\right) u = C_2 K_{1/2}(\beta r) \sin \frac{\theta}{2}. \tag{8.121}$$

Let

$$u(r, \theta) = u(r) \sin \frac{\theta}{2}. \tag{8.122}$$

Substitution of Equation (8.122) into (8.121) yields

$$\left(\frac{\partial^2}{\partial r^2} + \frac{\partial}{r \partial r} - \frac{1}{4r^2}\right) u = C_2 K_{1/2}(\beta r). \tag{8.123}$$

For Equation (8.123) the homogeneous solution can be obtained easily. It can be verified that the particular solution is proportional to $K_{1/2}$. Therefore the relevant solution is

$$u(r, \theta) = \left[C_3 \sqrt{r} + C_4 \frac{1}{\sqrt{r}} + \frac{C_2}{\beta^2} K_{1/2}(\beta r)\right] \sin \frac{\theta}{2}$$

$$= \left[C_3 \sqrt{r} + C_4 \frac{1}{\sqrt{r}} + \frac{C_2}{\beta^2} \sqrt{\frac{2}{\pi \beta r}} e^{-\beta r}\right] \sin \frac{\theta}{2}, \tag{8.124}$$

where C_3 and C_4 are undetermined constants. If we choose $C_4 = 0$ as in the classical solution, then u has a $1/\sqrt{r}$ singularity at the crack-tip. If we want u to be bounded at the crack-tip, we must have

$$C_4 + \frac{C_2}{\beta^2}\sqrt{\frac{2}{\pi\beta}} = 0. \tag{8.125}$$

Then

$$u(r,\theta) = \left[C_3\sqrt{r} + C_4\frac{1}{\sqrt{r}}(1 - e^{-\beta r}) \right]\sin\frac{\theta}{2}. \tag{8.126}$$

With Equation (8.126), from (8.62)$_2$ we find

$$\phi = \left[C_5\sqrt{r} - \frac{c}{e}C_4\frac{1}{\sqrt{r}}(1 - e^{-\beta r}) \right]\sin\frac{\theta}{2}, \tag{8.127}$$

where C_5 is an undetermined constant. The strain, electric field, stress, and electric displacement components are:

$$2S_{rz} = \frac{1}{2\sqrt{r}}\left[C_3 - C_4\frac{1}{r}(1 - e^{-\beta r}) + 2C_4\beta e^{-\beta r} \right]\sin\frac{\theta}{2},$$

$$2S_{\theta z} = \frac{1}{2\sqrt{r}}\left[C_3 + C_4\frac{1}{r}(1 - e^{-\beta r}) \right]\cos\frac{\theta}{2}, \tag{8.128}$$

$$E_r = -\frac{1}{2\sqrt{r}}\left[C_5 + \frac{c}{e}C_4\frac{1}{r}(1 - e^{-\beta r}) - 2\frac{c}{e}C_4\beta e^{-\beta r} \right]\sin\frac{\theta}{2},$$

$$E_\theta = -\frac{1}{2\sqrt{r}}\left[C_5 - \frac{c}{e}C_4\frac{1}{r}(1 - e^{-\beta r}) \right]\cos\frac{\theta}{2}, \tag{8.129}$$

$$T_{rz} = \frac{cC_3 + eC_5}{2\sqrt{r}}\sin\frac{\theta}{2}, \qquad T_{\theta z} = \frac{cC_3 + eC_5}{2\sqrt{r}}\cos\frac{\theta}{2}, \tag{8.130}$$

$$D_r = \left[\frac{eC_3 - \varepsilon C_5}{2\sqrt{r}} - \left(e + \frac{\varepsilon c}{e}\right)\frac{C_4}{2r\sqrt{r}} \right]\sin\frac{\theta}{2},$$

$$D_\theta = \left[\frac{eC_3 - \varepsilon C_5}{2\sqrt{r}} + \left(e + \frac{\varepsilon c}{e}\right)\frac{C_4}{2r\sqrt{r}} \right]\cos\frac{\theta}{2}. \tag{8.131}$$

Some observations can be made from Equations (8.126) through (8.131). When C_4 is equal to zero, the above solution reduces to the classical piezoelectric solution. This happens when $\alpha \to 0$, hence $\beta \to \infty$, and $C_4 \to 0$. The singularity of D_r and D_θ is of the order of $1/\sqrt{r^3}$. This is quite different from the classical theory. Electric field gradients effects in finite cracks were studied in [65, 66].

References

[1]   Landau DL, Lifshitz EM (1984) *Electrodynamics of Continuous Media*, 2nd edn. Butterworth-Heinemann, Oxford

[2]   Jackson JD (1990) *Classical Electrodynamics*. John Wiley & Sons, Singapore

[3]   Eringen AC, Maugin GA (1990), *Electrodynamics of Continua*, vol. II. Springer, New York

[4]   Eringen AC (1993), Vistas of nonlocal electrodynamics. In: Lee JS, Maugin GA, Shindo Y (ed) *Mechanics of Electromagnetic Materials and Structures*. American Society of Mechanical Engineers, New York

[5]   Eringen AC, Kim BS (1977) Relations between nonlocal elasticity and lattice dynamics. *Crystal Lattice Defects* 7:51–57

[6]   Maugin GA (1979) Nonlocal theories or gradient-type theories: a matter of convenience? *Arch Mech* 31:15–26

[7]   Eringen AC (1984) Theory of nonlocal piezoelectricity. *J Math Phys* 25:717–727

[8]   Yang JS (1997) Thin film capacitance in case of a nonlocal polarization law. *Int J Appl Electromag Mech* 8:307–314

[9]   Yang JS, Mao SX, Yan K et al (2006) Size effect on the electromechanical coupling factor of a thin piezoelectric film due to a nonlocal polarization law. *Scripta Materialia* 54:1281–1286

[10] Chopra KL (1969) *Thin Film Phenomena*. McGraw-Hill, New York

[11] Mindlin RD (1972) Elasticity, piezoelectricity and crystal lattice dynamics. *J Elasticity* 2:217–282

[12] Mindlin RD (1968), Polarization gradient in elastic dielectrics. *Int J Solids Struct* 4:637–642

[13] Mindlin RD (1969) Continuum and lattice theories of influence of electromechanical coupling on capacitance of thin dielectric films. *Int J Solids Struct* 5:1197–1208

[14] Askar A, Lee PCY, Cakmak AS (1970) A lattice dynamics approach to the theory of elastic dielectrics with polarization gradient. *Phys Rev B* 1:3525–3537

[15] Mindlin RD (1973) On the electrostatic potential of a point charge in a dielectric solid. *Int J Solids Struct* 9:233–235

[16] Mindlin RD (1971) Electromechanical vibrations of centrosymmetric cubic crystal plates. *PMM-J Mech Appl Mathe* 35:404–408

[17] Mindlin RD (1972) Coupled elastic and electromagnetic fields in a diatomic, electric continuum. *Int J Solids Struct* 8:401–408

[18] Mindlin RD (1974) Electromagnetic radiation from a vibrating, elastic sphere. *Int J Solids Struct* 10:1307–1314

[19] Askar A, Lee PCY, Cakmak AS (1971) The effect of surface curvature and discontinuity on the surface energy density and other induced fields in electric dielectrics with polarization gradient. *Int J Solids Struct* 7:523–537

[20] Schwartz J (1969) Solutions of the equations of equilibrium of elastic dielectrics: stress functions, concentrated force, surface energy. *Int J Solids Struct* 5:1209–1220

[21] Chowdhury KL, Glockner PG (1977) Point charge in the interior of an elastic dielectric half space. *Int J Engng Sci* 15:481–493

[22] Chowdhury KL, Glockner PG (1981) On a similarity solution of the Boussinesq problem of elastic dielectrics. *Arch Mech* 32:429–442

[23] Collet B (1981) One-dimensional acceleration waves in deformable dielectrics with polarization gradients. *Int J Engng Sci* 19: 389–407

[24] Dost S (1983) Acceleration waves in elastic dielectrics with polarization gradient effects. *Int J Engng Sci* 21:1305–1311

[25] Collet B (1982) Shock waves in deformable dielectrics with polarization gradients. *Int J Engng Sci* 20:1145–1160

[26] Yang JS, Batra RC (1995) Conservation laws in linear piezoelectricity. *Eng Fract Mech* 51:1041–1047

[27] Suhubi ES (1969) Elastic dielectrics with polarization gradients. *Int J Engng Sci* 7:993–997

[28] Chowdhury KL, Epstein M, Glockner PG (1979) On the thermodynamics of nonlinear elastic dielectrics. *Int J Non-Linear Mech* 13:311–322

[29] Chowdhury KL, Glockner PG (1976) Constitutive equations for elastic dielectrics. *Int J Non-Linear Mech* 11:315–324

[30] Chowdhury KL, Glockner PG (1977) On thermoelastic dielectrics. *Int J Solids Struct* 13:1173–1182

[31] Tiersten HF, Tsai CF (1972) On the interaction of the electromagnetic field with heat conducting deformable insulators. *J Math Phys* 13:361–378

[32] Maugin GA (1977) Deformable dielectrics II. Voigt's intramolecular force balance in elastic dielectrics. *Arch Mech* 29:143–151

[33] Maugin GA (1977) Deformable dielectrics III. A model of interactions. *Arch Mech* 29:251–258

[34] Maugin GA, Pouget J (1980) Electroacoustic equations for one-domain ferroelectric bodies. *J Acoust Soc Am* 68:575–587

[35] Askar A, Pouget J, Maugin GA (1984) Lattice model for elastic ferroelectrics and related continuum theories. In: Maugin GA (ed) *Mechanical Behavior of Electromagnetic Solid Continua*. Elsevier, North-Holland

[36] Pouget J, Askar A, Maugin GA (1986) Lattice model for elastic ferroelectric crystals: microscopic approximation. *Phys Rev B* 33:6304–6319

[37] Pouget J, Askar A, Maugin GA (1986), Lattice model for elastic ferroelectric crystals: continuum approximation. *Phys Rev B* 33:6320–6325

[38] Pouget J, Maugin GA (1980) Coupled acoustic-optic modes in deformable ferroelectrics. *J Acoust Soc Am* 68:588–601

[39] Pouget J, Maugin GA (1981) Bleustein-Gulyaev surface modes in elastic ferroelectrics. *J Acoust Soc Am* 69:1304–1318

[40] Pouget J, Maugin GA (1981) Piezoelectric Rayleigh waves in elastic ferroelectrics. *J Acoust Soc Am* 69:1319–1325

[41] Collet B (1984) Shock waves in deformable ferroelectric materials. In: Maugin GA (ed) *Mechanical Behavior of Electromagnetic Solid Continua*. Elsevier, North-Holland

[42] Sahin E, Dost S (1988) A strain-gradient theory of elastic dielectrics with spatial dispersion. *Int J Engng Sci* 26:1231–1245

[43] Demiray H, Dost S (1989) Diatomic elastic dielectrics with polarization gradient. *Int J Engng Sci* 27:1275–1284

[44] Askar A, Lee PCY (1974) Lattice dynamics approach to the theory of diatomic elastic dielectrics. *Phys Rev B* 9:5291–5299

[45] Maugin GA (1988) *Continuum Mechanics of Electromagnetic Bodies*. Elsevier, North-Holland

[46] Maugin GA, Pouget J, Drouot JR et al (1992) *Nonlinear Electromechanical Couplings*. John Wiley and Sons, Chichester

[47] Li JY (2003) Exchange coupling in P(VDF-TrFE) copolymer based all-organic composites with giant electrostriction. *Phys Rev Lett* 90:17601

[48] Kafadar CB (1971) Theory of multipoles in classical electromagnetism. *Int J Engng Sci* 9:831–853

[49] Demiray H, Eringen AC (1973) On the constitutive relations of polar elastic dielectrics. *Lett in Appl Engng Sci* 1:517–527

[50] Prechtl A (1980) Deformable bodies with electric and magnetic quadrupoles. *Int J Engng Sci* 18:665–680

[51] Nelson DF (1979) *Electric, Optic and Acoustic Interactions in Crystals*. Wiley, New York

[52] Kalpakides VK, Hadjigeorgiou EP, Massalas CV (1995) A variational principle for elastic dielectrics with quadruple polarization. *Int J Engng Sci* 33:793–801

[53] Kalpakides VK, Massalas CV (1993) Tiersten's theory of thermoelectroelasticity: An extension. *Int J Engng Sci* 31:157–164

[54] Hadjigeorgiou EP, Kalpakides VK, Massalas CV (1999) A general theory for elastic dielectrics. II. The variational approach. *Int J Non-Linear Mech* 34:967–980

[55] Kalpakides VK, Agiasofitou EK (2002) On material equations in second order gradient electroelasticity. *J Elasticity* 67:205–227

[56] Maugin GA (1980) The principle of virtual power: Application to coupled fields. *Acta Mech* 35:1–70

[57] Yang XM, Hu YT, Yang JS (2004) Electric field gradient effects in anti-plane problems of polarized ceramics. *Int J Solids Struct* 41:6801–6811

[58] Yang JS, Yang XM (2004), Electric field gradient effect and thin film capacitance. *World J Eng* 2:41–45

[59] Yang XM, Hu TY, Yang JS (2005) Electric field gradient effects in anti-plane problems of a circular cylindrical hole in piezoelectric materials of 6mm symmetry. *Acta Mech Solida Sinica* 18:29–36

[60] Li XF, Yang JS, Jiang Q (2005) Spatial dispersion of short surface acoustic waves in piezoelectric ceramics. *Acata Mechanica* 180:11–20

[61] Bleustein JL (1968) A new surface wave in piezoelectric materials. *Appl Phys Lett* 13:412–413

[62] Gulyaev YuV (1969) Electroacoustic surface waves in solids. *Sov Phys JETP Lett* 9:37–38

[63] Yang JS, Zhou HG, Li JY (2006) Electric field gradient effects in an anti-plane circular inclusion in polarized ceramics. *Proc Royal Soc London A* 462:3511–3522

[64] Yang JS (2004) Effects of electric field gradient on an anti-plane crack in piezoelectric ceramics. *Int J Fract* 127:L111–L116

[65] Zeng Y, Hu YT, Yang JS (2005) Electric field gradient effects in piezoelectric anti-plane crack problems. *J Huazhong Univ Sci Technol* 22:31–35

[66] Zeng Y (2005) Electric field gradient effects in anti-plane crack problems of piezoelectric ceramics. MS thesis, Huazhong University of Science and Technology

Appendix
Electroelastic Material Constants

Material constants for a few common piezoelectrics are summarized below.

Permittivity of free space $\varepsilon_0 = 8.854 \times 10^{-12}$ F/m.

Polarized ceramicss

The material matrices for PZT-5H are [1]

$$\rho = 7500\,\text{kg/m}^3,$$

$$[c_{pq}] = \begin{pmatrix} 12.6 & 7.95 & 8.41 & 0 & 0 & 0 \\ 7.95 & 12.6 & 8.41 & 0 & 0 & 0 \\ 8.41 & 8.41 & 11.7 & 0 & 0 & 0 \\ 0 & 0 & 0 & 2.3 & 0 & 0 \\ 0 & 0 & 0 & 0 & 2.3 & 0 \\ 0 & 0 & 0 & 0 & 0 & 2.325 \end{pmatrix} \times 10^{10}\,\text{N/m}^2,$$

$$[e_{ip}] = \begin{pmatrix} 0 & 0 & 0 & 0 & 17 & 0 \\ 0 & 0 & 0 & 17 & 0 & 0 \\ -6.5 & -6.5 & 23.3 & 0 & 0 & 0 \end{pmatrix}\,\text{C/m}^2,$$

$$[\varepsilon_{ij}] = \begin{pmatrix} 1700\varepsilon_0 & 0 & 0 \\ 0 & 1700\varepsilon_0 & 0 \\ 0 & 0 & 1700\varepsilon_0 \end{pmatrix}$$

$$= \begin{pmatrix} 1.505 & 0 & 0 \\ 0 & 1.505 & 0 \\ 0 & 0 & 1.302 \end{pmatrix} \times 10^{-8}\,\text{C/(V}-\text{m)}.$$

For PZT-5H, an equivalent set of material constants are [1]

$$s_{11} = 16.5, \qquad s_{33} = 20.7, \qquad s_{44} = 43.5,$$
$$s_{12} = -4.78, \qquad s_{13} = -8.45 \times 10^{-12}\,\mathrm{m^2/N},$$
$$d_{31} = -274, \qquad d_{15} = 741, \qquad d_{33} = 593 \times 10^{-12}\,\mathrm{C/N},$$
$$\varepsilon_{11} = 3130\varepsilon_0, \qquad \varepsilon_{33} = 3400\varepsilon_0.$$

When poling is along other directions, the material matrices can be obtained by tensor transformations. For PZT-5H, when poling is along the x_1-axis, we have

$$[c_{pq}] = \begin{pmatrix} 11.7 & 8.41 & 8.41 & 0 & 0 & 0 \\ 8.41 & 12.6 & 7.95 & 0 & 0 & 0 \\ 8.41 & 7.95 & 12.6 & 0 & 0 & 0 \\ 0 & 0 & 0 & 2.325 & 0 & 0 \\ 0 & 0 & 0 & 0 & 2.3 & 0 \\ 0 & 0 & 0 & 0 & 0 & 2.3 \end{pmatrix} \times 10^{10}\,\mathrm{N/m^2},$$

$$[e_{ip}] = \begin{pmatrix} 23.3 & -6.5 & -6.5 & 0 & 0 & 0 \\ 0 & 0 & 0 & 0 & 0 & 17 \\ 0 & 0 & 0 & 0 & 17 & 0 \end{pmatrix} \mathrm{C/m^2},$$

$$[\varepsilon_{ij}] = \begin{pmatrix} 1.302 & 0 & 0 \\ 0 & 1.505 & 0 \\ 0 & 0 & 1.505 \end{pmatrix} \times 10^{-8}\,\mathrm{C/Vm}.$$

When poling is along the x_2-axis,

$$[c_{pq}] = \begin{pmatrix} 12.6 & 8.41 & 7.95 & 0 & 0 & 0 \\ 8.41 & 11.7 & 8.41 & 0 & 0 & 0 \\ 7.95 & 8.41 & 12.6 & 0 & 0 & 0 \\ 0 & 0 & 0 & 2.3 & 0 & 0 \\ 0 & 0 & 0 & 0 & 2.325 & 0 \\ 0 & 0 & 0 & 0 & 0 & 2.3 \end{pmatrix} \times 10^{10}\,\mathrm{N/m^2},$$

$$[e_{ip}] = \begin{pmatrix} 0 & 0 & 0 & 0 & 0 & 17 \\ -6.5 & 23.3 & -6.5 & 0 & 0 & 0 \\ 0 & 0 & 0 & 17 & 0 & 0 \end{pmatrix} \mathrm{C/m^2},$$

$$[\varepsilon_{ij}] = \begin{pmatrix} 1.505 & 0 & 0 \\ 0 & 1.302 & 0 \\ 0 & 0 & 1.505 \end{pmatrix} \times 10^{-8}\,\mathrm{C/Vm}.$$

For PZT-G1195

$$\rho = 7500 \text{ kg/m}^3, \qquad c_{11}^E = c_{22}^E = 148, \qquad c_{33}^E = 131, \qquad c_{12}^E = 76.2,$$
$$c_{13}^E = c_{23}^E = 74.2, \qquad c_{44}^E = c_{55}^E = 25.4, \qquad c_{66}^E = 35.9 \text{ GPa},$$
$$e_{15} = 9.2, \qquad e_{31} = -2.1, \qquad e_{33} = 9.5 \text{ C/m}^2.$$

Material constants of a few other polarized ceramics are given in the following tables [2].

Material	c_{11}	c_{12}	c_{13}	c_{33}	c_{44}	c_{66}
PZT-4	13.9	7.78	7.40	11.5	2.56	3.06
PZT-5A	12.1	7.59	7.54	11.1	2.11	2.26
PZT-6B	16.8	8.47	8.42	16.3	3.55	4.17
PZT-5H	12.6	7.91	8.39	11.7	2.30	2.35
PZT-7A	14.8	7.61	8.13	13.1	2.53	3.60
PZT-8	13.7	6.99	7.11	12.3	3.13	3.36
BaTiO$_3$	15.0	6.53	6.62	14.6	4.39	4.24
	$\times 10^{10}$ N/m^2					

Material	e_{31}	e_{33}	e_{15}	ε_{11}	ε_{33}
PZT-4	−5.2	15.1	12.7	0.646	0.562
PZT-5A	−5.4	15.8	12.3	0.811	0.735
PZT-6B	−0.9	7.1	4.6	0.360	0.342
PZT-5H	−6.5	23.3	17.0	1.505	1.302
PZT-7A	−2.1	9.5	9.2	0.407	0.208
PZT-8	−4.0	13.2	10.4	0.797	0.514
BaTiO$_3$	4.3	17.5	11.4	0.987	1.116
	C/m^2			$\times 10^{-8}$ C/Vm	

Density	PZT-5H	PZT-5A	PZT-6B	PZT-4
kg/m^3	7500	7750	7550	7500

Dinsity	PZT-7A	PZT-8	BaTiO$_3$
kg/m^3	7600	7600	5700

Quartz

When referred to the crystal axes, the second-order material constants for left-hand quartz have the following values [3].

$$\rho = 2649 \, \mathrm{kg/m^3},$$

$$[c_{pq}] = \begin{pmatrix} 86.74 & 6.99 & 11.91 & -17.91 & 0 & 0 \\ 6.99 & 86.74 & 11.91 & 17.91 & 0 & 0 \\ 11.91 & 11.91 & 107.2 & 0 & 0 & 0 \\ -17.91 & 17.91 & 0 & 57.94 & 0 & 0 \\ 0 & 0 & 0 & 0 & 57.94 & -17.91 \\ 0 & 0 & 0 & 0 & -17.91 & 39.88 \end{pmatrix} \times 10^9 \, \mathrm{N/m^2},$$

$$[e_{ip}] = \begin{pmatrix} 0.171 & -0.171 & 0 & -0.0406 & 0 & 0 \\ 0 & 0 & 0 & 0 & 0.0406 & -0.171 \\ 0 & 0 & 0 & 0 & 0 & 0 \end{pmatrix} \mathrm{C/m^2},$$

$$[\varepsilon_{ij}] = \begin{pmatrix} 39.21 & 0 & 0 \\ 0 & 39.21 & 0 \\ 0 & 0 & 41.03 \end{pmatrix} \times 10^{-12} \, \mathrm{C/Vm}.$$

Temperature derivatives of the elastic constants of quartz at 25°C are [4]

pq	11	33	12	13
$(1/c_{pq}(dc_{pq}/dT)(10^{-6}/°C)$	18.16	−66.60	−1222	−178.6

pq	44	66	14
$(1/c_{pq}(dc_{pq}/dT)(10^{-6}/°C)$	−89.72	126.7	−49.21

For quartz there are 31 nonzero third-order elastic constants. Fourteen are given in the following table. These values, at 25°C, and based on a least squares fit, are all in $10^{11} \, \mathrm{N/m^2}$ [5].

Constant	Value	Standard Error
c_{111}	-2.10	0.07
c_{112}	-3.45	0.06
c_{113}	$+0.12$	0.06
c_{114}	-1.63	0.05
c_{123}	-2.94	0.05
c_{124}	-0.15	0.04
c_{133}	-3.12	0.07
c_{134}	$+0.02$	0.04
c_{144}	-1.34	0.07
c_{155}	-2.00	0.08
c_{222}	-3.32	0.08
c_{333}	-8.15	0.18
c_{344}	-1.10	0.07
c_{444}	-2.76	0.17

In addition, there are 17 relations among the third-order elastic constants of quartz [6].

$$c_{122} = c_{111} + c_{112} - c_{222}, \qquad c_{156} = \frac{1}{2}(c_{114} + 3c_{124}),$$

$$c_{166} = \frac{1}{4}(-2c_{111} - c_{112} + 3c_{222}),$$

$$c_{224} = -c_{114} - 2c_{124}, \qquad c_{256} = \frac{1}{2}(c_{114} - c_{124}),$$

$$c_{266} = \frac{1}{4}(2c_{111} - c_{112} - c_{222}),$$

$$c_{366} = \frac{1}{2}(c_{113} - c_{123}), \qquad c_{456} = \frac{1}{2}(-c_{144} + c_{155}),$$

$$c_{223} = c_{113}, \qquad c_{233} = c_{133}, \qquad c_{234} = -c_{134},$$

$$c_{244} = c_{155}, \qquad c_{225} = c_{144},$$

$$c_{355} = c_{344}, \qquad c_{356} = c_{134}, \qquad c_{455} = -c_{444}, \qquad c_{466} = c_{124}.$$

For the fourth-order elastic constants, there are 69 nonzero ones of which 23 are independent [7].

$$c_{1111}, \quad c_{3333}, \quad c_{4444}, \quad c_{6666}, \quad c_{1112}, \quad c_{1113}, \quad c_{1123}, \quad c_{2214}, \quad c_{3331},$$

$$c_{4456}, \quad c_{5524}, \quad c_{4443}, \quad c_{1133}, \quad c_{3344}, \quad c_{1456}, \quad c_{1155}, \quad c_{1134}, \quad c_{2356},$$

$$c_{4423}, \quad c_{4413}, \quad c_{3314}, \quad c_{6614}, \quad c_{6624}.$$

There are 46 relations [7]

$$c_{2222} = c_{1111}, \qquad c_{2266} = \frac{1}{6}(c_{1111} - c_{1112}), \qquad c_{2223} = c_{1113},$$

$$c_{2221} = c_{1112}, \qquad c_{6612} = \frac{1}{6}(c_{1111} - 4c_{6666} - c_{1112}), \qquad c_{2213} = c_{1123},$$

$$c_{1166} = c_{2266}, \qquad c_{1122} = \frac{1}{3}(-c_{1111} + 4c_{1112} + 8c_{6666}),$$

$$c_{6613} = \frac{1}{4}(c_{1113} - c_{1123}),$$

$$c_{5555} = c_{4444}, \qquad c_{4455} = \frac{1}{3}c_{4444}, \qquad c_{6623} = c_{6613},$$

$$c_{1124} = -c_{2214} + c_{6614} + c_{6624},$$

$$c_{3312} = -c_{1133}, \qquad c_{1114} = 3(-c_{2214} + 2c_{6614} - 2c_{6624}), \qquad c_{2233} = c_{1133},$$

$$c_{2256} = \frac{1}{2}(-2c_{2214} + 3c_{6614} - 5c_{6624}), \qquad c_{6633} = c_{1133},$$

$$c_{2224} = 3(c_{2214} - 3c_{6614} + c_{6624}),$$

$$c_{3355} = c_{3344}, \qquad c_{1156} = \frac{1}{2}(-2c_{2214} + 7c_{6614} - c_{6624}), \qquad c_{3332} = c_{3331},$$

$$c_{1256} = \frac{1}{2}(-2c_{2214} + 3c_{6614} - c_{6624}), \qquad c_{5534} = -c_{4443},$$

$$c_{6665} = \frac{3}{2}(c_{6614} - c_{6624}),$$

$$c_{4442} = -4c_{4456} - c_{5524}, \qquad c_{1234} = c_{1134} - 2c_{2356}, \qquad c_{2255} = c_{4412},$$

$$c_{5514} = 2c_{4456} + c_{5524}, \qquad c_{1356} = 2c_{1134} - 3c_{2356}, \qquad c_{5566} = c_{1456},$$

$$c_{5556} = 3c_{4456}, \qquad c_{2234} = 4c_{2356} - 3c_{1134}, \qquad c_{3324} = -c_{3314},$$

$$c_{4441} = 2c_{4456} - c_{5524}, \qquad c_{6634} = c_{1234}, \qquad c_{3356} = c_{3314},$$

$$c_{5512} = c_{4412} \qquad c_{1144} = c_{4412}, \qquad c_{5523} = c_{4413}, \qquad c_{2456} = c_{1456},$$

$$c_{2244} = c_{1155}, \qquad c_{5513} = c_{4423},$$

$$c_{4466} = c_{1456}, \qquad c_{4412} = c_{1155} - 4c_{1456}, \qquad c_{3456} = \frac{1}{2}(c_{4423} - c_{4413}).$$

The fourth-order elastic constants are usually unknown. Some scattered results are [7]

$$c_{1111} = 1.59 \times 10^{13} \text{ N/m}^2 \pm 20\%,$$

$$c_{3333} = 1.84 \times 10^{13} \text{ N/m}^2 \pm 20\%,$$

and [8]

$$c_{6666}^{E} = 77 \times 10^{11} \text{ N/m}^2.$$

AT-cut quartz is a special case of rotated Y-cut quartz ($\theta = 35.25°$) whose material constants are [9]

$$[c_{pq}] = \begin{pmatrix} 86.74 & -8.25 & 27.15 & -3.66 & 0 & 0 \\ -8.25 & 129.77 & -7.42 & 5.7 & 0 & 0 \\ 27.15 & -7.42 & 102.83 & 9.92 & 0 & 0 \\ -3.66 & 5.7 & 9.92 & 38.61 & 0 & 0 \\ 0 & 0 & 0 & 0 & 68.81 & 2.53 \\ 0 & 0 & 0 & 0 & 2.53 & 29.01 \end{pmatrix} \times 10^9 \text{ N/m}^2,$$

$$[e_{ip}] = \begin{pmatrix} 0.171 & -0.152 & -0.0187 & 0.067 & 0 & 0 \\ 0 & 0 & 0 & 0 & 0.108 & -0.095 \\ 0 & 0 & 0 & 0 & -0.0761 & 0.067 \end{pmatrix} \text{ C/m}^2,$$

$$[\varepsilon_{ij}] = \begin{pmatrix} 39.21 & 0 & 0 \\ 0 & 39.82 & 0.86 \\ 0 & 0.86 & 40.42 \end{pmatrix} \times 10^{-12} \text{ C/Vm.}$$

Langasite

The second-order material constants of $La_3Ga_5SiO_{14}$ are [10]

$$\rho = 5743 \text{ kg/m}^3,$$

$$[c_{pq}] = \begin{pmatrix} 18.875 & 10.475 & 9.589 & -1.412 & 0 & 0 \\ 10.475 & 18.875 & 9.589 & 1.412 & 0 & 0 \\ 9.586 & 9.586 & 26.14 & 0 & 0 & 0 \\ -1.412 & 1.412 & 0 & 5.35 & 0 & 0 \\ 0 & 0 & 0 & 0 & 5.35 & -1.412 \\ 0 & 0 & 0 & 0 & -1.412 & 4.2 \end{pmatrix} \times 10^{10} \text{ N/m}^2,$$

$$[e_{ip}] = \begin{pmatrix} -0.44 & 0.44 & 0 & -0.08 & 0 & 0 \\ 0 & 0 & 0 & 0 & 0.08 & 0.44 \\ 0 & 0 & 0 & 0 & 0 & 0 \end{pmatrix} \text{ C/m}^2,$$

$$[\varepsilon_{ij}] = \begin{pmatrix} 18.92\varepsilon_0 & 0 & 0 \\ 0 & 18.92\varepsilon_0 & 0 \\ 0 & 0 & 50.7\varepsilon_0 \end{pmatrix}$$

$$= \begin{pmatrix} 167.5 & 0 & 0 \\ 0 & 167.5 & 0 \\ 0 & 0 & 448.9 \end{pmatrix} \times 10^{-12} \text{ C/Vm.}$$

The third-order material constants of $La_3Ga_5SiO_{14}$ at 20°C are also given in [10]. The third-order elastic constants c_{pqr} (in 10^{10} N/m^2) are

c_{111}	-97.2	c_{134}	-4.1
c_{112}	0.7	c_{144}	-4.0
c_{113}	-11.6	c_{155}	-19.8
c_{114}	-2.2	c_{222}	-96.5
c_{123}	0.9	c_{333}	-183.4
c_{124}	-2.8	c_{344}	-38.9
c_{133}	-72.1	c_{444}	20.2

The third-order piezoelectric constants e_{ipq} (in C/m^2) are

e_{111}	9.3	e_{124}	-4.8
e_{113}	-3.5	e_{134}	6.9
e_{124}	1.0	e_{144}	-1.7
e_{122}	0.7	e_{315}	-4

The third-order electrostriction constants H_{pq} (in 10^{-9} N/V^2) are

H_{11}	-26	H_{31}	-24
H_{12}	65	H_{33}	-40
H_{13}	20	H_{41}	-170
H_{14}	-43	H_{44}	-44

The third-order dielectric permeability ε_{111} (in 10^{-20} F/V) are

ε_{111}	-0.5

Lithium Niobate

The second-order material constants for lithium niobate are [11]

$$\rho = 4700 \ \text{kg/m}^3,$$

$$[c_{pq}] = \begin{pmatrix} 2.03 & 0.53 & 0.75 & 0.09 & 0 & 0 \\ 0.53 & 2.03 & 0.75 & -0.09 & 0 & 0 \\ 0.75 & 0.75 & 2.45 & 0 & 0 & 0 \\ 0.09 & -0.09 & 0 & 0.60 & 0 & 0 \\ 0 & 0 & 0 & 0 & 0.60 & 0.09 \\ 0 & 0 & 0 & 0 & 0.09 & 0.75 \end{pmatrix} \times 10^{11}\,\mathrm{N/m^2},$$

$$[e_{ip}] = \begin{pmatrix} 0 & 0 & 0 & 0 & 3.70 & -2.50 \\ -2.50 & 2.50 & 0 & 3.70 & 0 & 0 \\ 0.20 & 0.20 & 1.30 & 0 & 0 & 0 \end{pmatrix} \mathrm{C/m^2},$$

$$[\varepsilon_{ij}] = \begin{pmatrix} 38.9 & 0 & 0 \\ 0 & 38.9 & 0 \\ 0 & 0 & 25.7 \end{pmatrix} \times 10^{-11}\,\mathrm{C/Vm}.$$

The third-order material constants of lithium niobate are given in [12]. The third-order elastic constants c_{pqr} (in 10^{11} N/m^2) are

Constant	Value	Standard Error
c_{111}	−21.2	4.0
c_{112}	−5.3	1.2
c_{113}	−5.7	1.5
c_{114}	2.0	0.8
c_{123}	−2.5	1.0
c_{124}	0.4	0.3
c_{133}	−7.8	1.9
c_{134}	1.5	0.3
c_{144}	−3.0	0.2
c_{155}	−6.7	0.3
c_{222}	−23.3	3.4
c_{333}	−29.6	7.2
c_{344}	−6.8	0.7
c_{444}	−0.3	0.4

The third-order piezoelectric constants $e_{ipq}(= -k_{1_{ipq}})$ are

Constant	Value	Standard Error
e_{115}	17.1	6.6
e_{116}	−4.7	6.4
e_{125}	19.9	2.1
e_{126}	−15.9	5.3
e_{135}	19.6	2.7
e_{136}	−0.9	2.7
e_{145}	20.3	5.7
e_{311}	14.7	6.0
e_{312}	13.0	11.4
e_{313}	−10.0	8.7
e_{314}	11.0	4.6
e_{333}	−17.3	5.9
e_{344}	−10.2	5.6
	C/m^2	

The third-order electrostrictive constants l_{pq} (compressed from $b_{ijkl} + \varepsilon_0\delta_{ij}\delta_{kl} - \varepsilon_0\delta_{ik}\delta_{jl} - \varepsilon_0\delta_{il}\delta_{kj}$) (in 10^{-9} F/m^2) are

Constant	Value	Standard Error
l_{11}	1.11	0.39
l_{12}	2.19	0.56
l_{13}	2.32	0.67
l_{31}	0.19	0.61
l_{33}	−2.76	0.41
l_{14}	1.51	0.17
l_{41}	1.85	0.17
l_{44}	−1.83	0.11

The third-order dielectric constants ε_{ip}(in 10^{-19} F/V) are

Constant	Value	Standard Error
ε_{31}	−2.81	0.06
ε_{22}	−2.40	0.09
ε_{33}	−2.91	0.06

Lithium Tantalate

The second-order material constants for lithium tantalate are [11]

$$\rho = 7450 \text{ kg/m}^3,$$

$$[c_{pq}] = \begin{pmatrix} 2.33 & 0.47 & 0.80 & -0.11 & 0 & 0 \\ 0.47 & 2.33 & 0.80 & 0.11 & 0 & 0 \\ 0.80 & 0.80 & 2.45 & 0 & 0 & 0 \\ -0.11 & -0.11 & 0 & 0.94 & 0 & 0 \\ 0 & 0 & 0 & 0 & 0.94 & -0.11 \\ 0 & 0 & 0 & 0 & -0.11 & 0.93 \end{pmatrix} \times 10^{11} \text{ N/m}^2,$$

$$[e_{ip}] = \begin{pmatrix} 0 & 0 & 0 & 0 & 2.6 & -1.6 \\ -1.6 & 1.6 & 0 & 2.6 & 0 & 0 \\ 0 & 0 & 1.9 & 0 & 0 & 0 \end{pmatrix} \text{ C/m}^2,$$

$$[\varepsilon_{ij}] = \begin{pmatrix} 36.3 & 0 & 0 \\ 0 & 36.3 & 0 \\ 0 & 0 & 38.2 \end{pmatrix} \times 10^{-11} \text{ C/Vm}.$$

References

[1] Auld, BA (1973) *Acoustic Fields and Waves in Solids*, vol. 1. John Wiley and Sons, New York

[2] Jaffe H, Berlincourt DA (1965) Piezoelectric transducer materials. *Proc IEEE* 53: 1372–1386

[3] Bechmann R (1958) Elastic and piezoelectric constants of alpha-quartz. *Phys Rev* 110: 1060–1061

[4] Sinha SH, Tiersten HF (1979) First temperature derivatives of the fundamental elastic constants of quartz. *J Appl Phys* 50: 2732–2739

[5] Thurston RN, McSkimin HJ, Andreatch P Jr. (1966) Third-order elastic constants of quartz. *J Appl Phys* 37:267–275

[6] Nelson DF (1979) *Electric, Optic and Acoustic Interactions in Crystals*. John Wiley and Sons, New York

[7] Gagnepain JJ, Besson R (1975) Nonlinear effects in piezoelectric quartz crystals. In: Mason WP, Thurston RN (ed) *Physical Acoustics, Vol.* XI. Academic Press, New York

[8] Tiersten HF (1975) Analysis of intermodulation in thickness-shear and trapped energy resonators. *J Acoust Soc* Am 57: 667–681

[9] Tiersten HF (1969) *Linear Piezoelectric Plate Vibrations*. Plenum, New York

[10] Sorokin BP, Turchin PP, Burkov SI et al (1996) Influence of static electric field, mechanical pressure and temperature on the propagation of acoustic waves in $La_3Ga_5SiO_{14}$ piezoelectric single crystals. In: *Proc IEEE Int Frequency Control Symp* 161–169

[11] Warner AW, Onoe M, Couqin GA (1967) Determination of elastic and piezoelectric constants for crystals in class (3m). *J Acoust Soc Am* 42: 1223–1231

[12] Cho Y, Yamanouchi K (1987), Nonlinear, elastic, piezoelectric, electrostrictive, and dielectric constants of lithium niobate. *J Appl Phys* 61: 875–887

LaVergne, TN USA
20 November 2009

164744LV00001B/55/P